Catalysis and the Mechanism of Methane Conversion to Chemicals

Toshihide Baba · Akimitsu Miyaji

Catalysis and the Mechanism of Methane Conversion to Chemicals

C–C and C–O Bonds Formation Using Heterogeneous, Homogenous, and Biological Catalysts

Springer

Toshihide Baba
Tokyo Institute of Technology
Yokohama, Japan

Akimitsu Miyaji
Tokyo Institute of Technology
Yokohama, Japan

ISBN 978-981-15-4131-5 ISBN 978-981-15-4132-2 (eBook)
https://doi.org/10.1007/978-981-15-4132-2

This Springer imprint is published by the registered company Springer Nature Singapore Pte Ltd.
The registered company address is: 152 Beach Road, #21-01/04 Gateway East, Singapore 189721, Singapore

Preface

Methane, which is a ubiquitous reservoir of chemical energy, is very thermodynamically stable, and its successful utilization in chemical synthesis requires quite significant kinetic barriers to be surmounted. Despite this, methane can be directly converted into high-value-added chemicals such as methanol, aromatic hydrocarbons, and olefins. This book deals principally with the direct conversion of methane to produce chemicals with C–O and C–C bonds, meaning that processes involving synthesis gas as an intermediate are not considered. Achieving the direct conversion of methane is a great challenge; however, the development of direct methane conversion processes is essential from both economic development and energy security points of view. Many researchers have sought to convert methane into chemicals via various catalytic systems involving heterogeneous, homogeneous, and biological catalysts by using advanced experimental and theoretical techniques together with various strategies. However, industrial processes for the direct conversion of methane to chemicals have not yet been developed. For scientists and engineers, the establishment of next-generation processes for direct methane conversion represents a grand challenge in chemistry, independent of trends in the larger world such as the discovery of shale gas and the drive to reduce carbon dioxide emissions. The authors are of the view that the development of direct methane conversion is crucial, and is highly preferable to the use of carbon dioxide, the most thermodynamically stable chemical, as a chemical feedstock, as carbon dioxide conversion inherently requires high energy consumption.

To convert methane into value-added chemicals using heterogeneous, homogeneous, or biological catalysts requires addressing questions such as:

1. How can catalysts activate methane for the formation of C–O and C–C bonds?
2. What determines the selective formation of C–O or C–C bonds?
3. Can the knowledge of methane chemistry be used to develop new catalytic processes suitable for industry?

The aim of this book is to discuss the highlights of the various investigations that have been conducted in this field, with an emphasis on the fundamental chemistry that has been found to be involved in heterogeneous, homogeneous, and biologically catalyzed methane activation. Thus, it is hoped that this book will provide a deep understanding of the work that has significantly impacted the field of the fundamental chemistry of heterogeneous, homogenous, and biological catalysis. Furthermore, by bridging insights from the various catalytic approaches together in each chapter, cross-disciplinary insights are obtained. Therefore, in this book, the use of heterogeneous, homogenous, and biological catalysis for direct methane conversion are often discussed together in a comparative fashion to connect the disciplines and generate new, unifying concepts.

This book, *Catalysis and the Mechanisms for Methane Conversion to Chemicals: C–O and C–C Bond Formation Using Heterogeneous, Homogenous, and Biological Catalysts* comprises seven chapters, among which Chaps. 2, 4, and 7 are quite unique compared to previous published books on the topic of methane conversion.

Chapter 1: Overview of Direct Methane Conversion to Chemicals with C–O and C–C Bonds

Chapter 2: Selective Production of Methanol from Methane and Molecular Oxygen at Atmospheric Temperature and Pressure Using Methane Monooxygenases

Chapter 3: Heterogeneous and Homogeneous Catalytic Partial Oxidations of Methane to Methanol and Its Derivatives

Chapter 4: Application of Biocatalysts for the Production of Methanol from Methane

Chapter 5: C–C Bond Formation via the Condensation of Methane in the Presence or Absence of Oxygen

Chapter 6: Conversion of Methane to Aromatic Hydrocarbons

Chapter 7: C–C Bond Formation via Carbocations in the Methane Conversion Under Non-Oxidative Conditions

Both authors would like to express their deep gratitude to Prof. Yoshio Ono and Prof. Ichiro Okura, who opened the vistas of heterogeneous, homogeneous, and biological catalysis to the authors and always inspired them through their enthusiasm for the science of catalysis and long-term encouragement. We also sincerely thank Mr. Shinichi Koizumi and Ms. Takeko Sato, who handled the editing and publication.

Yokohama, Japan

Toshihide Baba
Akimitsu Miyaji

Contents

Chapter 1
Overview of Direct Methane Conversion to Chemicals with C–O and C–C Bonds

1.1 Social Considerations and Global Circumstances Surrounding Methane Consumption

The consumption of energy and the production of materials are essential for human beings to live a comfortable lifestyle; accordingly, both primary energy consumption and materials production have continually increased [1]. In 1965, the total primary energy consumption worldwide was about 38 billion tons as oil equivalent; this figure had increased to 153 billion tons in 2017 [2]. That is, in approximately 50 years, primary energy consumption has increased nearly four times. In 2017, more than 80% of primary energy was derived from fossil resources with the following breakdown: 32.4% oil, 27.6% coal, 23.4% natural gas, 6.9% hydroelectricity, and 3.6% renewables. Because of their high consumption rate, fossil energy resources are dwindling; oil will probably be depleted first, followed by natural gas and finally coal. Under these circumstances, dramatic changes to the energy system will take place over the coming decades. Additionally, chemical products are often produced from petroleum feedstocks; nearly 11% of oil is used for chemical manufacture [3]. Therefore, in addition to the primary energy system, chemical production processes that currently rely on fossil resources will have to be changed.

In this context, natural gas has recently attracted increasing attention due to the discovery of shale gas. Natural gas contains 70–90% methane [4], which is an ideal energy source, having the highest energy value (~56 kJ g^{-1}) among hydrocarbons. Furthermore, since methane has the greatest hydrogen-to-carbon ratio among all hydrocarbons, it also provides the maximum amount of energy per mole CO_2 produced during combustion [5]. In view of the abovementioned properties, ~94% of globally produced natural gas is used as a direct fuel for heating and power generation in both household and industrial settings [6]. However, the use of methane as a raw material for the production of chemicals accounts for only ~5% of its total consumption.

T. Baba and A. Miyaji, *Catalysis and the Mechanism of Methane Conversion to Chemicals*,
https://doi.org/10.1007/978-981-15-4132-2_1

Although fossil fuel reserves are dwindling, the use of fossil resources such as natural gas to obtain primary energy and chemical products cannot be replaced until new technologies such as solar light and artificial photosynthesis have been developed and implemented. Thus, to maintain a comfortable lifestyle, human beings will have to continue to consume finite oil, coal, and natural gas resources for the foreseeable future. The use of the abundant and inexpensive gas methane as a raw material for energy and chemical products has been proposed as a means of transitioning away from the current dependence on crude oil, which is localized in a limited number of geographic areas. Thus, methane is important in terms of establishing energy security, continuing economic development, and maintaining a comfortable lifestyle for humans.

However, industrial processes for the direct conversion of methane to chemicals have not yet been developed. For scientists and engineers, the establishment of next-generation processes for direct methane conversion represents a grand challenge in chemistry, independent of trends in the larger world such as the discovery of shale gas and the drive to reduce carbon dioxide emissions. The authors are of the view that the development of direct methane conversion is crucial, and is highly preferable to the use of CO_2, the most thermodynamically stable chemical, as a chemical feedstock, as CO_2 conversion inherently requires high energy consumption. Therefore, we herein focus on the conversion of methane to value-added chemicals, especially on the direct conversion of methane to methanol and higher hydrocarbons (such as propylene) bypassing the stage of synthesis gas (CO and H_2) production.

1.2 Industrial Methane Conversion for the Production of Chemicals

Although natural gas can be converted into various chemicals such as methanol (Fig. 1.1), the industrial processes of methane conversion to chemicals are limited (Fig. 1.2) and classified into two categories. The first category is the indirect conversion via synthesis gas (mixture of CO and H_2), the production of which accounts for ~96% of total methane consumption for the production of chemicals (right side of Fig. 1.2), and the second one is direct conversion to produce chemicals (left side of Fig. 1.2). Herein, the term "direct methane conversion" refers to processes that do not involve the production of synthesis gas as an intermediate. In other words, "direct methane conversion" implies that reforming agents, i.e., H_2O and CO_2, are not added.

In indirect methane conversion for chemical production, methane is first converted to synthesis gas. Technologies for the production of synthesis gas are well established and in current industrial use [7]. Synthesis gas, produced by the steam reforming and partial oxidation of methane, is used for the production of methanol/hydrocarbons via Fischer–Tropsch synthesis [8] and NH_3. In direct methane conversion, ~70% of methane is used for ammonia synthesis as a hydrogen source, ~20% is used for methanol synthesis, and ~10% is used for Fischer–Tropsch synthesis.

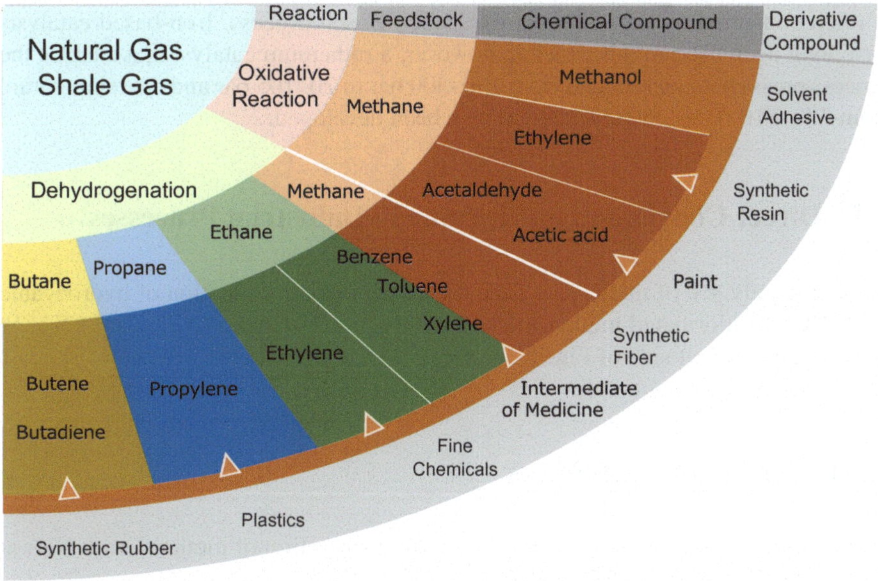

Fig. 1.1 Production of chemicals from natural gas

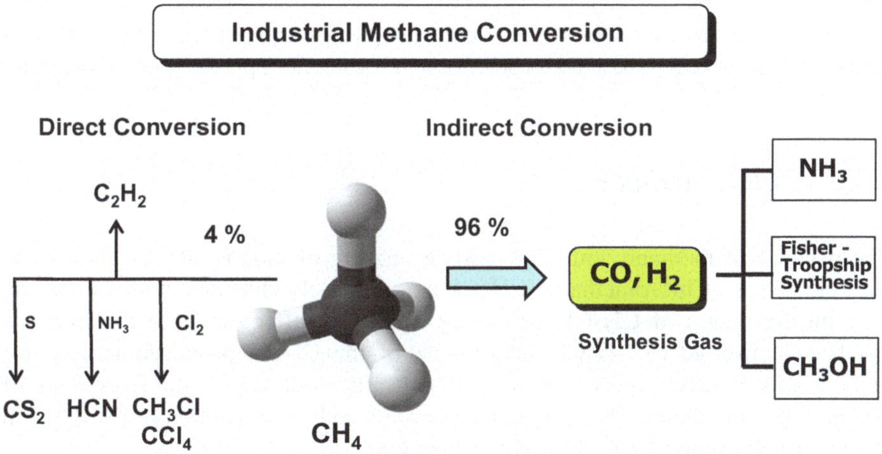

Fig. 1.2 Industrial processes of methane conversion for the production of chemicals

Methanol is a large volume commodity chemical whose production is almost entirely based on synthesis gas. The conversion of synthesis gas to methanol can be achieved in high selectivity using Cu–ZnO-based catalysts. Synthesis gas is also used in Fischer–Tropsch synthesis, mainly to produce straight-chain liquid hydrocarbons with carbon numbers from 5 to 27 (C5–C27), although methane, ethylene, and propylene are also produced. Additionally, the hydrogen (H_2) in synthesis gas is often converted to ammonia via the Haber–Bosch process. Iron-based catalysts have long been used in this process; however, a ruthenium catalyst that enables the process pressure to be reduced from 150–300 bar to 70–105 bar and the temperature from 370–510 °C to 350–470 °C has also been developed.

1.3 Direct Conversion of Methane in Industrial Processes

Approximately 4% of methane is used for the industrial production of hydrocyanic acid (HCN), chlorinated methane derivatives (e.g., CH_3Cl and CCl_4), and CS_2 [7], as shown on the left side of Fig. 1.2.

1.3.1 Hydrocyanic Acid

In the Degussa process, HCN is produced via the reaction of methane with NH_3 at about 1700 K using a Pt–Rh catalyst.

$$CH_4 + NH_3 \longrightarrow HCN + 3\,H_2 \tag{1.1}$$

This process represents one of the few examples of an industrial intermolecular dehydrogenation reaction, and is made possible by the high processing temperature.

1.3.2 Chloromethane

The reaction of methane with Cl_2 yields a mixture of mono-, di-, tri-, and tetra-chloromethanes. Chloromethane (CH_3Cl) is more easily chlorinated than methane, with the formation of CH_2Cl_2 becoming significant after ~20% of the methane has been converted to CH_3Cl. About 80% of the CH_3Cl produced is used for making silicon resins (polysiloxanes), which are produced by the hydrolysis of dimethyldichlorosilane. This compound is produced by the reaction of CH_3Cl with silicon, usually in the form of the copper alloy Si(Cu).

$$CH_3Cl + Si(Cu) \longrightarrow (CH_3)_2SiCl_2 \tag{1.2}$$

1.3.3 *Carbon Disulfide*

Carbon disulfide (CS_2) is produced by the reaction of methane with sulfur vapor.

$$CH_4 + \alpha S_x \longrightarrow CS_2 + 2H_2S \tag{1.3}$$

Above, x is between 2 and 8, because αS_x is an equilibrium mixture of S_2, S_6, and S_8. The main use of CS_2 is the production of regenerated cellulose for rayon, cellophane, and synthetic sponges.

1.3.4 *Acetylene*

In the oldest process for the synthesis of acetylene, coke and lime are heated to 2000 °C in an electric furnace to produce calcium carbide, which is hydrolyzed by H_2O to produce acetylene. The more modern Wulff process utilizes natural gas (primarily methane) and produces a 1:2 mixture of acetylene and ethylene via the pyrolysis of methane at high temperatures (~1600 K). Acetylene can be further reacted with formaldehyde to produce 1,4-butynediol, which can then be hydrogenated stepwise to butenediol and butanediol.

1.4 Direct Conversion of Methane for C–O and C–C Bond Formation: Methanol, Methanol Derivatives, and Higher Hydrocarbons

Great effort has been invested in making direct natural gas conversion processes economically competitive with indirect processes, especially for the production methanol and ethylene [6, 9–14]. Many researchers have reported methods for the control of methane conversion based on investigations of the role of heterogeneous [15–20], homogeneous [21–30], and biological [31–37] catalysts, determination of reaction mechanisms using spectroscopic analysis [38–40] and computational chemistry [41–44], and optimization of the reaction environment [45–47].

An overall classification scheme for chemical production using methane is shown in Fig. 1.3. This classification is based mainly on whether C–O or C–C bonds are formed via the direct conversion of methane. The products are methanol, methanol derivatives such as methyl sulfate (CH_3OSO_3H), and higher hydrocarbons with a carbon number of two or greater. In this book, these higher hydrocarbons are referred to as C_2^+ hydrocarbons.

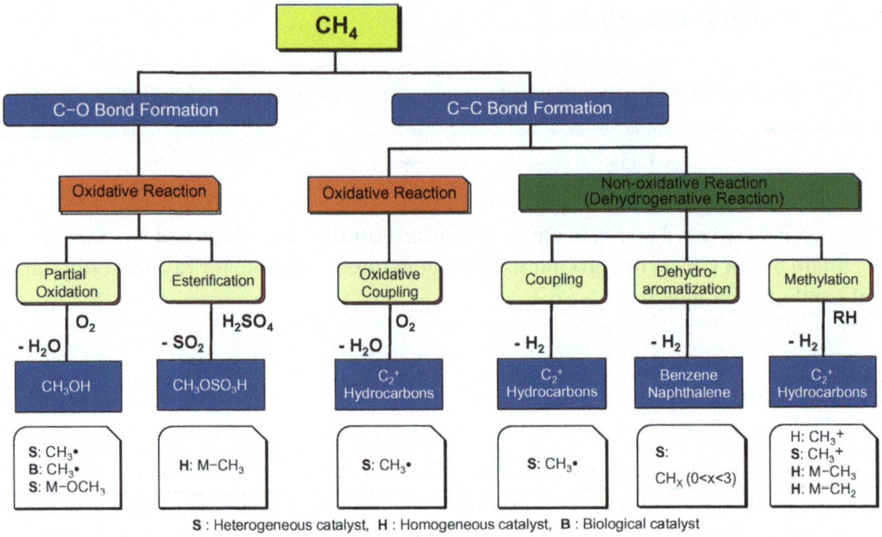

Fig. 1.3 Classification of direct methane conversion for the C–C and C–O bonds formation and corresponding key reaction intermediates

1.4.1 C–O Bond Formation: Production of Methanol and Its Derivatives

C–O bond formation reactions can be classified into two categories as described below.

1.4.1.1 Direct Oxidation of Methane to Methanol

The first type of C–O bond formation reaction is the direct production of methanol via the reaction of methane with oxygen. The overall reaction is shown in reaction (1.4) below.

$$CH_4 + 1/2\, O_2 \longrightarrow CH_3OH \quad (1.4)$$

This reaction is usually referred to as the partial oxidation of methane to methanol or the hydroxylation of methane to methanol. The free energy change (ΔG) of reaction (1.4) at 473 K is -103 kJ mol^{-1}. Thus, the partial oxidation reaction is thermodynamically favorable. Some bacteria contain enzymes known as methane monooxygenases (MMO) that catalyze reaction (1.4) at ambient conditions with 100% selectivity for methanol, and the application of MMOs as biocatalysts has been investigated [31–37]. Additionally, some heterogeneous catalysts for the selective production of methanol from methane, such as FeO_2/SiO_2 [48], inverse $CeO_2/Cu_2O/Cu(111)$ [49], Ni/CeO_2 (111) [50], and copper-exchanged zeolites [51], have recently been developed; however, their catalytic performance cannot yet match that of MMO. The mechanism of the conversion of methane to methanol is discussed in detail in Chapter 2, and the practical use of MMO as a biocatalyst for methane production is described in Chapter 4.

The methyl radical ($\bullet CH_3$) is considered a key reaction intermediate in the partial oxidation of methane. This intermediate has not been observed experimentally under an oxygen atmosphere, but was detected via electron paramagnetic resonance (EPR) spectroscopy at 77 K using molybdenum supported on silica as a heterogeneous catalyst and N_2O as the oxidant [52]. Furthermore, several theoretical calculations have suggested the involvement of $\bullet CH_3$ [31, 33]. Based on these experimental and theoretical results, reaction (1.4) is believed to proceed via $\bullet CH_3$ in the presence of oxygen when catalyzed by heterogeneous or biological catalysts.

Additionally, theoretical investigations have predicted that this reaction could be catalyzed by homogeneous organometallic catalysts [53, 54]. A theoretical investigation of the oxidation of methane to methanol on organometallic compounds such as copper and nickel complexes suggested that metal–methoxy species ($M–OCH_3$, M: metal cations) could act as the key intermediate in reaction (1.4) [19, 54]. The theoretical reaction of $M–OCH_3$ with methane via the oxidative addition pathway is shown in reaction (1.5) below.

$$L_nM{-}OCH_3 + CH_4 \longrightarrow L_nM{-}CH_3 + CH_3OH \tag{1.5}$$

In reaction (1.5), L represents a ligand. However, such catalytic reactions have not been experimentally observed yet.

1.4.1.2 Conversion of Methane to Methanol Derivatives

The second category of C–O bond formation reactions involves the production of methanol derivatives from methane. For example, the reaction of methane with strong acids such as H_2SO_4 has been reported.

$$CH_4 + 2\,H_2SO_4 \longrightarrow CH_3OSO_3H + 2\,H_2O + SO_2 \tag{1.6}$$

In this reaction, the produced methanol derivative, methyl sulfate, must be hydrolyzed as below to obtain methanol [24].

$$CH_3OSO_3H + H_2O \longrightarrow CH_3OH + H_2SO_4 \tag{1.7}$$

The formation of CH_3OSO_3H via the reaction of methane with H_2SO_4 is shown below.

$$CH_4 + 4\,H_2SO_4 \longrightarrow CH_3OSO_3H + 2\,H_2O{\cdot}H_2SO_4 + SO_2 \quad (1.8)$$

Above, $H_2O{\cdot}H_2SO_4$ represents H_2SO_4 hydrated with H_2O. The ΔG value of the reaction was estimated to be –25 kJ mol^{-1}. Therefore, the reaction is thermodynamically downhill.

On the other hand, the direct oxidation of methane to methanol with H_2SO_4 shown in reaction (1.9) is thermodynamically unfavorable.

$$CH_4 + H_2SO_4 \longrightarrow CH_3OH + H_2O + SO_2 \quad (1.9)$$

The ΔG of reaction (1.9) was estimated to be nearly 79 kJ mol^{-1} at 298 K based on the standard electrode potential [10]. That is, SO_4^{2-} is too weak of an oxidant for reaction (1.9) to proceed. Thus, the direct production of methanol by methane conversion via the intermediate M–CH_3 is considered to be difficult under mild conditions, while the strategy of converting methane to a methanol derivative is useful for avoiding the further oxidation of the product.

In this book, the direct formation of methanol from methane with oxygen (reaction (1.4)) is clearly distinguished from the formation of methanol derivatives such as methyl sulfate. The latter topic is discussed in further detail in Chapters 2–4.

1.4.2 C–C Bond Formation: Production of Higher Hydrocarbons

The formation of C–C bonds to produce higher hydrocarbons (C_2^+) can be classified into two further categories based on whether the C_2^+ hydrocarbons are produced in the presence or the absence of oxygen. In this book, the former is referred to as the oxidative coupling of methane, while reactions of the latter type are referred to as non-oxidative reactions of methane, and are further classified into three kinds of reactions as described below.

1.4.2.1 C–C Bond Formation via the Oxidative Coupling of Methane

The oxidative coupling of methane produces mainly ethane and ethylene, accompanied by the simultaneous production of H_2O. Reaction temperatures higher than 800 K and heterogeneous catalysts are typically used in such processes. The key reaction intermediate may be •CH_3, which is often generated on the surface of heterogeneous catalysts. For example, an EPR signal corresponding to •CH_3 was observed on a catalyst that was formed by subjecting MgO and Li_2CO_3 on an MgO support to pretreatment at 732 K under oxygen (Li/MgO) [55]. Further details are given in Chapter 5.

1.4.2.2 C–C Bond Formation via Non-Oxidative Reactions of Methane

The formation of C–C bonds to produce C_2^+ hydrocarbons can also proceed in the absence of oxygen. These non-oxidative reactions are considered to be dehydrogenation reactions, since they produce hydrogen and C_2^+ hydrocarbons simultaneously. In this book, non-oxidative reactions of methane are classified into the following three categories based on their products:

(1) Non-oxidative coupling of methane
(2) Dehydroaromatization of methane
(3) Methylation of hydrocarbons with methane

As mentioned above, hydrogen is commonly produced along with the C_2^+ hydrocarbons in these reactions. The formation of hydrogen makes the non-oxidative reactions less thermodynamically favorable than oxidative reactions such as the oxidative coupling of methane.

(1) Non-oxidative coupling of methane

In the non-oxidative coupling of methane, a heterogeneous or homogeneous catalyst generates $\bullet CH_3$ or $M\text{–}CH_3$ as the key reaction intermediate. Typically, ethane and ethylene are produced together with hydrogen. However, heterogeneous catalysts often yield aromatic hydrocarbons at high reaction temperatures. Further details are given in Chapter 5.

(2) Dehydroaromatization of methane

The dehydroaromatization of methane is typically carried out at around 900 K using transition metal ions such as Mo and W supported on and/or exchanged into zeolites [16, 19] or supported metal catalysts [56]. The main products are benzene and naphthalene, with lower molecular weight hydrocarbons such as ethylene also being formed as initial reaction products. The key reaction intermediates for C–C bond formation and the reaction mechanism for the formation of aromatic hydrocarbons are described in Chapter 6.

(3) Methylation of hydrocarbons with methane

C–C bond formation can also proceed via the reaction of methane with any hydrocarbon, including methane. For example, superacids can act as homogenous catalysts to catalyze the self-coupling of methane at near room temperature. Such reactions produce low-molecular-weight hydrocarbons in solution, such as C_2H_6 in the solution of $FSO_3H\text{–}SbF_5$. The hydrogen produced by the reaction escapes into the gas phase [57], as shown in reaction (1.10).

$$2\,CH_4 \longrightarrow C_2H_6 + H_2 \tag{1.10}$$

Although Ag^{+}-exchanged zeolites (Ag-zeolites) can catalyze the reaction of methane with alkenes (e.g., C_2H_4) and aromatic hydrocarbons (e.g., benzene), as shown in reactions (1.11) and (1.12), respectively, they cannot catalyze the self-coupling of methane shown in reaction (1.10) [58, 59].

$$CH_4 + C_2H_4 \longrightarrow C_3H_6 + H_2 \quad (1.11)$$

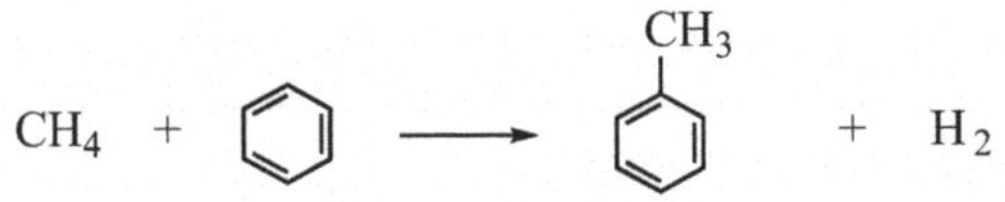

(1.12)

When Ag-zeolites were used as catalysts in reactions (1.11) and (1.12), the key reaction intermediate was the methyl carbenium ions ($^{+}CH_3$). Methylation of hydrocarbons with methane according to reactions (1.11) and (1.12) is more thermodynamically favorable than the self-coupling of methane to produce C_2H_6 and hydrogen according to reaction (1.10) [60].

Superacids such as HF-SbF_5 catalyze reaction (1.10) to produce $^{+}CH_3$ from methyl carbonium ions ($^{+}CH_5$) as a key reaction intermediate. However, when the reaction of C_2H_4 with CH_4 is carried out in the presence of superacids as catalysts, propane (C_3H_8) is produced, while propylene (C_3H_6) is not obtained, i.e., $^{+}CH_3$ is not a reaction intermediate in this case. Furthermore, superacids do not catalyze reaction (1.12), i.e., their catalytic properties are not always identical to those of Ag-zeolites, with further details provided in Chapter 7.

M–CH_3 and M–CH_2 (M: metal cation) are also key reaction intermediates in C–C bond formation. M–CH_3 takes part in the formation of C_2H_6 [22], while M–CH_2 induces metathesis reactions, such as the reaction shown in Eq. 1.13 [61].

$$CH_4 + C_3H_8 \longrightarrow 2\,C_2H_6 \quad (1.13)$$

Additionally, the oxidative methylation of ethylene [62] and aromatics [63, 64] by methane can proceed in the presence of oxygen. These reactions may proceed via $\bullet CH_3$ as a key reaction intermediate, since $\bullet CH_3$ was observed in the EPR spectra of CH_4 adsorbed on a Li_2O/MgO catalyst [62].

Further details of the methylation of hydrocarbons with methane are discussed in Chapter 7.

1.5 Types of Key Reaction Intermediates

As briefly discussed in the preceding sections, the generation of the key reaction intermediate by the cleavage of a C–H bond of methane or the insertion of metal ions into a C–H bond is crucial (Fig. 1.3). These key reaction intermediates are $\bullet CH_3$,

CH_x ($0 < x < 3$) including the carbene methylene ($:CH_2$), $^+CH_3$, M–CH_3, M–CH_2, and M–OCH_3. The key reaction intermediate depends on both the type of methane conversion reaction and the type of used catalyst.

On the other hand, the key reaction intermediates shown in Fig. 1.3 can be generated by methods other than the use of catalysts to accomplish the direct conversion of methane to chemicals. In particular, methane conversion can be achieved using an electric field [65–67] or plasma [68, 69]; the combination of these methods with catalysts [67–75] including photocatalysts [74, 75] is also effective.

It should be noted that methyl carbanions ($^-CH_3$) do not take part in the formation of C–O and C–H bonds to produce methanol or higher hydrocarbons in methane conversion. In other words, $^-CH_3$ is not an intermediate in the formation of C–O and C–H bonds in the methane conversion, even though in deuterium exchange reactions, the C–H bond of CH_4 is cleaved and H–D exchange occurs between D_2 and CH_4, as shown in reaction (1.14).

$$D_2 + CH_4 \longrightarrow CH_3D + HD \tag{1.14}$$

For example, deuterium exchange proceeds over strong base catalysts, such as KNH_2 supported on Al_2O_3 (KNH_2/Al_2O_3), and the reaction between C_2H_6 and D_2 over KNH_2/Al_2O_3 exhibits a faster exchange rate than that of CH_4 and D_2 [76]. These results indicate that the H–D exchange reaction between C_2H_6 with D_2 in the presence of a catalyst proceeds via $^-CH_3$ reaction intermediates. Prior to this exchange reaction, the cleavage of a C–H bond of CH_4 should occur.

However, $^-CH_3$ does not proceed to form C–O and C–H bonds in methane conversion. Thus, methanol and C_2^+ hydrocarbons are not produced via $^-CH_3$ reaction intermediates, and C–H bond cleavage does not always result in the formation of a new C–H bond. This means that the rate of the H–D exchange reaction does not always correspond to the activity of the catalyst in the methane conversion reaction.

1.6 Generation of Key Reaction Intermediates

As shown in Fig. 1.3, in order for the direct conversion of methane to proceed, key reaction intermediates must be generated from methane using various catalysts. However, a given catalyst may generate more than one kind of reaction intermediate. For example, CH_x ($0 < x < 3$) intermediates were generated over Mo-zeolites in dehydroaromatization [13, 16, 17], while some reaction intermediates with metal–carbon bonds have been observed in the reaction of metal ions with CH_4 [43, 77, 78]. These results demonstrate the importance of investigating the reaction intermediates responsible for the production of C–O and C–H bonds in the direct conversion of methane. Therefore, in this section, the methods by which various reaction intermediates are generated are briefly illustrated via some main examples; the individual reactions are further discussed in their respective chapters.

1.6.1 Methyl Radical (•CH_3) Formation

The methyl radical (•CH_3), which is generated by the hemolytic scission of a C–H bond in CH_4, is an important reaction intermediate for the production of methanol and/or C_2^+ hydrocarbons both in the presence and absence of oxygen, especially when heterogeneous or biological catalysts are used. For example, over metal oxides or metal–oxo groups, the formation of •CH_3 often proceeds as shown in reaction (1.15).

$$\text{M-O} + CH_4 \longrightarrow \text{M-OH} + \bullet CH_3 \tag{1.15}$$

Using biological catalysts such as methane monooxygenase, •CH_3 may be generated and then participate in methanol formation. This overall reaction is shown below.

$$CH_4 + O_2 + 2\,H^+ + 2\,e^- \longrightarrow CH_3OH + H_2O \tag{1.16}$$

In this reaction, one oxygen atom of the oxygen molecule is incorporated into the methanol molecule, while the other forms H_2O, and the supply of H^+ and e^- is essential for the production of methanol. The mechanism of biological catalysts seems to be essentially different from that of heterogeneous catalysts. Chapters 2 and 3 discuss the mechanisms of methanol formation via •CH_3 in biological and heterogeneous catalysis, respectively, and the different mechanisms of these catalysts are explained in terms of their properties.

1.6.2 Formation of Carbene-like Species CH_x ($0 < x < 3$) for Dehydroaromatization

Carbene-like species, CH_x ($0 < x < 3$), are key reaction intermediates in dehydroaromatization reactions involving metal ions, such as Mo and W ions loaded on or exchanged into various zeolites, as mentioned previously. For example, CH_x species can be generated at the pores and surface of a zeolite via the reaction of Mo_2O_3 with CH_4 followed by calcination; these species act as active sites. During the initial stage of reaction, ethylene is formed; the ethylene is then converted into aromatic hydrocarbons in a stepwise fashion. The details of this reaction and the properties of the catalysts are discussed in Chapter 6.

1.6.3 *Formation of Methyl Carbenium Ions ($^+CH_3$)*

Olah et al. reported that self-coupling of methane to produce C_2H_6 and hydrogen proceeds in the presence of superacid catalysts. Superacids such as FSO_3H-SbF_5 generate $^+CH_3$, which is expressed in the following reactions as $^+CH_5$.

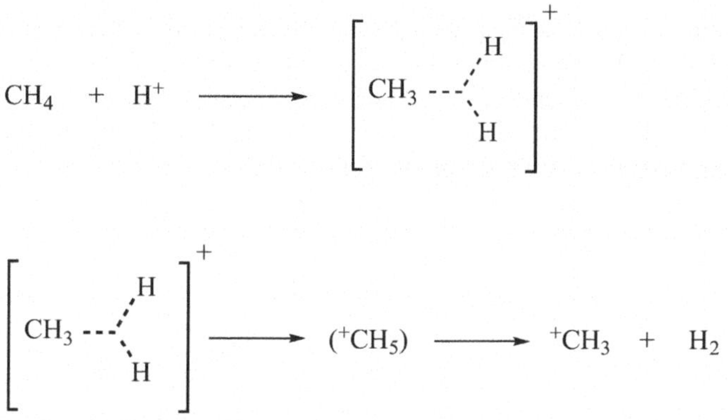

(1.17)

In superacid solutions, CH_4 is protonated to form the $^+CH_5$ ion. The $^+CH_5$ ion then undergoes deprotonation related to the loss of hydrogen to form the $^+CH_3$, which subsequently reacts with methane to form C_2H_6 as shown in reaction (1.18).

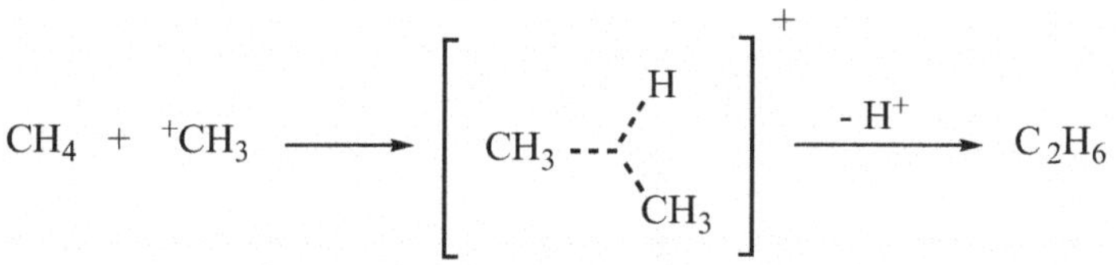

(1.18)

$^+CH_3$ is also generated by other catalysts such as Ag^+-exchanged zeolites and plays an important role in electrophilic reactions for C–C bond formation, as shown in reactions (1.11) and (1.12). This means that catalysts other than superacids possibly offer a potential means of C–C bond formation. Further details are discussed in Chapter 7.

1.6.4 *Classification of M–CH_3 Generation Methods*

As shown in Fig. 1.3, M–CH_3 reaction intermediates are mainly important to the production of methanol derivatives, not methanol. Among the various $^+CH_3$ intermediates, Pd–CH_3 species have been shown to take part in the formation of C–C bonds to produce C_2H_6 [22].

M–CH_3 formation has been aggressively researched, and many reviews have been published on this topic [27, 44, 79–83]. Labinger et al. conveniently classified M–CH_3 formation reactions into five types according to their overall stoichiometry [79]: Class 1: Oxidative addition, Class 2: σ-bond metathesis, Class 3: 1,2-addition, Class 4: Metalloradical activation, and Class 5: Electrophilic activation.

Class 1: Oxidative addition

Oxidative addition reactions are typical for electron-rich, low-valence complexes of the late transition metals, which are found toward the right side of the periodic table, such as Re, Fe, Ru, Os, In, and Pt. As shown in reaction (1.19), the species L_nM^x (L_n: ligand, x: number of valence electrons) is generated in situ by the thermal or photochemical decomposition of a suitable precursor.

$$L_nM^x + CH_4 \longrightarrow L_nM^{(x+2)}(H)(CH_3) \tag{1.19}$$

The reaction of methane with L_nM^x changes the oxidation state of the metal ions.

Class 2: σ-bond metathesis

Alkyl (R) or hydride complexes of early transition metals with a d^0 electric configuration may undergo the reversible reaction shown in reaction (1.20).

$$L_nM_x{-}R + CH_4 \longrightarrow \underset{\textbf{(I)}}{[L_nM\cdots CH_3\cdots H\cdots R]} \longrightarrow L_nM_x{-}CH_3 + RH \tag{1.20}$$

In the above reaction, R is an alkyl group or hydride. In the case of the hydride complexes (R = H), hydrogen (H_2) is eliminated. In this reaction, the intermediate **I** involves a four-center transition state with no change in the metal oxidation state.

Class 3: 1,2-addition

1,2-addition reactions differ from σ-bond metathesis (Class 2) reactions in that they involve the addition of CH_4 to a metal–nonmetal double bond, as shown in reaction (1.21).

$$L_nM^x{=}E + CH_4 \longrightarrow L_nM^x(CH_3){-}E(H) \tag{1.21}$$

Furthermore, the oxidation state of the metal ion is unchanged, which differentiates these reactions from oxidative addition (Class 1) reactions. For example, methane addition via a $Zr = N$ double bond has been reported [83].

$$(RNH)_3ZrR \xrightarrow[-RH]{\Delta} (RNH)_2Zr{=}NR \xrightarrow{CH_4} (RNH)_2Zr(CH_3)-N(H)-R \tag{1.22}$$

Class 4: Metalloradical activation
Sherry et al. [84, 85] reported that rhodium (II) porphyrin complexes exist in a monomer–dimer equilibrium and can reversibly cleave the C–H bond of CH_4 via the attachment of the monomeric rhodium (II)–porphyrin fragments, as shown in Eq. (1.23). Methane is the most reactive hydrocarbon in metalloradical activation.

$$\left[(Por)Rh^{II}\right]_2 \rightleftharpoons 2\,(Por)Rh^{II} \xrightleftharpoons{CH_4} (Por)Rh^{III}-CH_3 + (Por)Rh^{III}-H \tag{1.23}$$

Class 5: Electrophilic activation
In Ref. [86], certain reactions that lead directly to the production of functionalized alkanes from CH_4 rather than observable organometallic species are classified as electrophilic activation reactions. This kind of reaction is shown in reaction (1.24).

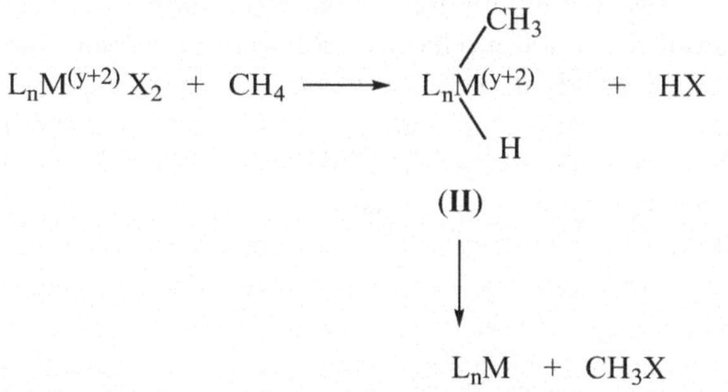

(1.24)

Above, $M^{(y+2)}$ is a late or a post-transition metal (Pd^{2+}, Pt^{2+} and/or Pt^{4+}, Hg^{2+}, and Tl^{3+}) in a strong acid such as H_2SO_4 [86]. The presumed reaction intermediates involved in the transformation are formed through the exchange of a given metal cation for a proton, which is called "electrophilic activation".

1.6.5 Formation of Metal Carbene Complexes (M–CH_2)

The remaining category of intermediates is metal carbene species (M–CH_2, where M is a metal ion, metal atom, or metal cluster), which are generated by the dehydrogenation of CH_4 by the M species, which is typically Ta, W, Ir, Pt, or Os [87, 88]. The intermediate M–CH_2 then takes part in C–C bond formation.

1.7 Why This Book Was Written and How It Is Structured

Many reports of methods for the conversion of methane to produce oxygenates, especially methanol and methanol derivatives, and higher hydrocarbons (C_2^+), such as C_2H_4, by controlling the generation of the reaction intermediates have been published over the past century; most of these methods involve the use of catalysts and careful control of reaction conditions. However, very few commercial processes for the direct conversion of methane have been developed. This is due to the fact that methane is the most inert of the hydrocarbons. CH_4 lacks low-energy empty orbitals and high-energy filled orbitals; thus, it does not readily participate in chemical reactions. Its negligible electron affinity, large ionization energy ($^+CH_3/CH_4 = 13.5 \pm 0.1$ eV), and high pK_a (40–46) also impede the reaction of CH_4. Furthermore, the C–H bond in CH_4 is stronger than those in the products; that is, products such as methanol and ethylene are more reactive than methane. Based on these chemical properties, the selectivity toward the reaction products would seem to be more important than the reactivity of CH_4, especially in oxidative reactions of methane.

As discussed at the beginning of this chapter, the development of processes based on the direct conversion of methane is essential from both an economic development and an energy security point of view. However, for scientists and engineers, the development of direct methane conversion as a next-generation process represents a "grand challenge" in chemistry independent of trends in the larger world, such as the discovery of shale gas and the drive to reduce carbon dioxide emissions.

In this book, the authors have consciously focused on providing various examples of the direct conversion of methane using heterogeneous, homogeneous, and biological catalysts that illustrate their mechanistic and functional aspects. The main aim is to provide an understanding of the way in which these catalysts and their reaction environments control the formation of reaction intermediates from methane and contribute to the formation of C–O or C–C bonds to produce methanol, methanol derivatives, or hydrocarbons via the direct conversion of methane.

To achieve the objectives mentioned above, this book consists of the following chapters.

Chapter 1: Overview of direct methane conversion to chemicals with C–O and C–C bonds

Chapter 2: Selective production of methanol from methane and molecular oxygen at atmospheric temperature and pressure using methane monooxygenases

Chapter 3: Heterogeneous and homogeneous catalytic partial oxidations of methane to methanol and its derivatives
Chapter 4: Application of biocatalysts for the production of methanol from methane
Chapter 5: C–C bond formation via the condensation of methane in the presence or absence of oxygen
Chapter 6: Conversion of methane to aromatic hydrocarbons
Chapter 7: C–C bond formation via carbocations in the methane conversion under non-oxidative conditions

References

1. BP Energy Economics (2018) BP Energy Outlook, 2018 edition. https://www.bp.com/en/global/corporate/energy-economics/energy-outlook.html
2. British Petroleum, BP (2018) Statistical review of world energy, June 2018, 67th edition. https://www.bp.com/en/global/corporate/energy-economics/statistical-review-of-world-energy.html
3. IEA (2018) World energy balance 2018
4. Bullin KA, Kroskop PE (2009) Compositional variety complicates processing plant for US shale gas. Oil Gas J 107:50–55
5. Schüeth F (2011) Chemical compounds for energy storage. Chemical Ingenieur Technik 83:1984–1993
6. Huang K, Miller JB, Huber GW, Dumescic JA, Maravelias CT (2018) A general framework for the evaluation of direct nonoxidative methane conversion strategies. Joule 2:349–365
7. Rostrup-Nielsen JR (2008) Organic reactions. In: Ertl G, Knözinger H, Schüth F, Weitkamp J (eds) Handbook of heterogeneous catalysis, vol 6. Wiley-VCH, Mannheim, p 2882
8. Li J, He Y, Tan L, Zhang P, Peng X, Oruganti A, Yang G, Abe H, Wang Y, Tsubaki N (2018) Integrated tunable synthesis of liquid fuels *via* Fischer-Tropsch technology. Nature Catal 1:787–793
9. Schwach P, Pan X, Bao X (2017) Direct conversion of methane to value-added chemicals over heterogeneous catalysts: challenges and prospects. Chem Rev 117:8497–8520
10. Gunsalus NJ, Koppaka A, Park SH, Bischot SM, Hashiguchi BG, Perianna RA (2017) Homogeneous functionalization of methane. Chem Rev 117:8521–8573
11. Wang B, Albarracín-Suazo S, Pagán-Torres Y, Nikolla E (2017) Advances in methane conversion processes. Catal Today 285:147–158
12. Ravi M, Ranocchiari M, von Bokhoven JA (2017) The direct catalytic oxidation of methane—a critical assessment. Angew Chem Int Ed 56:16464–16483
13. Olivos-Suarez AI, Szécsényi Á, Hensen EJM, Ruiz-Maritinez J, Pidko EA, Gascon J (2016) Strategies for the direct catalytic valorization of methane using heterogeneous catalysis: challenges and opportunities. ACS Catal 6:2965–2981
14. Karakaya C, Kee RJ (2016) Progress in the direct catalytic conversion of methane to fuels and chemicals. Prog Energy Combust Sci 55:60–97
15. Horn R, Schlögl R (2015) Methane activation by heterogeneous catalysis. Catal Lett 145:23–39
16. Vollmer I, Yarulina I, Kapateijn F, Gascon J (2019) Progress in developing a structure-activity relationship for the direct aromatization of methane. ChemCatChem 11:39–52
17. Gabrienko AA, Arzumanov SS, Tokarev AV, Danilova IG, Prosvirin IP, Kriventov VV, Zaikovski VI, Freude D, Stepanov AG (2017) Different efficiency on Zn^{2+} and ZnO species for methane activation on Zn-modified zeolites. ACS Catal 7:1818–1830
18. Abate S, Barbera K, Centi G, Lazafame P, Perathoner S (2016) Disruptive catalysis by zeolites. Catal Sci Technol 6:2485–2501
19. Ma S, Guo X, Zhao L, Scott S, Bao X (2013) Recent progress in methane dehydroaromatization: from laboratory curiosities to promising technology. J Energy Chem 22:10–20

20. Tang P, Zlau Q, Wu Z, Ma D (2014) Methane activation: the past and future. Energy Environ Sci 7:2580–2591
21. Shtenman AA (2016) Coordinational activation of methane and other alkanes by metal complexes. Russ Chem Bull Int Ed 65:1930–1944
22. Lotz MD, Remy MS, Lao DB, Ariafard A, Yates BF, Canty AJ, Mayer JM, Sanford MS (2014) Formation of ethane from mono-methyl palladium (II) complexes. J Am Chem Soc 136:8237–8242
23. Citek C, Herres-Pawlis S, Daniel T, Stack P (2015) Low temperature synthesis and reactivity of Cu_2O_2 models. Acc Chem Res 48:2424–2433
24. Perianna RA, Bhalla G, Tenn WT III, Yong KJH, Kiu XY, Mironov O, Jones CJ, Ziatdinov VR (2004) Perspective on some challenges and approaches for developing the next generation of selective low temperature, oxidation catalysts for alkane hydroxylation based on the C-H activation reaction. J Mol Catal A: Chem 220:7–25
25. Cavaliere VN, Mindiola DJ (2012) Methane: a new frontier in organometallic chemistry. Chem Sci 3:3356–3365
26. Schwarz H (2014) How and why do cluster size, charge state and ligands affect the course of metal mediated gas-phase activation of methane. Isr J Chem 54:1413–1431
27. Golisz SR, Gunnoe TB, Goddard WA III, Groves JT, Perianna RA (2011) Chemistry in the center for catalytic hydrocarbon functionalization: an energy frontier research center. Catal Lett 141:213–221
28. Olah GA, Surya Prakash GK, Sommer J (1985) Superacids. Wiley, Wiley Interscience Publication
29. Olah GA, Surya Prakash GK, Sommer J (1979) Superacids. Science 206:13–20
30. Olah GA (1973) Carbocations and electrophilic reactions. Angew Chem Inter Ed 12:173–254
31. Jansiewski AJ, Que L (2018) Dioxygen activation by nonheme diiron enzymes: diverse dioxygen adducts, high-valent intermediate, and related model complexes. Chem Rev 118:2554–2592
32. Zhang S, Karthikeyan R, Fernando SD (2017) Low-temperature biological activation of methane: structure, function and molecular interactions of soluble and particulate methane monooxygenases. Rev Environ Sci Biotechnol 16:611–623
33. Wang VCC, Maji S, Chen PPY, Lee HK, Yu SSF, Chan SI (2017) Alkane oxidation: methane monooxygenases, related enzymes, and their biomimetics. Chem Rev 117:8574–8621
34. Wang W, Liang AD, Lippard SJ (2015) Coupling oxygen consumption with hydrocarbon oxidation in bacterial multicomponent monooxygenases. Acc Chem Res 48:2632–2639
35. Haynes CA, Gonzalez R (2014) Rethinking biological activation of methane and conversion to liquid fuels. Nat Chem Biol 10:331–339
36. Sirajuddin S, Rosenzweig AC (2015) Enzymatic oxidation of methane. Biochemistry 54:2283–2294
37. Bjorck CE, Dobson PD, Pandhal J (2018) Biotechnological conversion of methane to methanol: evaluation of progress and potential. AIMS Bioeng 5:1–38
38. Xu J, Zheng A, Wang X, Qi G, Su J, Du J, Gan Z, Wu J, Wang W, Deng F (2012) Room temperature activation of methane over Zn modified H-ZSM-5 zeolites: insight from solid-state NMR and theoretical calculations. Chem Soc 3:2932–2940
39. Bango A, Saielli G (2011) Relative DFT calculations of the NMR properties and reactivity of transition metal σ-complexes insight on C-H bond activation. Phys Chem Chem Phys 13:4285–4291
40. Gabrienko AA, Arzumanov SS, Morzo IB, Toktarev AV, Wang W, Stepanov AG (2013) Methane activation and transformation on Ag/H-ZSM-5zeolite studied with solid-state NMR. J Phys Chem C 117:7690–7702

41. Yoshizawa K (2013) Quantum chemical studies on dioxygen activation and methane hydroxylation by diiron and dicopper species as well as related metal-oxo species. Bull Chem Soc Jpn 86:1083–1116
42. Latimer AA, Kulkarini AR, Aljama H, Montoya JH, Yoo JS, Tsai C, Abild-Pedersen F, Studt F, Norskov JK (2017) Understanding trends in C-H bond activation in heterogeneous catalysis. Nat Mater 16:225–229
43. Armentrout PB (2017) Methane activation by 5d transition metals: energetics mechanism and periodic trends. Chem Eur J 23:10–18
44. Ess DH, Goddard WA III, Perianna RA (2010) Electric, ambiphilic and nucleophilic C-H activation through transition-state and reaction pathway interaction energy decompositions. Organometallics 29:6459–6472
45. Cruellas A, Melchiori T, Gallucci F, van Sint Annaland M (2017) Advanced reactor concept for oxidative coupling of methane. Catal Rev Sci Technol 59:234–294
46. Mohammad Y, Penlidis A (2018) Optimization in chemical reaction engineering: oxidative coupling of methane as a case study. Ind Eng Chem Res 57:8664–8678
47. Kondratenko EV, Peppel T, Seeburg D, Kobndratenko VA, Kalevaru N, Martin A, Wohlrab S (2017) Catal Sci Technol 7:366–381
48. Fajardo CAG, Niznansky D, N'Guyen Y, Courson C, Roger AC (2008) Methane selective oxidation to formaldehyde with Fe-catalysts supported on silica or incorporated into the support. Catal Commun 9:864–869
49. Zuo Z, Ramírez PJ, Senanayake SD, Liu P, Rodriguez JA (2016) Low-temperature conversion of methane to methanol on CeO_x/Cu_2O catalysts: water controlled activation of the C-H Bond. J Am Chem Soc 138:13810–13813
50. Lustemberg PG, Palomino RM, Gutiérrez RA, Grinter DC, Vorokhta M, Liu Z, Ramírez PJ, Matolín V, Ganduglia-Pirovano MV, Senanayake SD, Rodriguez JA (2018) Direct conversion of methane to methanol on Ni-Ceria surfaces: metal–support interactions and water-enabled catalytic conversion by site blocking. J Am Chem Soc 140:7681–7687
51. Narsimhan K, Iyoki K, Dinh K, Román-Leshkov Y (2016) Catalytic oxidation of methane into methanol over copper-exchanged zeolites with oxygen at low temperature. ACS Cent Sci 2:424–429
52. Liu HF, Liu RS, Liew KY, Jonson RE, Lusnford JH (1984) Partial oxidation of methane by nitrous oxide over molybdenum on silica. J Am Chem Soc 106:4117–4121
53. Najafian A, Cundari TR (2017) Methane C-H activation via 3d metal oxide complexes with potentially redox-noninnocent pincer ligands: density functional theory study. Inorg Chem 56:12282–12290
54. Najafian A, Cundari TR (2018) C-H activation methane by nickel methoxide complexes: A density functional theory study. Organometallics 37:3111–3121
55. Driscoll DJ, Martir W, Wang JW, Lusnford JH (1985) Formation of gas-phase methyl radicals over MgO. J Am Chem Soc 107:58–63
56. Grecker D, Motagamwala AH, Rivera-Dones KR, Miller JB, Hiuber GW, Mavrikakis M, Dumesic JA (2017) Methane conversion to ethylene and aromatics on PtSn catalysts. ACS Catal 7:2088–2100
57. Olah GA, Sclosberg RH (1968) Chemistry in super acids. I. Hydrogen exchange and polycondensation of methane and alkane in FSO_3H-SbF_5 ("magic acid") solution. Protonation of alkanes and the intermediacy of CH_5^+ and related hydrocarbon ions. The high chemical reactivity of "paraffins" in ionic solution reaction. J Am Chem Soc 90:2726–2727
58. Baba T, Inazu K (2006) Heterolytic dissociation of C-H bond of methane over Ag+-exchanged zeolites and conversion of methane into higher hydrocarbons in the presence of ethylene or benzene. Chem Lett 35:142–147
59. Baba T, Komatsu N, Sawada H, Yamaguchi Y, Takahashi T, Sugisawa H, Ono Y (1999) ^1H magic angle spinning NMR evidence for dissociate adsorption of hydrogen on Ag^+-exchanged A- and Y-zeolites. Langmuir 15:7894–7896
60. Guo J, Lou H, Zheng X (2009) Energy-efficient coaromatization of methane and propane. J Natural Gas Chem 18:260–272

61. Soulivong D, Copéret C, Thivolle-Cazat J, Basset JM (2004) Cross-metathesis of propane and methane: a catalytic reaction of C-C bond cleavage of a higher alkane by methane. Angew Chem Int Ed 43:5366–5369
62. Lunsford JH, Qiu P, Rosynek MP, Yu Z (1998) Catalytic conversion methane and ethylene to propylene. J Phys Chem B 102:167–173
63. Adebajo MO (2007) Green chemistry perspectives of methane conversion *via* oxidative methylation of aromatics. Green Chem 9:526–539
64. Kim H, Suh HM, Paik H (1992) Oxidative methylation of toluene with methane over lead-lithum-magnesium oxide catalysts. Appl Catal A General 87:115–127
65. Stoukides M (1995) Electrochemical studies of methane activation. J Appl Electrochem 25:899–912
66. Spinner N, Mustain WE (2013) Electrochemical methane activation and conversion to oxygenates at room temperature. ECS Trans 53:1–20
67. Arnason L, Schmidt PS, Pandey M, Bagger A, Thygesen KS, Steprens IEL, Rossmeisl J (2018) Fundamental limitation of electrocatalytic methane conversion to methanol. Phys Chem Chem Phys 20:11152–11159
68. Minea T, vanden Bekerom DCM, Peeters FJJ, Zoethout E, Graswinckel MF, van de Sanden MCM, Cents T, Lefferts, van Rooij LGJ (2018) Non-oxidative methane coupling to C_2 hydrocarbons in a microwave plasma reactor. Plasma Process Polym 15:e1800087 (1–16)
69. Yang Y (2002) Methane conversion and reforming by nonthermal plasma on pins. Ind Eng Chem Res 41:5198–5926
70. Ogo S, Sekine Y (2017) Catalytic reaction assisted by plasma or electronic field. Chem Rec 17:726–738
71. Ogo S, Nakatsubo H, Iwasaki K, Sato A, Murakami K, Yabe T, Ishikawa A, Nakai H, Seine Y (2018) Electron-hopping brings lattice strain and high catalytic activity in the low-temperature oxidative coupling of methane in an electric filed. J Phys Chem C 122:2089–2096
72. Delikonsntis E, Scapinello M, Stefanidis GD (2018) Low energy cost conversion of methane to ethylene in a hybrid plasma-catalytic reactor system. Fuel Process Technol 176:33–42
73. Park S, Lee M, Bae J, Hong DY, Park YH, Hwang YK, Jeong MG, Kim YD (2017) Plasma-assisted no-oxidative conversion of methane over Mo/H-ZSM-5 catalyst in DBP reactor. Top Catal 60:735–742
74. Baltrusaitis J, Jansen I, Christus JDS (2014) Renewable energy based catalytic CH_4 conversion to fuels. Catal Sci Technol 4:2397–2411
75. Shimura S, Yoshida H (2014) Semiconductor photocatalysts for non-oxidative coupling, dry reforming, and steam reforming of methane. Catal Surv Asia 18:24–33
76. Handa H, Baba T, Ono Y (1998) H-D exchange between methane and deuteriated potassium amide supported on alumina. J Chem Soc, Faraday Trans 94:451–454
77. Eller K, Schwarz H (1991) Organometallic chemistry in the gas phase. Chem Rev 91:1121–1177
78. Schwarz H, Shaik S, Li J (2017) Electronic effect on room-temperatures, Gas-phase C-H bond activation by cluster oxides and metal carbide: the methane challenges. J Am Chem Soc 139:17201–17212
79. Labinger JA, Bercaw JE (2002) Understanding and exploiting C-H bond activation. Nature 417:507–514
80. Armold PL, Mcnullon MW, Rieb Kühn JFE (2015) C-H bond activation by f-block complexes. Angew Chem Int Ed 54:82–100
81. Cavaliere VN, Wicker BF, Mindiola DJ (2012) Homogeneous organometallic chemistry of methane. Adv Organomet Chem 60:1–47
82. Schwarz H, Schröder D (2000) Concepts of metal-mediated methane functionalization. An interaction of experiment and theory. Pure Appl Chem 72:2319–2332
83. Cummins CC, Baxter SM, Wolczanski PT (1988) Methane and benzene activation *via* transient $(^{t}BuSiNH)_2Zr = NSi^{t}Bu_3$. J Am Chem Soc 110:8731–8733
84. Sherry AE, Wayland BB (1990) Metalloradical activation of methane. J Am Chem Soc 112:1259–1261

85. Wayland BB, Sherry AE (1991) Activation of methane and toluene by rhodium (II) porphyrin complexes. J Am Chem Soc 113:5305–5311
86. Stahl SS, Labinger JA, Bercaw JE (1998) Homogeneous oxidation of alkanes by electrophilic late transition metals. Angew Chem Int Ed 37:2180–2192
87. Zhou S, Li J, Schlangen M, Schwarz H (2016) Bond activation by metal-carbene complexes in the gas phase. Acc Chem Res 49:494–502
88. Lapoutre VJF, Redlich B, van der Meer AFG, Oomens J, Bakker JM, Sweeney A, Mookherjee A, Armentrout PB (2013) Structure of the dehydrogenation products of methane activation by 5d transition metal cations. J Phys Chem A 117:4115–4126

Chapter 2
Selective Production of Methanol from Methane and Molecular Oxygen at Atmospheric Temperature and Pressure Using Methane Monooxygenases

2.1 Introduction

Methane monooxygenase (MMO; EC 1.14.13.25) is an enzyme that is biosynthesized by methane-oxidizing bacteria (methane-utilizing bacteria), which are microorganisms that utilize methane as a carbon and energy source [1, 2]. In bacterial cells, MMO oxidizes methane to methanol using molecular oxygen and electron donors at atmospheric temperature and pressure (20–40 °C, 1 bar) as follows [1, 2]:

$$CH_4 + O_2 + e^- + H^+ \rightarrow CH_3OH + H_2O \quad (2.1)$$

Methane conversion via MMO is attractive as a methane-utilization technique. Firstly, MMO converts methane into methanol at atmospheric temperature and pressure. In principle, the cleavage of a C–H bond in methane requires a large amount of energy, *i.e.*, high-temperature conditions, because of the high energy of homolytic C–H bond dissociation (435 kJ mol^{-1}). For example, the production of synthesis gas (carbon monoxide and hydrogen) from methane requires temperatures of 850 °C or higher in the presence of a nickel catalyst [2, 3]. The ability of MMO to cleave the C–H bond of methane under mild conditions is believed to originate from the generation of extremely reactive oxygen species at the catalytic site of MMO. Therefore, the structure and properties of the catalytic site in MMO and the active oxygen species involved in the catalysis have been investigated.

The other attractive feature of methane conversion via MMO is its 100% selectivity toward methanol; the produced methanol is not further oxidized by the reactive oxygen species. As mentioned in the introduction, methanol is more reactive than methane itself. The C–H bond energy of methanol is 393–402 kJ mol^{-1} [4]; as a result, the C–H bond of methanol is typically cleaved under methane activation conditions. That is, in circumstances in which methanol can be produced via the oxidation of methane, further oxidation of the produced methanol would also be expected to occur. Thus, the selective production of methanol by MMO indicates that MMO can control the oxidation reaction in some way. Generally, the protein

T. Baba and A. Miyaji, *Catalysis and the Mechanism of Methane Conversion to Chemicals*,
https://doi.org/10.1007/978-981-15-4132-2_2

structure of an enzyme plays a role in the control of enzymatic reactions. Therefore, to understand the mechanism of the conversion of methane to methanol via MMO, not only the catalytic site of MMO, but also its protein structure and dynamics, must be understood. To this end, Extensive efforts have been made to obtain the protein structure of MMO using X-ray crystallography and NMR analysis.
To analyze the protein structure and function of MMO, it has been isolated from the bacterial cells of methane-oxidizing bacteria. Two types of MMO have been isolated: soluble MMO (sMMO) and membrane-bound MMO (particulate MMO, pMMO) [1, 2]. sMMO is located in the cytoplasm of methane-oxidizing bacteria, while pMMO is embedded in the cellular membrane. sMMO and pMMO have completely different protein structures, but both catalyze the reaction shown in reaction (2.1).

sMMO has been extensively studied, as it is relatively stable under conditions other from those of biological surroundings (hereafter referred to as in vitro conditions); studies of sMMO are steadily approaching the elucidation of its mechanism of methane oxidation. On the other hand, pMMO is very unstable in the presence of molecular oxygen and at room temperature [5–7]. This instability complicates its structural and functional analysis, and has led to inconsistent experimental results. Although the methane oxidation mechanism of pMMO remains less well understood than that of sMMO, the progress in the investigation of pMMO in the twenty-first century has been remarkable.

This chapter describes the catalytic cycles for the conversion of methane to methanol by the two MMOs, as well as structural studies of the two MMOs. These catalytic cycles involve tremendously complex molecular mechanisms. Each enzyme undergoes changes in its protein scaffold in order to interact with four substrates (methane, molecular oxygen, electrons, and protons), deliver the substrates to the catalytic site, and control each stage of the catalytic cycle. Thus, some knowledge of the protein structure and dynamics of these enzymes is necessary to understand their reaction mechanisms.

2.2 sMMO

2.2.1 Overall Protein Structure

The protein structure and reaction mechanism of sMMO have been thoroughly studied using sMMO from two bacterial strains, *Methylococcus capsulatus* (Bath) and *Methylosinus trichosporium* OB3b. Although the protein structure and reaction behavior of the two sMMOs are almost the same, some differences have been found.

sMMO is a protein complex composed of three proteins [8, 9]: Hydroxylase (MMOH), reductase (MMOR), and component B (MMOB). All three proteins are required for efficient methane oxidation. The structures of these three proteins are shown in Fig. 2.1, and their structural properties are briefly summarized in Table 2.1.

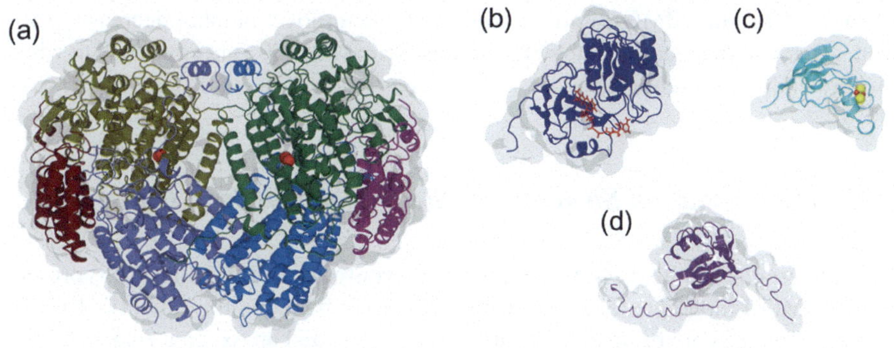

Fig. 2.1 Protein structures of MMOH (PDB ID: 1MTY) (**a**) ferredoxin domain of MMOR (PDB ID: 1TVC) (**b**) 2Fe-2S domain of MMOR (PDB ID:1JQ4) (**c**) and MMOB (PDB ID: 1CKV) (**d**) from *Methylococcus capsulatus* (Bath)

Table 2.1 Structural properties of the sMMO components (*Methylococcus capsulatus* (Bath))

	MMOH	MMOR	MMOB
Protein quaternary structure	Homodimer	Single peptide	Single peptide
Subunits of monomer	α, β, γ	–	–
Molecular weight	α: 60,600	~39,000	~16,000
	β: 45,100		
	γ: 19,800		
Cofactor	2 Fe	[2Fe-2S]	No cofactor
		FAD	

These proteins are water-soluble with polar amino acid side chains distributed on their surfaces.

MMOH (Fig. 2.1a) exists as a dimer, with a monomeric unit consisting of three subunits (α, β, and γ). The di-nuclear iron center present in the α subunit is the catalytic site for the oxidation of methane to methanol [10]. The iron atoms of this center are coordinated by the nitrogen atoms of histidine and glutamic acid residues and the oxygen atom of a water molecule.

MMOR (Fig. 2.1b) plays a role in the transfer of electrons from nicotinamide adenine dinucleotide (NADH) to MMOH [11]. MMOR has two domains: A FAD-binding domain and a ferredoxin domain, which contains a [2Fe–2S] cluster. The former domain accepts electrons from NADH, which are subsequently transferred to the ferredoxin domain. The ferredoxin domain donates the electrons to the di-nuclear iron center of MMOH.

MMOB (Fig. 2.1c) is a small protein compared to MMOH and MMOR, but plays an important role in the oxidation of methane to methanol at the catalytic site of

MMOH. When the MMOB unit was removed, the turnover number decreased 150-fold, and the initial reaction rate with oxygen decreased 1000-fold compared to when this protein was present [12].

2.2.2 Coordination Structure of the Di-Nuclear Iron Cluster

Crystallographic studies of MMOH indicate that only one iron-binding site is present in the MMOH monomer, and that two iron ions are bound at this site [10, 13–15]. The two iron ions are bridged by a carboxylate group and an oxygen. As mentioned above, this di-nuclear iron site is the catalytic center for the oxidation of methane to methanol.

The crystal structures of the reduced and oxidized forms of MMOH from *Methylococcus capsulatus* (Bath) (Fig. 2.2a, b) [13] and the oxidized form of MMOH from *Methylosinus trichosporium* [16] have been reported. The overall structure of the two enzymes is very similar. The di-nuclear iron cluster is located in the α-subunit, approximately 12 Å inside the protein surface; thus, it cannot directly access the bulk water [10, 16]. The amino acid residues coordinated to the di-nuclear iron cluster are identical or highly conserved between the two strains, and the coordination structures can be superimposed.

The reduced form of MMOH can exhibit various conformations, which demonstrates the flexibility of the protein structure and the diversity in the number of water molecules coordinated to the iron ions. In the structure of the reduced form of MMOH from *Methylococcus capsulatus* (Bath) (Fig. 2.2a), two carboxylates from Glu bridge the iron ions: a μ-1,3 carboxylate from Glu144 and a μ-1,1 carboxylate from Glu243 (Glu243 coordinates Fe_2 in a bidentate manner) [13]. His246 and monodentate Glu209 coordinate with one of the iron ions. The other iron ion is coordinated by His147, Glu114 (in a monodentate fashion), and at least one water

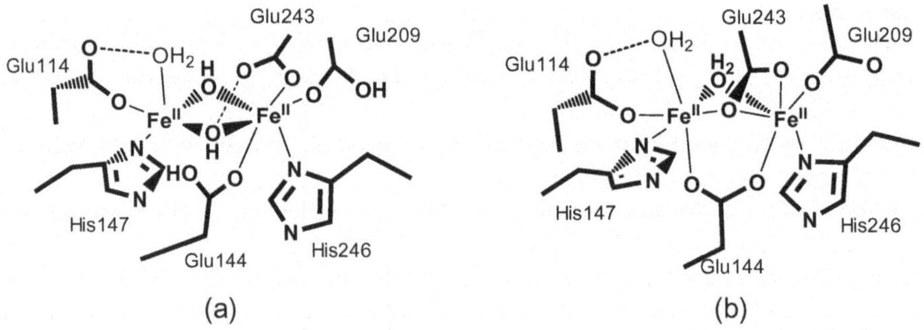

Fig. 2.2 Coordination structure of the di-iron site of MMOH. **a** $MMOH_{ox}$ from *Methylococcus capsulatus* (Bath) (PDB ID 1MTY) **b** $MMOH_{red}$ from *Methylococcus capsulatus* (Bath) (PDB ID 1FYZ)

molecule to complete the di-iron coordination sphere. Two water molecules are coordinated in another conformation of reduced MMOH, in which one water molecule bridges to iron ion. The pentacoordinate geometry of both iron ions was supported by magnetic circular dichroism (MCD) spectroscopy of MMOH [17, 18]. On the other hand, in the crystal structure of oxidized MMOH (Fig. 2.2b) [13, 16], both iron atoms exhibit hexacoordinate geometry and are bridged by a μ-1,3 carboxylate from Glu144, by three or by one hydroxo and one water molecule.

2.2.3 Catalytic Cycle of the Oxidation of Methane to Methanol

The catalytic cycle of oxygen activation followed by methane hydroxylation via MMOH is shown in Fig. 2.3. Several intermediates formed during sMMO catalysis

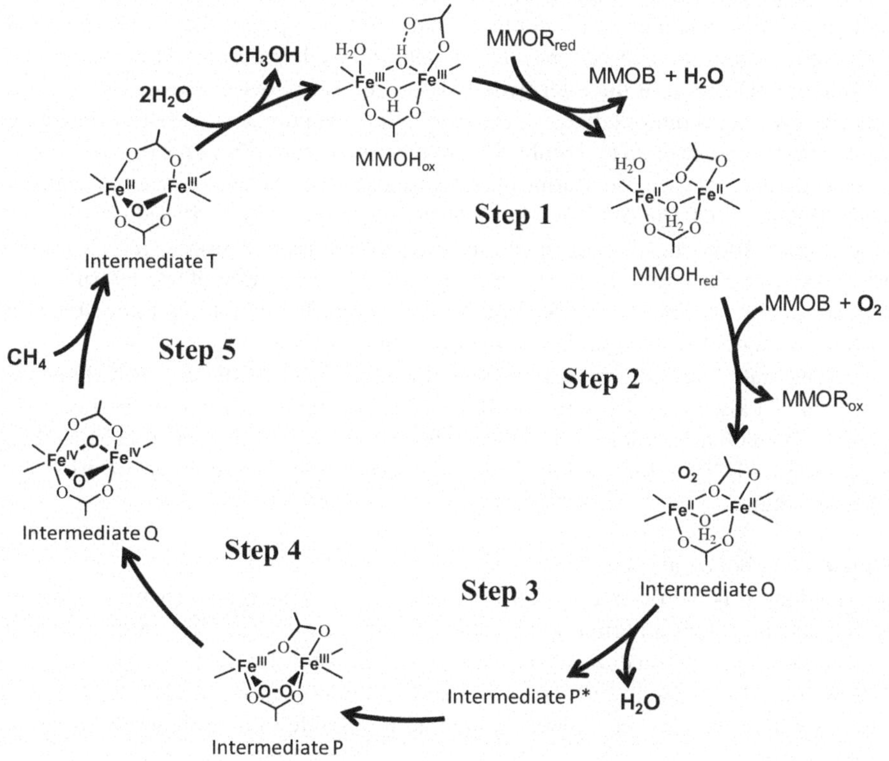

Fig. 2.3 Catalytic cycle for the oxidation of methane to methanol at the di-iron site of MMOH using molecular oxygen or hydrogen peroxide as the oxidant

(Fig. 2.3) have been characterized using various kinetic and spectroscopic methods [19–24], from which the following catalytic cycle has been developed.

Step 1: Reduction of MMOH
The oxidized form of MMOH ($MMOH_{ox}$), in which the oxidation state of the di-iron center is $Fe^{II}Fe^{II}$, accepts two electrons from NADH via MMOR to give $MMOH_{red}$, which has a high-spin weakly ferromagnetically coupled $Fe^{II}Fe^{II}$ center. Only $MMOH_{red}$ can react with molecular oxygen to initiate the catalytic cycle.

Step 2: Binding of molecular oxygen to MMOH
In the presence of two equivalents of MMOB, $MMOH_{red}$ reacts with molecular oxygen to form intermediate O. In this intermediate, molecular oxygen binds to MMOH, but not directly to the di-nuclear iron center.

Step 3: Formation of the peroxo intermediates
Intermediate O transforms into intermediate P*, which has a peroxo di-iron center. The binding mode of O_2 to the di-nuclear iron center is unknown, and the oxidation state of the di-iron center is still controversial. $Fe^{III}Fe^{III}$ species that differ in ligand arrangement from the subsequent intermediate P have been found. However, when the rate of conversion of the intermediate P* into P was slowed, an $Fe^{II}Fe^{II}$ species was observed. This intermediate is proposed to form via replacement of the bridging H_2O ligand by the oxygen molecule without electron transfer from the di-iron center.

Intermediate P* then transforms into a peroxo di-iron species referred to as intermediate P, in which the oxidation state of the two irons is Fe^{III}. Based on spectroscopic observations and density functional theory (DFT) calculations, a μ-1,2-peroxo bridge between the two irons has been suggested. The formation of this intermediate depends on the pH, and shows a kinetic solvent isotope effect in D_2O, indicating that proton transfer is involved in the formation of intermediate P.

Intermediate P can also be generated by the reaction of $MMOH_{ox}$ with H_2O_2 via a mechanism known as the peroxide shunt. The apparent K_m value of the peroxide shunt is about 66 mM, while that of the reaction of molecular oxygen with $MMOH_{ox}$ is about 3 μM. Therefore, a high concentration of H_2O_2 is required to drive the peroxide shunt.

Step 4: O–O bond cleavage
Intermediate P is then converted into intermediate Q. This transformation depends on the pH and shows a normal kinetic solvent isotope effect in D_2O, indicating that the formation of intermediate Q requires proton transfer. Intermediate Q exhibits a yellow color due to its absorption at 430 and 330 nm ($\varepsilon_{430,\ 330} = 7500\ M^{-1}\ cm^{-1}$) and has a lifetime of several seconds, making it possible to investigate its structure and properties.

Step 5: Reaction with methane
Intermediate Q reacts with methane, resulting in C–H bond cleavage and in turn, the formation of methanol. The structure and properties of this intermediate have been extensively investigated, and are described in detail in Sect. 2.2.4.

After the oxidation of methane, intermediate Q is converted to intermediate T, which is a complex of MMOH and the product. In this intermediate, the di-iron center is proposed to have a mono-μ-oxo bridge. MMOH returns to its resting state ($MMOH_{ox}$) via the release of methanol from the intermediate.

All of the transformations in the catalytic cycle above depend on the molecular dynamics of the MMOH protein. The formation of $MMOH_{red}$ from $MMOH_{ox}$ via the electrons supplied by NADH involves interaction between the MMOH and MMOR proteins. The activation of molecular oxygen occurs through the transformation from $MMOH_{red}$ to intermediate Q; interaction between the MMOH and MMOB proteins is also involved in this transformation. Methane activation and methanol production occur via reaction with intermediate Q. The release of methanol from MMOH occurs through the transformation of intermediate T to $MMOH_{ox}$. Clearly, the mechanism of the oxidation of methane to methanol by sMMO cannot be understood without understanding the protein structure of sMMO and its dynamics.

2.2.4 Intermediate Q: Methane Activation and Methanol Production

As shown in the catalytic cycle, intermediate Q is an important reaction intermediate that is responsible for the cleavage of the C–H bond of methane. Thus, its geometrical and electronic structures have been studied in detail.

(1) Structure of intermediate Q

The iron atoms of intermediate Q are believed to form a $Fe_2^{IV}(\mu - O)_2$ diamond core structure as shown in Fig. 2.4, although, as will be discussed later, the structure of this intermediate is still controversial.

This structure has been determined and supported mainly using the following analyses.

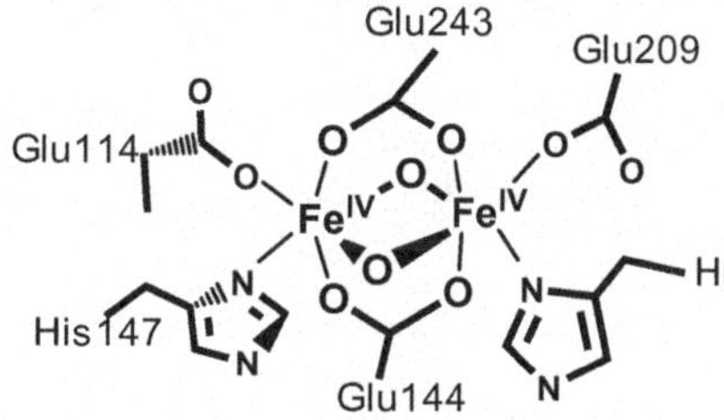

Fig. 2.4 Proposed $Fe_2^{IV}(\mu - O)_2$ diamond core structure

(a) Mössbauer spectroscopic study [25]: Mössbauer analysis revealed that the two iron atoms were high-spin Fe^{IV} ions and antiferromagnetically coupled in nearly identical environments.
(b) Extended X-ray absorption fine structure (EXAFS) analysis [26, 27]: A sample of intermediate Q was prepared by rapid freeze quenching. The results showed that the two Fe–O bonds were asymmetric (1.77 and 2.05 Å). Both Fe–O bonds were longer than typical terminal $Fe^{IV} = O$ bonds (about 1.65 Å). The distance between the two iron ions was short (2.46 Å) The coordination number of the two iron ions was likely five or less.
(c) Time-resolved resonance Raman spectroscopy at 4 °C [28]: The results revealed that the two equivalent oxygen atoms in the diamond core of intermediate Q originated from molecular oxygen, providing support for the homolytic cleavage of the O–O bond of molecular oxygen. Additionally, the vibration modes of Fe = O and Fe-OH were not observed, although theoretical calculations of these frequencies have not been conducted yet.

However, the structure of the diamond core remains controversial, as the spectroscopic results are not totally consistent with some computational chemistry and coordination chemistry results. First, the short distance between the two iron atoms in the diamond core structure (Fig. 2.5a) has not been reproduced in various computational studies. In these computational studies, the optimized diamond core structures have iron–iron distances of 2.6–2.8 Å [29–35], suggesting that the short distance observed spectroscopically does not correspond to an optimum structure. The theoretical results are consistent with the experimental geometry of a model complex of the diamond core, $[Fe_2(\mu\text{-}O)_2(TPA^*)_2](ClO_4)_4$ (TPA*: tris(3,5-dimethyl-4-methoxypyridyl-2-methyl)amine), which has an iron–iron distance of 2.72 Å [36]. Second, the model complex of the diamond core is insufficiently reactive to abstract hydrogen atoms from methane and isomerizes to the open ring form (Fig. 2.5b).

In view of this ongoing debate, a study using FeKα high energy resolution fluorescence detection X-ray absorption spectroscopy (HERFD XAS), which has 2–4 times greater experimental resolution compared to standard XAS, was recently carried out [37]. The HERFD XAS analyses could differentiate the HERFD XAS pre-edge energies and strengths of the high-valent diamond core $\left[Fe_2^{IV}(\mu - O)_2(TPA*)_2\right](ClO_4)_4$ di-iron model complex and the open-core $[(O = Fe^{IV}\text{-}O\text{-}Fe^{IV}(OH)(TPA^*)_2](ClO_4)_3$. The data for intermediate Q in sMMO were the same as those of the open-core structure model complex, suggesting that the intermediate Q can be rationalized by invoking an open-core structure with a terminal $Fe^{IV} = O$ motif. This analysis did

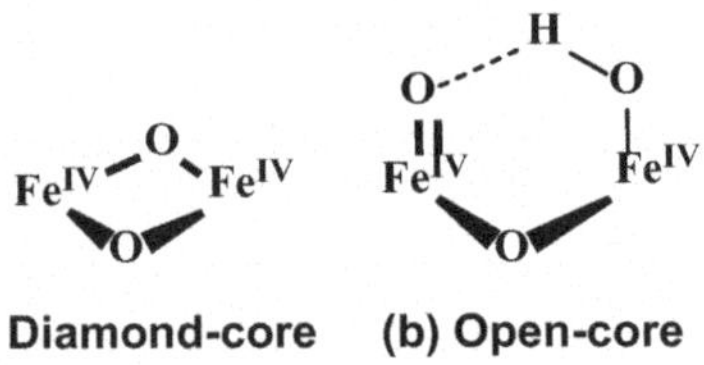

Fig. 2.5 Diamond-core and open-core structures of the di-iron center

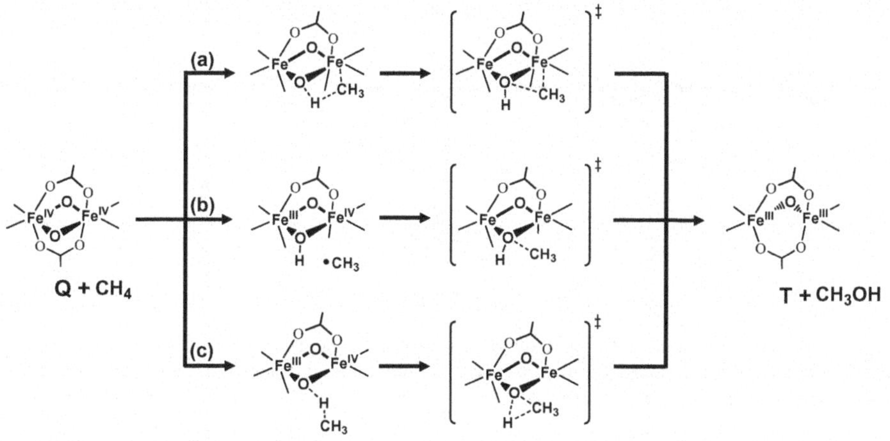

Fig. 2.6 Three proposed mechanisms for the transformation of methane to methanol at the diamond core di-nuclear iron center of sMMO: **a** Non-radical mechanism, **b** asynchronous mechanism, and **c** radical mechanism

not consider the regulation of the core structure by the protein environment. Thus, additional experimental and theoretical studies to confirm the conformation of the active site of intermediate Q are needed.

(2) Mechanism of the reaction of intermediate Q with methane

Based on various approaches, such as analysis of the enantioselectivity of the product using chiral alkanes [38, 39] and radical clock substrates [40–43], measurement of kinetic isotope effects [21, 44–46], and theoretical calculations [35, 47–51], several mechanisms for the reaction between methane and intermediate Q have been proposed [52, 53]: A radical mechanism, a non-radical mechanism, and an asynchronous concerted mechanism, as shown in Fig. 2.6. However, the experimental results have not been able to provide conclusive data regarding the reaction mechanism.

Theoretical chemistry has provided great insight into the details of these reaction mechanisms. Most theoretical results have supported the radical mechanism, although some have supported the non-radical mechanism. A representative reaction mechanism is shown in Fig. 2.7 [35].

The reaction mechanism was developed using the model structure shown in Fig. 2.7 (Q), which was constructed based on the crystal structure of $MMOH_{red}$. This putative species binds a methane molecule and abstracts a hydrogen atom from methane with a barrier of 60 kJ mol^{-1} to produce a methyl radical and an OH group. The methyl radical rebinds to the OH group via a CH_3–M–OH transition structure to form a methanol molecule with a barrier of 35 kJ mol^{-1}. Initially, the high-spin antiferromagnetic state is favorable, while the high-spin ferromagnetic spin state is more favorable for the intermediate. The shift from an antiferromagnetic state to

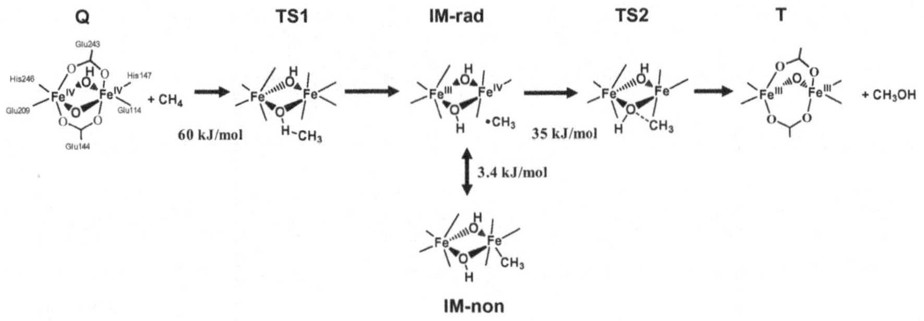

Fig. 2.7 Reaction mechanism for the transformation of methane into methanol at the diamond core di-nuclear iron center of sMMO as determined using DFT calculations taking the spin state of the iron ions into account

a ferromagnetic state was also observed in a previous study, which reported that this so-called spin-crossover makes the reaction favorable. The overall reaction of methane with the diamond core species is exothermic, and the activation barriers are similar to those determined experimentally (58–71 kJ mol^{-1}).

However, computational chemistry research is limited by the fact that the theoretical results depend on the molecular model used for the calculation and the calculation methods. Calculations based on the protein structures of MMOH in the various states described above would be expected to elucidate the reaction mechanism with higher accuracy.

In summary, despite great progress toward its elucidation, the structure of intermediate Q remains controversial, and is still being revised. Therefore, the associated reaction mechanisms will likely be revised as well in the future.

2.2.5 *Binding of the Substrate to MMOH*

The di-nuclear iron site exists within the protein MMOH. Therefore, the reaction cannot proceed unless oxygen molecules and methane molecules can reach the di-nuclear iron site from outside MMOH. Here, the binding site for the reactive molecule, which is located close to the di-nuclear iron site, will be described.

The crystallographic structures of various sMMO hydroxylases that bind analogs of the products and substrates have been clarified. Through these structural studies, three hydrophobic cavities, which are referred to as cavities 1–3, were found in the α subunit of MMOH (Fig. 2.8) [54–56]. These cavities are considered to form a route by which methane and oxygen molecules can access the di-iron center of MMOH.

Cavity 1 is the space adjacent to the di-iron center of MMOH. In $MMOH_{ox}$, this cavity is connected to a pore region that opens to the surface of MMOH. The pore region, which is defined by the residues Glu 240, Thr 213, and Asn 214, is the shortest access route from the protein surface to the di-iron center. This pore has

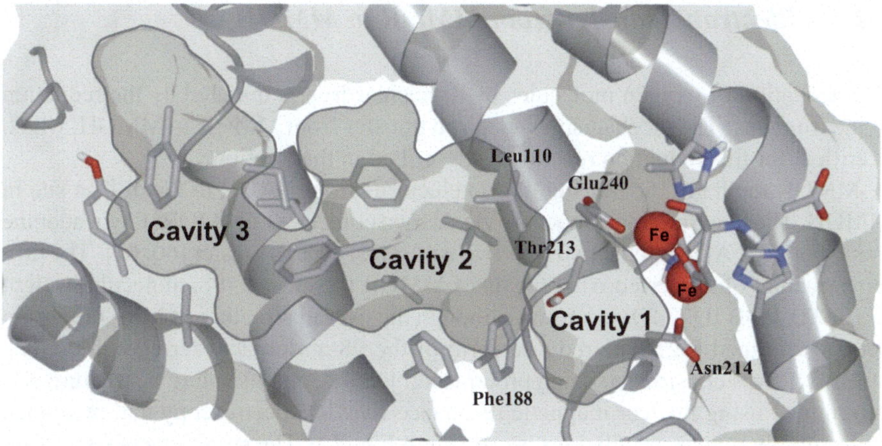

Fig. 2.8 The hydrophobic cavities linked to the di-iron center of MMOH from *Methylococcus capsulatus* (Bath) in which the substrate and product analogs can bind

been proposed to play a role in proton transfer [57], as well as to provide a route for methanol release [54].

Cavity 2 is located next to cavity 1 and is linked to cavity 3. Cavity 1 and cavity 2 are not connected in $MMOH_{ox}$ because of the two amino acid residues, Leu110 and Phe 188 [10, 54]. However, when MMOH forms a complex with MMOB, the orientation of Phe 188 changes in a gate-like fashion, connecting the two cavities. Simultaneously, the pore region of cavity 1 is closed by a conformational change in Thr 213, which forms hydrogen bonds with Glu 240 [58]. Double electron–electron resonance (DEER) spectroscopy suggests that the conformation of $MMOH_{red}$ changes upon binding to MMOB, while that of $MMOH_{ox}$ does not [59]. Based on these structural investigations, the following mechanism is proposed to explain how the substrate accesses the catalytic site of sMMO and how the product is released from this site:

(1) Initially, methane and oxygen molecules can bind to MMOH but cannot access the di-iron center because the gate at the entrance of cavity 1 is closed.
(2) When MMOB binds to MMOH, the gate is opened, allowing methane and oxygen molecules to access the di-iron center.
(3) Methane is oxidized to methanol, and the di-iron center is oxidized ($MMOH_{ox}$).
(4) MMOB dissociates from MMOH, causing the gate of cavity 1 to close and the pore region to open. The produced methanol is released through the opened pore.

2.2.6 *Electron Transfer from MMOR to MMOH*

The activation of oxygen molecules by di-nuclear iron is initiated by the reduction of the iron ions. The mechanism of electron transfer from MMOR to MMOH, which is critical for this reduction reaction, is described in this section.

MMOR mediates the transfer of two electrons from NADH to the di-iron site in MMOH. The electron transfer occurs via the two cofactors of MMOR: flavin adenine dinucleotide (FAD) and the two iron/two sulfur ([2Fe–2S]) cluster [60, 61]. The protein structure of the ferredoxin domain of MMOR (residues 1–98, molecular weight = 10,900, PDB ID 1J4Q) is known, although the overall structure of MMOR has not been determined [62]. The reduced form of the 98-amino-acid ferredoxin domain of MMOR was constructed using protein engineering, and its solution structure was determined using nuclear magnetic resonance (NMR) spectroscopy and restrained molecular dynamics calculations, as shown in Fig. 2.1. The structure consists of six β strands arranged into two β sheets, along with three α helices. Two of these helices form a helix–proline–helix motif, which is unprecedented among [2Fe–2S] cluster proteins. The [2Fe–2S] cluster is coordinated by the sulfur atoms of three cysteine residues. The domain-binding FAD accepts electrons from NADH; the domain carrying the [2Fe–2S] cofactor then accepts the electrons from the FAD domain and donates the electrons to the di-iron center of MMOH.

MMOR plays a role in the conversion of the hydride from NADH into single-electron equivalents and then transfers them sequentially to the di-nuclear iron center of MMOH. This electron transfer was investigated using stopped-flow spectroscopy [63]. According to the experimental data, NADH binds to and forms a charge-transfer complex with the oxidized form of the FAD moiety of MMOR. Then, NADH transfers a hydride, leading to the formation of a charge-transfer complex between the reduced form of the FAD moiety and NAD^+. Sequential electron transfer from the FAD moiety to [2Fe–2S] occurs in the same kinetic phase as the release of NAD^+ from MMOR. The intramolecular electron transfer results in the formation of a FAD semiquinone and the reduced [2Fe–2S] state in MMOR. Thus, the [2Fe–2S] center serves to transfer the electrons from MMOR to MMOH.

During the electron transfer, MMOR interacts with MMOH. This interaction affects the redox potential of the di-nuclear iron center of MMOH, increasing it from +48 to around +100 mV (vs the standard hydrogen electrode (SHE)) [64]. The positive shift makes it easier for the di-nuclear iron center of MMOH to accept electrons from MMOR. The structure of the two-protein complex (MMOH–MMOR) has not yet been obtained, but the formation of the complex has been investigated using various biochemical experiments and computational simulations. These investigations have revealed that MMOR binds to the surface of MMOH in a so-called canyon-like region, which is a polar center surrounded by hydrophobic residues (Fig. 2.9).

Some of the amino acid residues of MMOR that interact with MMOH have been identified. Residues 24–31 in α helix 1 and residues 67–69 in α helix 2 of the ferredoxin domain of MMOR were identified via NMR spectroscopy [62]. Additionally, the residues Glu-56 and Glu-91 of the ferredoxin domain of MMOR were identified

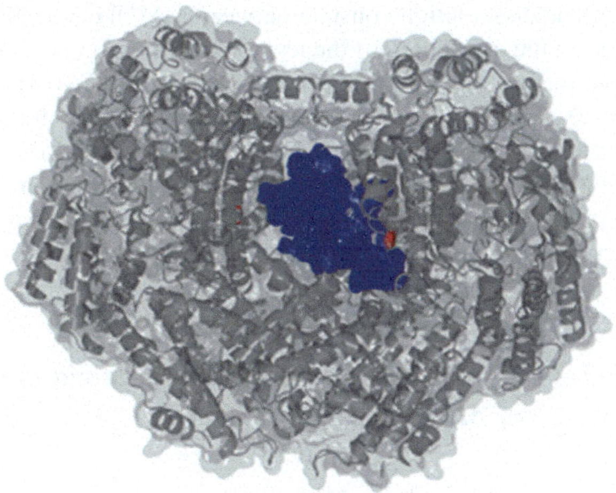

Fig. 2.9 Canyon region of MMOH where MMOR binds (shown as blue). Di-iron center is shown as red balls

using matrix-assisted laser desorption/ionization time-of-flight mass spectrometry (MALDI-TOF-MS) and capillary high-performance liquid chromatography–mass spectrometry (capillary HPLC/MS) of a proteolytic sample in which MMOH and the ferredoxin domain of MMOR were cross-linked by a carbodiimide reagent [65]. Glu-56 and Glu-91 are cross-linked to the N-terminal amino group of the α-subunit of MMOH. The two Glu are located near one another on the surface of the ferredoxin domain of MMOR and >25 Å from the [2Fe–2S] cluster. Site-directed mutagenesis at Glu56 and Glu91 decreased the activity of sMMO to about 65% that of the wild-type MMOR, while the mutant MMOR retained more than 80% NADH oxidase activity [65], indicating that these residues are involved in the formation of a complex between MMOH and MMOR. Other chemical cross-linking experiments also have indicated that the binding sites of MMOR are located on the β subunits of MMOH [66]. Additionally, H–D exchange coupled with mass spectrometry (HDX-MS) and computational docking revealed that the ferredoxin domain of MMOR binds to the canyon region of MMOH [59, 67]. According to the simulation, the ferredoxin domain of MMOR covers part of the α-and β-subunits of MMOH, which is consistent with the results of the cross-linking experiments described above. Based on these results, the binding of MMOR to the canyon region of MMOH is highly probable.

A structure for the MMOH–MMOR complex was proposed based on the results of a computational docking study. According to the putative protein structure, the [2Fe-2S] cluster of MMOR is positioned 14 Å away from the di-iron center of MMOH. This distance is not too large to preclude electron transfer reactions, as these reactions within and between proteins often involve prosthetic groups separated by distances greater than 10 Å. The gating residue of the pore region of MMOH, Glu240, is situated between the [2Fe–2S] cluster and the di-nuclear iron center. This residue may be involved in the proton-coupled electron transfer [58].

The dissociation constant between MMOH and MMOR was measured by monitoring the quenching of the tryptophan fluorescence spectrum. Based on these measurements, two distinct dissociation constants (K_d) of approximately 10 nM and 8 μM were estimated [66]. The larger K_d value was reproduced in a fluorescence anisotropy titration of the formation of the complex between MMOH and the ferredoxin domain of MMOR in the presence of MMOB [67], while a smaller value (0.9 μM) was measured in the absence of MMOB [63]. The results imply that MMOR and MMOB share the same binding site on the surface of MMOH.

2.2.7 Control of the Catalytic Cycle via the Interaction of MMOB with MMOH

Electron transfer from MMOR, access of oxygen and methane molecules to the di-nuclear iron site, and control of the dissociation of the product methanol from MMOH are achieved by the interaction of the small protein MMOB with MMOH and MMOR. The mechanisms by which MMOB controls the oxidation of methane to methanol are described in this section.

MMOB has been reported to have the following effects in the catalytic cycle of methane hydroxylation in MMOH:

(a) Increasing the turnover number of MMOH 150-fold and the rate of the initial reaction with O_2 by a factor of 1000 [9, 12].
(b) Decreasing the redox potential of the di-nuclear iron center of MMOH [15, 68, 69].
(c) Diminishing the rate of intermolecular electron transfer between MMOR and MMOH [70].
(d) Altering the conversion rate of methane and C2–C8 *n*-alkanes and changing product regioselectivity [22, 71, 72].

These effects of MMOB on the enzymatic properties of MMOH arise from specific features of the interaction of MMOB with MMOH. Therefore, the structure and function of MMOB have been investigated intensively in order to obtain a better understanding of the mechanism of methane hydroxylation by sMMO.

The results of various experiments, including saturation-recovery EPR spectroscopy of spin-labeled MMOB [73], matrix-assisted laser desorption/ionization time-of-flight (MALDI-TOF) peptide identification of chemical cross-linking between MMOH and MMOB [74, 75], and crystallographic analysis of the MMOH–MMOB complex [58, 76] have shown that MMOB interacts with MMOH at the canyon-like cavity, as shown in Fig. 2.10.

In particular, crystallographic data derived from the MMOH–MMOB complex provided detailed information regarding the conformational changes in the protein structure of MMOH after the binding of MMOB. MMOB has an unstructured N-terminal peptide chain (Met1–Ser35), a well-folded core region (Asp36–Leu129) with seven β-strands and three α-helices, and a short C-terminal sequence

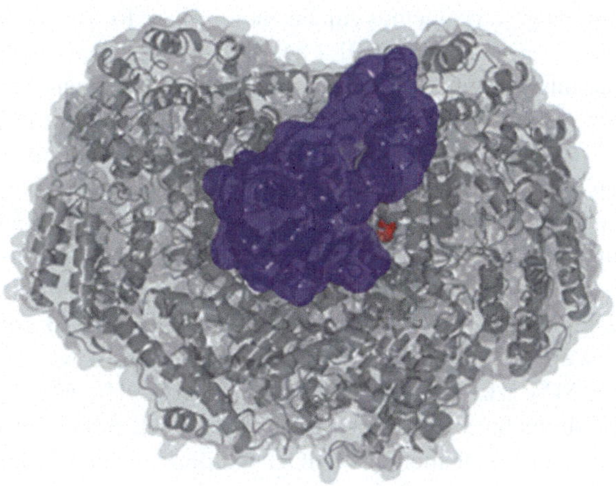

Fig. 2.10 Protein structure of the MMOH (gray)–MMOB (purple) complex from *M. capsulatus* Bath. Di-iron center is shown as red balls

(Met130–Ala141). The N-terminus of MMOB has eight hydrogen bonds, and shows hydrophobic interactions with the MMOH helices H and 4. This interaction makes the unstructured N-terminal peptide chain of MMOB change into an ordered conformation on the surface of MMOH. The N-terminal chain also has a small α-helix (Gly17–Phe25) that forms a hydrophobic interaction with MMOH. The aromatic residues Phe20, Phe24, and Phe 25 of the peptide chain exhibit hydrophobic interactions with Val 302, Lys303, and Tyr 340 of MMOH. The hydrophilic residues Lys18, Asp19, Asp22, and Gln23 of MMOB are oriented toward the other side of the α-helix, opposite of the hydrophobic helix face, and endow the MMOH–MMOB complex with solubility in aqueous media.

The core region (Asp36-Leu129) of MMOB binds to the surface of the α-subunit near the di-nuclear iron site of MMOH. The pore region constructed by Thr213, Asn214, and Glu240 (see Sect. 2.2.4) is located between MMOB and the di-nuclear iron center. In the absence of MMOB, the pore is opened, and the side chains of Glu240 and Asn214 are oriented toward the solvent accessible area. This position allows water molecules or hydronium ions (H_3O^+) to access the di-nuclear iron site for proton transfer [77]. When the core region of MMOB is bound close to the pore region, the orientations of the amino acids in the pore of MMOH are altered. The carboxylate of Glu240 becomes oriented toward the interior of MMOH. Because of this change in orientation, protons required for O_2 activation (i.e., the formation of intermediate P) are delivered. Asn214 of MMOH forms a hydrogen bond with Ser111 of MMOB. This hydrogen bond induces a major allosteric effect at the active site of MMOH [45]. The change in the orientation of Asn214 in helix E of MMOH further induces the transformation of this helix region from a π-helix to an α-helix. In turn, the structural change in the π-helix alters the orientation of Thr213, which then forms a hydrogen bond with the carboxylate of Glu240, stabilizing the pore blocking.

The binding of MMOB to MMOH also significantly changes the positions of other key amino acid residues of MMOH. The cavities where substrates such as methane and oxygen molecules can be enclosed are located close to the di-nuclear iron center (see Sect. 2.2.4) [78, 79]. Phe 188 and Ile 110 act as gates between the exterior and the interior, where the di-iron active sites are located. These conformational changes are responsible for the effects described in points (a)–(d) above.

The binding of MMOB to MMOH also affects the conformation of MMOB. DEER spectroscopic data suggest that the conformation of the N-terminal peptide chain of MMOB becomes more rigid when MMOH is reduced [59]. The key role of the MMOB N-terminus is underscored by data showing that mutants of MMOB in which the N-terminus is disrupted display significantly reduced activity [58, 80]. The dissociation constant of the MMOH–MMOB complex was estimated by fluorescence anisotropy experiments, and MMOB was found to show a higher affinity for $MMOH_{red}$ ($K_d = 0.17\ \mu M$) than $MMOH_{ox}$ ($K_d = 0.55\ \mu M$) [59]. The different affinities indicate that MMOB binds to MMOH after its reduction by MMOR. The core region of mutant MMOB with a truncated N-terminus is still able to bind to MMOH, but with lower affinity ($K_d = 2.67\ \mu M$ [67]). Therefore, the N-terminal peptide chain serves to anchor MMOB to MMOH.

Based on the findings that the binding site of MMOH is shared between MMOR and MMOB and that the affinity of MMOR and MMOB for MMOH depends on the redox state of the di-nuclear iron center, the following functional model has been proposed [67]:

(1) First, MMOR binds to $MMOH_{ox}$ by replacing MMOB; the di-nuclear iron center of $MMOH_{ox}$ is then reduced by MMOR as described in Sect. 2.2.5.
(2) MMOB rebinds to the $MMOH_{red}$–MMOR complex competitively to form the $MMOH_{red}$–MMOB complex. Oxygen and methane molecules can then access the di-iron center of MMOH, as described in Sect. 2.2.4, as the binding of MMOB to MMOH triggers simultaneous conformational changes in the α-subunit of MMOH that open the pore region to O_2 and CH_4 and enable proton delivery to the di-iron center.
(3) After the methane oxidation reaction described in Sect. 2.2.3, MMOB dissociates from $MMOH_{ox}$ and the produced methanol is released.
(4) MMOR binds to $MMOH_{ox}$ again, initiating the catalytic cycle shown in Fig. 2.3.

This scenario is still speculative because of the absence of experimental evidence for the displacement of MMOB by MMOR at the oxidized MMOH. However, it is obvious that MMOB precisely controls the routes of the substrates, electrons, and protons to the di-nuclear iron center in MMOH. Without MMOB, oxidation of NADH to NAD^+, rather than the oxidation of methane to methanol, is dominant in sMMO [63].

2.3 pMMO

Particulate methane monooxygenases (pMMOs) are enzymes embedded in the membranes of methane-utilizing bacteria (methanotrophs) [1, 2, 81]. As the use of pMMOs represents a promising avenue for the establishment of catalytic processes for the conversion of methane to useful chemicals, their protein structure and enzymatic properties are currently being investigated. At present, the changes in the protein structure, the structure of the catalytic site, and the reaction mechanism of pMMO are much less well understood than those of sMMO.

2.3.1 Overall Protein Structure

Protein crystallization and X-ray analysis of pMMO isolated from some methanotrophs have revealed its protein structure [82–85], which is shown in Fig. 2.11. pMMO consists of a trimerized tertiary structure, with each monomer being composed of three polypeptides: PmoB (α-subunit), PmoA (β-subunit), and PmoC (γ-subunit). In methane-oxidizing bacteria, pMMO is embedded in the intracytoplasmic membrane, which is the inner membrane of double cell membranes, as shown schematically in Fig. 2.11. The protein molecule contains a domain composed of a large number of hydrophobic amino acid side chains. This domain is integrated into the intracytoplasmic membrane. Other domains are composed of many hydrophilic amino acid side chains, and are exposed and oriented toward the outside of the intracytoplasmic membrane [82, 86].

The enzymatic hydroxylation of methane by pMMO requires copper ions [5, 87, 88]. pMMO contains multiple metal-binding sites, as shown in Fig. 2.11 [89–91]. These metal-binding sites are thought to be involved in the function of the enzyme, such as the catalytic site, electron and proton transfer, and stabilization of the protein structure.

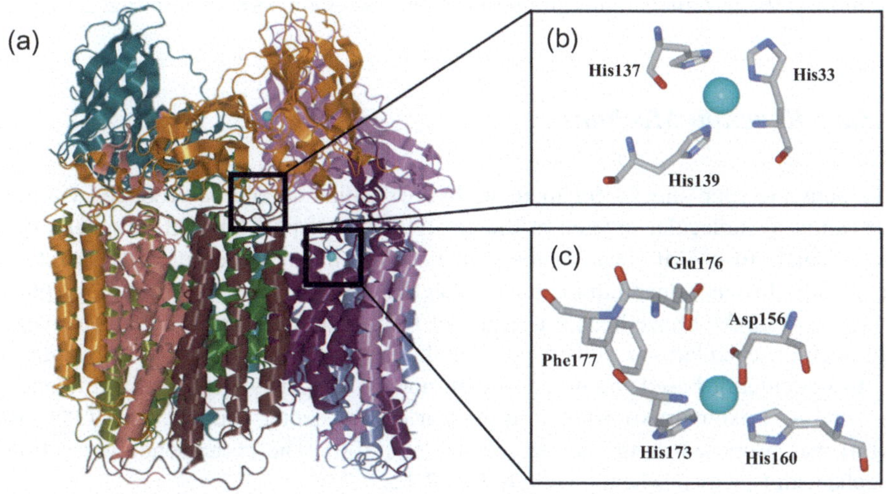

Fig. 2.11 Protein structure of pMMO from *Methylosinus trichosporium* OB3b and its metal-binding sites (PDB ID: 3CHX)

A metal-binding site housing one copper ion has been found in the hydrophilic N-terminal region of the polypeptide PmoB [82–85] and was first modeled as a di-nuclear copper site [82]. However, a recent study using quantum mechanical refinement of X-ray crystallographic data suggested that the data were better fith by a single copper center rather than two copper ions [92]. In fact, pMMO containing only two mono-nuclear copper ions has been purified with retaining its activity [93]. In this book, this copper site is referred to as the PmoB–N copper site. An additional copper-binding site is located in the transmembrane region of the PmoC polypeptide. Copper ions bind to this site with lower affinity than to the PmoB–N copper site; thus, the copper can be replaced by a zinc ion when pMMO is isolated from bacterial cells [85]. Thus, this site is referred to as a Cu/Zn site in this book. These two copper-binding sites have been identified in a number of X-ray analyses of pMMO protein crystals [82–85].

Several additional copper-binding sites other than the PmoB–N copper site and Cu/Zn site have been identified via X-ray analysis of pMMO protein crystals; one is located in the transmembrane region of the PmoA polypeptide [94–96], and another is found in the C-terminal hydrophilic region of PmoB [97].

2.3.2 Catalytic Site

Among the metal-binding sites that have been identified in pMMO, the PmoB–N site is considered the most probable catalytic site for the hydroxylation of methane, because recombinant pMMO subdomain proteins containing only the PmoB–N copper site have shown the ability to oxidize methane to methanol [84, 90, 98]. At this site, one or two copper ions are coordinated with a histidine residue and a nitrogen atom from the terminal amino group of the peptide chain, as shown in Fig. 2.11.

2.3.3 Reaction Mechanism

The active oxygen species for methane hydroxylation should be produced via the activation of molecular oxygen at the catalytic site of pMMO. Unfortunately, the intermediate of pMMO bearing the active species has not been detected spectroscopically. Instead, quantum chemistry calculations based on the crystallographic analysis of pMMO have been a useful tool for exploring the reaction mechanism, although the calculation results depend on the experimental crystallographic data.

In a calculation based on a model with di-nuclear copper as site A, a mixed valence (+2 and +3) oxohydroxo-bridged di-nuclear copper center ($Cu^{II}Cu^{III}$-(μ-O^-) (μ-OH)) was proposed as the active species [99, 100]. The corresponding reaction mechanism is shown schematically in Fig. 2.12.

In this reaction mechanism, the oxygen cross-linked to the metal ion is a radical and extracts a hydrogen radical from the C–H bond of methane. OH and methyl radicals cross-linked to metal ions are believed to be generated as a result, with C–O bonds being formed via a "radical rebound mechanism" and "asynchronous

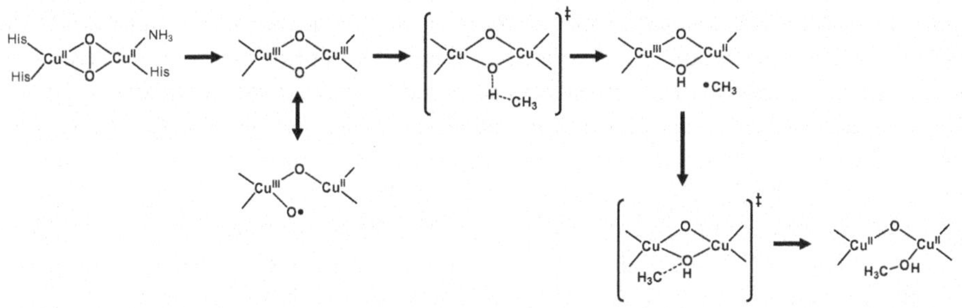

Fig. 2.12 Proposed reaction mechanism for the hydroxylation of methane by pMMO using a model involving a di-nuclear copper catalytic site

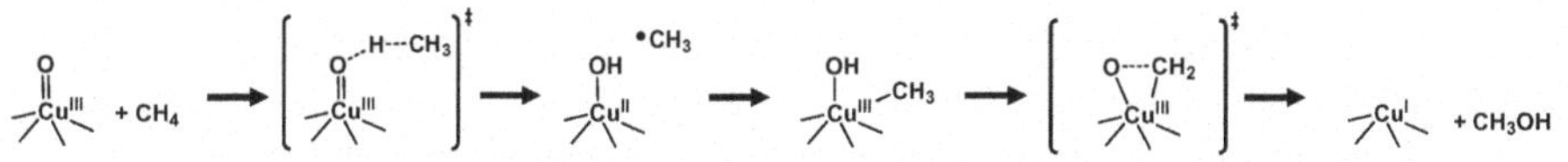

Fig. 2.13 Proposed reaction mechanism for the hydroxylation of methane by pMMO based on a model involving a mono-nuclear copper catalytic site

cooperative mechanism" [101]. The possibility of methanol production via a non-radical mechanism has also been reported based on a quantum chemical calculation [102].

On the other hand, in the mono-nuclear copper site model, a Cu^{III}–oxo species is assumed to be an active species [92], although this species has not been observed experimentally. The corresponding reaction mechanism is shown schematically in Fig. 2.13. The Cu^{III}–oxo species binds a methane molecule and abstracts a hydrogen atom from the methane with a barrier of 22 kJ mol^{-1}. A methyl radical and an OH group are produced by the hydrogen abstraction, followed by the rebound of the OH group to the methyl radical to produce methanol. The barrier for the rebound of the methyl radical to the OH group was calculated to be 53 kJ mol^{-1}. At the beginning of the reaction, the triplet spin state is favorable, while the singlet spin state is more favorable for the intermediate. Therefore, the spin state shifts from the triplet state to the singlet state in this reaction mechanism; this so-called spin-crossover was also found to make the reaction favorable in a previous study. Overall, the reaction of methane with the Cu^{III}–oxo species is exothermic, and the activation barriers are lower than those in the reaction mechanism based on a di-nuclear copper site (107 kJ mol^{-1}).

2.3.4 Substrate-Binding Cavity

As no crystal structures of pMMO containing analogs of the substrate or product have been obtained yet, the substrate-binding site of pMMO has not been identified but is expected to be located in the cavity or pocket adjacent to the catalytic site in pMMO. The properties of the substrate-binding cavity of pMMO have been analyzed using structural information for pMMO obtained from *Methylosinus trichosporium* OB3b and *Methylococcus capsulatus* (Bath), the substrate specificity of pMMO, and the general mechanism of host–guest molecular recognition [103–105]. In terms of specificity, pMMO can oxidize not only methane, but also C2–C5 *n*-alkanes. However, *n*-hexane cannot be oxidized by pMMO. This substrate specificity implies that the greatest volume that the substrate-binding site of pMMO can accommodate is that of *n*-pentane (96.4 $Å^3$), while larger substrates such as *n*-hexane (120.4 $Å^3$) cannot be accommodated. In host–guest chemistry, it is generally accepted that the maximum guest molecule to host cavity volume ratio (packing coefficient) is 0.7 [104]. By assuming that the packing coefficient of *n*-pentane in the substrate-binding site of pMMO is ~0.7 (v/v), the volume of the pMMO substrate-binding site can be estimated as ~138 $Å^3$. The predicted cavity volume is very large compared to the molecular volume of methane (28.4 $Å^3$). However, the volume of cavity 1 adjacent to the di-nuclear iron center of MMOH (Fig. 2.8) is even larger at 225 $Å^3$.

In pMMO, the putative cavities where the substrate could bind are located adjacent to the metal-binding centers (Fig. 2.14). Comparison of the amino acid residues comprising the surfaces of the cavities of the two pMMOs and ammonia monooxygenase (AMO), a protein from the same family as pMMO, (Fig. 2.14) shows that three amino acid residues of AMO (His 141 and Ala 156 of AmoB, and Met 56 of AmoC) adjacent to the PmoB-N copper site are substituted for Trp, Ile, and Leu, respectively, in the two pMMOs. On the other hand, all amino acid residues that make up the cavities adjacent to the Cu/Zn site are completely conserved among the three enzymes. Amino acid residues proximal to the substrate-binding site tend to affect the substrate-binding ability more strongly than those distal from the site [106]. Therefore, the distinct product selectivities of AMO and the pMMOs suggest that one or more of the amino acid residues comprising the substrate-binding site of AMO

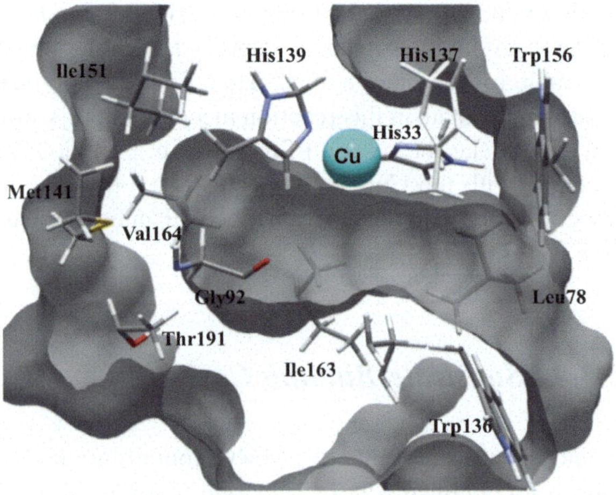

Fig. 2.14 Putative substrate-binding cavity of pMMO from *Methylosinus trichosporium* OB3b

should be different from those of the binding sites of the two pMMOs; additionally, these amino acid residues likely affect the ability of pMMO to control the orientation of *n*-butane and *n*-pentane. Thus, comparison of amino acids in the cavities of the two pMMOs and AMO implies that the cavity adjacent to the di-copper center is most likely the substrate-binding site in these three enzymes.

2.3.5 Electron Transfer from the Electron Donor

Unlike in the case of sMMO, several electron donors for pMMO in vitro have been identified, namely hydroquinone derivatives, NADH, methanol, succinate, and hydrogen, as shown in Fig. 2.15.

(1) *Duroquinol*

pMMO shows activity when hydroquinone derivatives are used as the electron donor. Alternatively, 2,3,5,6-tetramethyl hydroquinone (duroquinol) can be used as a reductant [107]. Duroquinol is an analog of hydrophobic hydroquinone derivatives present within bacterial membranes in the quinone pool and is believed to donate electrons directly to pMMO, as duroquinol-driven methane hydroxylation was observed even when pMMO was solubilized from the hydrophobic bacterial membrane and purified [108, 109].

However, the PmoB copper center is located in the hydrophilic region of pMMO. Therefore, it is unclear how the hydrophobic hydroquinone derivatives in the quinone

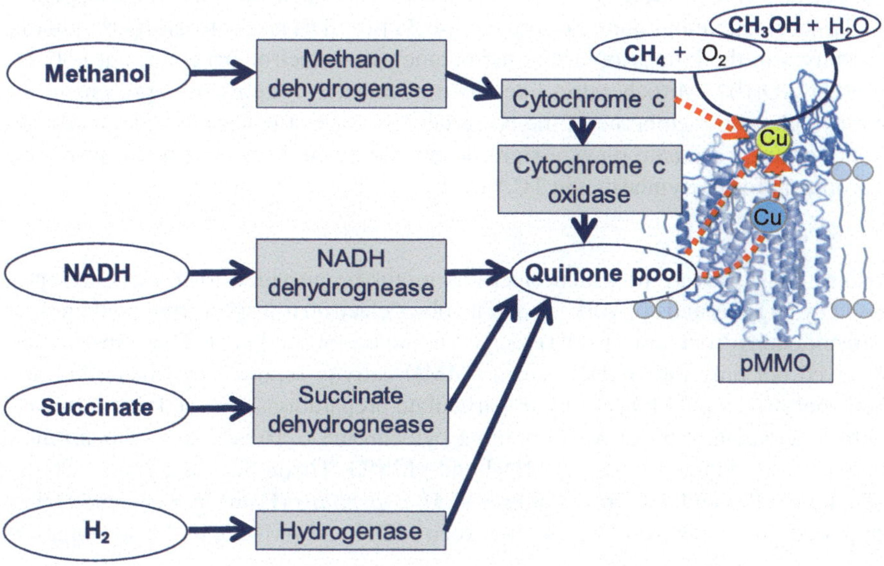

Fig. 2.15 Proposed pathways of electron transfer to pMMO

pool can donate electrons to the PmoB copper center. One possibility is that copper ions other than those in the PmoB copper center mediate electron transfer from the quinone pool to the PmoB copper center [110].

(2) *NADH*

At least three models of electron donation to pMMO have been proposed.

It is generally accepted that NADH can be used in vitro as an electron source for pMMO, although in an early study, Ribbons suggested that NADH may not be the immediate electron donor for pMMO [111]. Electron transfer proteins in methane-oxidizing bacterial cells are presumed to be involved in electron transfer from NADH to the protein, as shown in Fig. 2.15 [88, 107, 112, 113], although a site that accepts electrons directly from NADH has also been proposed [82].

NADH donates electrons to pMMO via other membrane components within the bacterial membrane, such as NADH dehydrogenase and the quinone pool [88, 112, 113]. Therefore, in most cases, no NADH-driven activity was observed when pMMO was isolated from the bacterial membrane.

It was subsequently shown that exogenous quinols can act as electron donors for pMMO, and that the resulting quinones can be reduced in vitro via an NADH-dependent reaction involving a type 2 NADH-quinone reductase [113]. The latter enzyme has been isolated and purified from *M. capsulatus* (Bath) as a 36.6 kDa monomer with a single non-covalently associated FAD cofactor. However, experiments with inhibitors of the enzyme yielded contradictory results [113].

(3) *Succinate*

Cornish et al. proposed another electron donation pathway in which succinate provides electrons to pMMO by way of membrane-associated succinate dehydrogenase (SDH, succinate:ubiquinone oxidoreductase) [114]. SDH is able to use hydroquinone derivatives in the intracytoplasmic membrane as its electron acceptor, but not able to use $NAD(P)^+$. Mitochondria-like electron transfer systems may present in the intracytoplasmic membrane of the bacterial. However, it is not physiologically relevant because succinate plays a signifficant role as the intermediate for providing precurosors for biosynthesis via TCA cycle.

(4) *Methanol*

A third possible way in which electrons might be provided to pMMO was proposed by Higgins and co-workers, and involves electron transport from methanol via methanol dehydrogenase (MDH) located in the periplasm [115]. These researchers observed not only the NADH-driven pMMO activity reported by others, but also methanol-driven pMMO activity in particulate preparations; this activity correlated with the concentration of a CO-binding cytochrome *c*. Hence, this cytochrome *c* may act as a mediator between MDH and pMMO. Theoretical analysis of experimental growth yield data from cultures of *M. capsulatus* (Bath) by Leak and Dalton supported direct electron coupling between MDH and pMMO, and also suggested multiple pathways of electron transfer between these two enzymes [116]. These authors presented an alternative model for electron transport in which electrons from methanol are transferred to pMMO not only via MDH and cytochrome *c*, but also via low-potential respiratory chain components such as cytochrome *b* and quinone along a pathway that requires reverse electron flow. However, the proteins involved in this pathway have not been identified biochemically. Although in vivo studies have revealed the effect of methanol on methane oxidation in methanotrophs [117, 118], the electron transfer pathway from methanol to the hydroxylase component of pMMO (pMMO-H) in methanotrophic bacteria remains controversial because of issues with the reproducibility of methanol-driven pMMO activity in vitro.

(5) *Hydrogen*

Two dehydrogenases are present in the bacterial membrane of *M. capsulatus* Bath, and hydrogen-driven methane hydroxylation has been observed [119].

2.3.6 Reaction Control

Regarding the control of the enzymatic reaction of pMMO, a pMMO–methanol dehydrogenase complex that is believed to be involved in reaction control has been isolated. This enzyme shows higher activity than pMMO alone and highly stable enzyme activity. Low-temperature electron microscope image analysis indicated that methanol dehydrogenase is bound to the hydrophilic domain in which the active site is located [91]. On the other hand, since pMMO is a membrane protein, interaction with the electron transfer proteins of the cell membrane would also seem to be involved in its control [113, 120].

2.4 Summary

In this chapter, the protein structure of methane monooxygenase, an enzyme that selectively synthesizes methanol from methane using oxygen molecules as oxidants under atmospheric temperature/pressure, and the structural changes and protein interactions involved in the oxidation of methane to methanol were reviewed. The molecular mechanism of catalysis by methane monooxygenase is extremely complex, and mimicking all its functions with inorganic materials and complex organic molecules would seem to be difficult. Therefore, catalysts for the oxidation of methane to methanol should not necessarily imitate the molecular structure and function of methane monooxygenase. In the next chapter, we will review the development of heterogeneous and homogeneous catalysts that enable the methane-to-methanol oxidation reaction.

References

1. Hanson RS, Hanson TE (1996) Methanotrophic bacteria. Microbiol Rev 60:439–471
2. Dalton H (2005) The Leeuwenhoek Lecture 2000 The natural and unnatural history of methane-oxidizing bacteria. Philos Trans R Soc Lond B Biol Sci 360:1207–1222
3. Enger BC, Lødeng R, Holmen A (2008) A review of catalytic partial oxidation of methane to synthesis gas with emphasis on reaction mechanisms over transition metal catalysts. Appl Catal A 346:1–27
4. Cruickshank FR, Benson SW (1969) Carbon-hydrogen bond dissociation energy in methanol. J Phys Chem 733:733–737
5. Nguyen HH, Elliott SJ, Yip JH, Chan SI (1998) The particulate methane monooxygenase from *Methylococcus capsulatus* (Bath) is a novel copper-containing three-subunit enzyme. Isolation and characterization. J Biol Chem 273:7957–7966
6. Lieberman RL, Rosenzweig AC (2004) Biological methane oxidation: regulation, biochemistry, and active site structure of particulate methane monooxygenase. Crit Rev Biochem Mol Biol 39:147–164
7. Miyaji A (2011) Particulate methane monooxygenase from *Methylosinus trichosporium* OB3b. Methods Enzymol 495:211–225
8. Colby J, Dalton H (1978) Resolution of the methane monooxygenase of *Methylococcus capsulatus* (Bath) into three components. Biochem J 171:461–468
9. Fox BG, Froland WA, Dege JE, Lipscomb JD (1989) Methane monooxygenase from *Methylosinus trichosporium* OB3b. Purification and properties of a three component system with high specific activity from a type II methanotroph. J Biol Chem 264:10023–10033
10. Rosenzweig AC, Frederick CA, Lippard SJ, Nordlund P (1993) Crystal structure of a bacterial non-heme iron hydroxylase that catalyzes the biological oxidation of methane. Nature 366:537–543
11. Colby J, Dalton H (1979) Characterization of the second prosthetic group of the flavoenzyme NADH-acceptor reductase (component C) of the methane mono-oxygenase from Methylococcus capsulatus (Bath). Biochem J 177:903–908
12. Green J, Dalton H (1985) Protein B of soluble methane monooxygenase from *Methylococcus capsulatus* (Bath). A novel regulatory protein of enzyme activity. J Biol Chem 260:15795–15801
13. Rosenzweig AC, Nordlund P, Takahara PM, Frederick CA, Lippard SJ (1995) Geometry of the soluble methane monooxygenase catalytic diiron center in two oxidation states. Chem Biol 2:409–418
14. Rosenzweig AC, Brandstetter H, Whittington DA, Nordlund P, Lippard SJ, Frederick CA (1997) Crystal structure of the methane monooxygenase hydroxylase from *Methylococcus capsulatus* (Bath): implications for substrate gating and component interactions. Proteins 29:141–152
15. Whittington DA, Lippard SJ (2001) Crystal structures of the soluble methane monooxygenase hydroxylase from *Methylococcus capsulatus* (Bath) demonstrating geometric variability at the dinuclear iron active site. J Am Chem Soc 123:827–838
16. Elango N, Radhakrishnan R, Froland WA, Wallar BJ, Earhart CA, Lipscomb JD, Ohlendorf DH (1997) Crystal structure of the hydroxylase component of methane monooxygenase from *Methylosinus trichosporium* OB3b. Protein Sci 6:556–568
17. Pulver SC, Froland WA, Fox BG, Lipscomb JD, Solomon EI (1993) Spectroscopic studies of the coupled binuclear non-heme iron active site in the fully reduced hydroxylase component of methane monooxygenase: comparison to deoxy and deoxyazide hemerythrin. J Am Chem Soc 115:12409–12422
18. Pulver SC, Froland WA, Lipscomb JD, Solomon EI (1997) Ligand field circular dichroism and magnetic circular dichroism studies of component B and substrate binding to the hydroxylase component of methane monooxygenase. J Am Chem Soc 119:387–395
19. Lee SK, Nesheim JC, Lipscomb JD (1993) Transient intermediates of the methane monooxygenase catalytic cycle. J Biol Chem 268:21569–21577

20. Lee SK, Fox BG, Froland WA, Lipscomb JD, Münck E (1993) A transient intermediate of the methane monooxygenase catalytic cycle containing an Fe(IV)-Fe(IV) cluster. J Am Chem Soc 115:6450–6451
21. Lee SY, Lipscomb JD (1999) Oxygen activation catalyzed by methane monooxygenase hydroxylase component: proton delivery during the O-O bond cleavage steps. Biochemistry 38:4423–4432
22. Liu KE, Valentine AM, Wang DL, Huynh BH, Edmondson DE, Salifoglou A, Lippard SJ (1995) Kinetic and spectroscopic characterization of intermediates and component interactions in reactions of methane monooxygenase from *Methylococcus capsulatus* (Bath). J Am Chem Soc 117:10174–10185
23. Shu LJ, Nesheim JC, Kauffmann K, Münck E, Lipscomb JD, Que L Jr (1997) An Fe(IV)-O_2 diamond core structure for the key intermediate Q of methane monooxygenase. Science 275:515–518
24. Brazeau BJ, Lipscomb JD (2000) Kinetics and activation thermodynamics of methane monooxygenase hydroxylase compound Q formation and reaction with substrates. Biochemistry 39:13503–13515
25. DeWitt JG, Bentsen JG, Rosenzweig AC, Hedman B, Green J, Pilkington S, Papaefthymiou GC, Dalton H, Hodgson KO, Lippard SJ (1991) X-ray absorption, Moessbauer, and EPR studies of the dinuclear iron center in the hydroxylase component of methane monooxygenase. J Am Chem Soc 113:9219–9235
26. Woodland MP, Patil DS, Cammack R, Dalton H (1986) ESR studies of protein A of the soluble methane monooxygenase from *Methylococcus capsulatus* (Bath). Biochim Biophys Acta 873:237–242
27. Ericson A, Hedman B, Hodgson KO, Green J, Dalton H, Bentsen JG, Beer RH, Lippard SJ (1988) J Am Chem Soc 110:2330–2332
28. Banerjee R, Proshlyakov Y, Lipscomb JD, Proshlyakov DA (2015) Structure of the key species in the enzymatic oxidation of methane to methanol. Nature 518:431–434
29. Rinaldo D, Philipp DM, Lippard SJ, Friesner RAJ (2007) Intermediates in dioxygen activation by methane monooxygenase: a QM/MM study. J Am Chem Soc 129:3135–3147
30. Han WG, Noodleman L (2008) Structural model studies for the high-valent intermediate Q of methane monooxygenase from broken-symmetry density functional calculations. Inorg Chim Acta 361:973–986
31. Siegbahn PEM, Crabtree RH, Nordlund P (1998) Mechanism of methane monooxygenase—a structural and quantum chemical perspective. J Biol Inorg Chem 3:314–317
32. Siegbahn PEM (1999) Theoretical model studies of the iron dimer complex of MMO and RNR. Inorg Chem 38:2880–2889
33. Lovell T, Han WG, Liu TQ, Noodleman L (2002) A structural model for the high-valent intermediate Q of methane monooxygenase from broken-symmetry density functional and electrostatics calculations. J Am Chem Soc 124:5890–5894
34. Friedle S, Reisner E, Lippard S (2010) Current challenges of modeling diiron enzyme active sites for dioxygen activation by biomimetic synthetic complexes. J Chem Soc Rev 39:2768–2779
35. Huang SP, Shiota Y, Yoshizawa K (2013) DFT study of the mechanism for methane hydroxylation by soluble methane monooxygenase (sMMO): effects of oxidation state, spin state, and coordination number. Dalton Trans 42:1011–1023
36. Xue G, Wang D, De Hont R, Fiedler AT, Shan X, Münck E, Que L Jr (2007) A synthetic precedent for the $Fe_2^{IV}(\mu-O)_2]$ diamond core proposed for methane monooxygenase intermediate Q. Proc Natl Acad Sci USA 104:20713–20718
37. Cutsail GE 3rd, Banerjee R, Zhou A, Que L Jr, Lipscomb JD, DeBeer S (2018) High-resolution extended X-ray absorption fine structure analysis provides evidence for a longer Fe···Fe distance in the Q intermediate of methane monooxygenase. J Am Chem Soc 140:16807–16820
38. Priestley ND, Floss HG, Froland WA, Lipscomb JD, Williams PG, Morimoto H (1992) Cryptic stereospecificity of methane monooxygenase. J Am Chem Soc 114:7561–7562

39. Valentine AM, Wilkinson B, Liu KE, Komar-Panicucci S, Priestley ND, Williams PG, Morimoto H, Floss HG, Lippard SJ (1997) Tritiated chiral alkanes as substrates for soluble methane monooxygenase from *Methylococcus capsulatus* (Bath): probes for the mechanism of hydroxylation. J Am Chem Soc 119:1818–1827
40. Ruzicka F, Huang DS, Donnelly MI, Frey PA (1990) Methane monooxygenase catalyzed oxygenation of 1,1-dimethylcyclopropane. Evidence for radical and carbocationic intermediates. Biochemistry 29:1696–1700
41. Choi SY, Eaton PE, Hollenberg PF, Liu KE, Lippard SJ, Newcomb M, Putt DA, Upadhyaya SP, Xiong Y (1996) Regiochemical variations in reactions of methylcubane with tert-butoxyl radical, cytochrome P-450 enzymes, and a methane monooxygenase system. J Am Chem Soc 118:6547–6555
42. Liu KE, Johnson CC, Newcomb M, Lippard SJ (1993) Radical clock substrate probes and kinetic isotope effect studies of the hydroxylation of hydrocarbons by methane monooxygenase. J Am Chem Soc 115:939–947
43. Valentine AM, LeTadic-Biadatti MH, Toy PH, Newcomb M, Lippard SJ (1999) Oxidation of ultrafast radical clock substrate probes by the soluble methane monooxygenase from *Methylococcus capsulatus* (Bath). J Biol Chem 274:10771–10776
44. Wilkins PC, Dalton H, Samuel CJ, Green J (1994) Further evidence for multiple pathways in soluble methane-monooxygenase-catalysed oxidations from the measurement of deuterium kinetic isotope effects. Eur J Biochem 226:555–560
45. Stahl SS, Francisco WA, Merkx M, Klinman JP, Lippard SJ (2001) Oxygen kinetic isotope effects in soluble methane monooxygenase. J Biol Chem 276:4549–4553
46. Tinberg CE, Lippard SJ (2009) Revisiting the mechanism of dioxygen activation in soluble methane monooxygenase from *M. capsulatus* (Bath): evidence for a multi-step, proton-dependent reaction pathway. Biochemistry 48:12145–12158
47. Yoshizawa K, Yumura T (2003) A non-radical mechanism for methane hydroxylation at the diiron active site of soluble methane monooxygenase. Chemistry 9:2347–2358
48. Mai BK, Kim Y (2014) Substrate-dependent H/D kinetic isotope effects and the role of the di(μ-oxo)diiron(IV) core in soluble methane monooxygenase: a theoretical study. Chemistry 20:6532–6541
49. Yoshizawa K (2000) Two-step concerted mechanism for methane hydroxylation on the diiron active site of soluble methane monooxygenase. J Inorg Biochem 78:23–34
50. Xue G, Pokutsa A, Que L Jr (2011) Substrate-triggered activation of a synthetic [$Fe_2(\mu\text{-}O)_2$] diamond core for C-H bond cleavage. J Am Chem Soc 133:16657–16667
51. Saito T, Shoji M, Isobe H, Yamanaka S, Kitagawa Y, Yamada S, Kawakami T, Okumura M, Yamaguchi K (2010) Theory of chemical bonds in metalloenzymes. XIV. Correspondence between magnetic coupling mode and radical coupling mechanism in hydroxylations with methane monooxygenase and related species. Int J Quant Chem 110:2955–2981
52. Tinberg CE, Lippard SJ (2011) Dioxygen activation in soluble methane monooxygenase. Acc Chem Res 44:280–288
53. Banerjee R, Jones JC, Lipscomb JD (2019) Soluble methane monooxygenase. Annu Rev Biochem 88:409–431
54. Whittington DA, Rosenzweig AC, Frederick CA, Lippard SJ (2001) Xenon and halogenated alkanes track putative substrate binding cavities in the soluble methane monooxygenase hydroxylase. Biochemistry 40:3476–3482
55. Whittington DA, Sazinsky MH, Lippard SJ (2001) X-ray crystal structure of alcohol products bound at the active site of soluble methane monooxygenase hydroxylase. J Am Chem Soc 123:1794–1795
56. Sazinsky MH, Lippard SJ (2005) Product bound structures of the soluble methane monooxygenase hydroxylase from *Methylococcus capsulatus* (Bath): protein motion in the α-subunit. J Am Chem Soc 127:5814–5825
57. McCormick MS, Lippard SJ (2011) Analysis of substrate access to active sites in bacterial multicomponent monooxygenase hydroxylases: X-ray crystal structure of xenon-pressurized phenol hydroxylase from *Pseudomonas sp.* OX1. Biochemistry 50:11058–11069

58. Lee SJ, McCormick MS, Lippard SJ, Cho US (2013) Control of substrate access to the active site in methane monooxygenase. Nature 494:380–384
59. Wang W, Lippard SJ (2014) Diiron oxidation state control of substrate access to the active site of soluble methane monooxygenase mediated by the regulatory component. J Am Chem Soc 136:2244–2247
60. Lund J, Woodland MP, Dalton H (1985) Electron transfer reactions in the soluble methane monooxygenase of *Methylococcus capsulatus* (Bath). Eur J Biochem 147:297–305
61. Blazyk JL, Lippard SJ (2002) Expression and characterization of ferredoxin and flavin adenine dinucleotide binding domains of the reductase component of soluble methane monooxygenase from *Methylococcus capsulatus* (Bath). Biochemistry 41:15780–15794
62. Müller J, Lugovskoy AA, Wagner G, Lippard SJ (2002) NMR structure of the [2Fe-2S] ferredoxin domain from soluble methane monooxygenase reductase and interaction with its hydroxylase. Biochemistry 41:42–51
63. Gassner GT, Lippard SJ (1999) Component interactions in the soluble methane monooxygenase system from *Methylococcus capsulatus* (Bath). Biochemistry 38:12768–12785
64. Liu Y, Nesheim JC, Paulsen KE, Stankovich MT, Lipscomb JD (1997) Roles of the methane monooxygenase reductase component in the regulation of catalysis. Biochemistry 36:5223–5233
65. Kopp DA, Berg EA, Costello CE, Lippard SJ (2003) Structural features of covalently cross-linked hydroxylase and reductase proteins of soluble methane monooxygenase as revealed by mass spectrometric analysis. J Biol Chem 278:20939–20945
66. Fox BG, Liu Y, Dege JE, Lipscomb JD (1991) Complex formation between the protein components of methane monooxygenase from *Methylosinus trichosporium* OB3b. Identification of sites of component interaction. J Biol Chem 266:540–550
67. Wang W, Iacob RE, Luoh RP, Engen JR, Lippard SJ (2014) Electron transfer control in soluble methane monooxygenase. J Am Chem Soc 136:9754–9762
68. Paulsen KE, Liu Y, Fox BG, Lipscomb JD, Münck E, Stankovich MT (1994) Oxidation-reduction potentials of the methane monooxygenase hydroxylase component from *Methylosinus trichosporium* OB3b. Biochemistry 33:713–722
69. Liu KE, Lippard SJ (1991) Redox properties of the hydroxylase component of methane monooxygenase from *Methylococcus capsulatus* (Bath). Effects of protein B, reductase, and substrate. J Biol Chem 266:12836–12839
70. lazyk JL, Gassner GT, Lippard SJ (2005) Intermolecular electron-transfer reactions in soluble methane monooxygenase: a role for hysteresis in protein function. J Am Chem Soc 127:17364–17376
71. Froland WA, Andersson KK, Lee SK, Liu Y, Lipscomb JD (1992) Methane monooxygenase component B and reductase alter the regioselectivity of the hydroxylase component-catalyzed reactions. J Biol Chem 267:17588–17597
72. Liu Y, Nesheim JC, Lee SK, Lipscomb JD (1995) Gating effects of component B on oxygen activation by the methane monooxygenase hydroxylase component. J Biol Chem 270:24662–24665
73. MacArthur R, Sazinsky MH, Kühne H, Whittington DA, Lippard SJ, Brudvig GW (2002) Component B binding to the soluble methane monooxygenase hydroxylase by saturation-recovery EPR spectroscopy of spin-labeled MMOB. J Am Chem Soc 124:13392–13393
74. Brazeau BJ, Wallar BJ, Lipscomb JD (2003) Effector proteins from $P450_{(cam)}$ and methane monooxygenase: lessons in tuning nature's powerful reagents. Biochem Biophys Res Commun 312:143–148
75. Zhang JY, Wallar BJ, Popescu CV, Renner DB, Thomas DD, Lipscomb JD (2006) Methane monooxygenase hydroxylase and B component interactions. Biochemistry 45:2913–2926
76. Lee SJ (2016) Hydroxylation of methane through component interactions in soluble methane monooxygenases. J Microbiol 54:277–282
77. Wang W, Liang AD, Lippard SJ (2015) Coupling oxygen consumption with hydrocarbon oxidation in bacterial multicomponent monooxygenases. Acc Chem Res 48:2632–2639

78. Merkx M, Kopp DA, Sazinsky MH, Blazyk JL, Müller J, Lippard SJ (2001) Dioxygen activation and methane hydroxylation by soluble methane monooxygenase: a tale of two irons and three proteins. Angew Chem Int Ed 40:2782–2807
79. Murray LJ, Lippard SJ (2007) Substrate trafficking and dioxygen activation in bacterial multicomponent monooxygenases. Acc Chem Res 40:466–474
80. Chang SL, Wallar BJ, Lipscomb JD, Mayo KH (2001) Residues in *Methylosinus trichosporium* OB3b methane monooxygenase component B involved in molecular interactions with reduced- and oxidized-hydroxylase component: a role for the N-terminus. Biochemistry 40:9539–9551
81. Conrado RJ, Gonzalez R (2014) Chemistry. Envisioning the bioconversion of methane to liquid fuels. Science 343:621–623
82. Lieberman LL, Rosenzweig AC (2005) Crystal structure of a membrane-bound metalloenzyme that catalyses the biological oxidation of methane. Nature 434:177–182
83. Hakemian AS, Kondapalli KC, Telser J, Hoffman BM, Stemmler TL, Rosenzweig AC (2008) The metal centers of particulate methane monooxygenase from *Methylosinus trichosporium* OB3b. Biochemistry 47:6793–6801
84. Smith SM, Rawat S, Telser J, Hoffman BM, Stemmler TL, Rosenzweig AC (2011) Crystal structure and characterization of particulate methane monooxygenase from *Methylocystis species* strain M. Biochemistry 50:10231–10240
85. Sirajuddin S, Barupala D, Helling S, Marcus K, Stemmler TL, Rosenzweig AC (2014) Effects of zinc on particulate methane monooxygenase activity and structure. J Biol Chem 289:21782–21794
86. Myronova N, Kitmitto A, Collins RF, Miyaji A, Dalton H (2006) Three-dimensional structure determination of a protein supercomplex that oxidizes methane to formaldehyde in *Methylococcus capsulatus* (Bath). Biochemistry 45:11905–11914
87. Prior SD, Dalton H (1985) The effect of copper ions on membrane content and methane monooxygenase activity in methanol-grown cells of *Methylococcus capsulatus* (Bath). J Gen Microbiol 131:155–163
88. Cook SA, Shiemke AK (2002) Evidence that a type-2 NADH:quinone oxidoreductase mediates electron transfer to particulate methane monooxygenase in *Methylococcus capsulatus*. Arch Biochem Biophys 398:32–40
89. Rosenzweig AC (2008) The metal centres of particulate methane monooxygenase. Biochem Soc Trans 36:1134–1137
90. Culpepper MA, Rosenzweig AC (2012) Architecture and active site of particulate methane monooxygenase. Crit Rev Biochem Mol Biol 47:483–492
91. Wang VC, Maji S, Chen PP, Lee HK, Yu SS, Chan SI (2017) Alkane oxidation: methane monooxygenases, related enzymes, and their biomimetics. Chem Rev 117:8574–8621
92. Cao L, Caldararu O, Rosenzweig AC, Ryde U (2018) Quantum refinement does not support dinuclear copper sites in crystal structures of particulate methane monooxygenase. Angew Chem Int Ed 57:162–166
93. Miyaji A, Kamachi T, Okura I (2002) Improvement of the purification method for retaining the activity of the particulate methane monooxygenase from *Methylosinus trichosporium* OB3b. Biotech Lett 24:1883–1887
94. Chan SI, Wang VCC, Lai JCHL, Yu SSF, Chen PPY, Chen KHC, Chen CL, Chan MK (2007) Redox potentiometry studies of particulate methane monooxygenase: support for a trinuclear copper cluster active site. Angew Chem Int Ed 46:1992–1994
95. Chan SI, Yu SS (2008) Controlled oxidation of hydrocarbons by the membrane-bound methane monooxygenase: the case for a tricopper cluster. Acc Chem Res 41:969–979
96. Chan SI, Chen KH, Yu SS, Chen CL, Kuo SS (2004) Toward delineating the structure and function of the particulate methane monooxygenase from methanotrophic bacteria. Biochemistry 43:4421–4430
97. Yu SS, Ji CZ, Wu YP, Lee TL, Lai CH, Lin SC, Yang ZL, Wang VC, Chen KH, Chan SI (2007) The C-terminal aqueous-exposed domain of the 45 kDa subunit of the particulate methane monooxygenase in *Methylococcus capsulatus* (Bath) is a Cu(I) sponge. Biochemistry 46:13762–13774

98. Balasubramanian R, Smith SM, Rawat S, Yatsunyk LA, Stemmler TL, Rosenzweig AC (2010) Oxidation of methane by a biological dicopper centre. Nature 465:115–119
99. Yoshizawa K, Shiota Y (2006) Conversion of methane to methanol at the mononuclear and dinuclear copper sites of particulate methane monooxygenase (pMMO): a DFT and QM/MM study. J Am Chem Soc 128:9873–9881
100. Shiota Y, Yoshizawa K (2009) Comparison of the reactivity of bis(μ-oxo)Cu(II)Cu(III) and Cu(III)Cu(III) species to methane. Inorg Chem 48:838–845
101. Da Silva JC, Pennifold RC, Harvey JN, Rocha WR (2015) A radical rebound mechanism for the methane oxidation reaction promoted by the dicopper center of a pMMO enzyme: a computational perspective. Dalton Trans 45:2492–2504
102. Wilkinson B, Zhu M, Priestley ND, Nguyen HHT, Morimoto H, Williams PG, Chan SI, Floss HG (1996) A concerted mechanism for ethane hydroxylation by the particulate methane monooxygenase from *Methylococcus capsulatus* (Bath). J Am Chem Soc 118:921–922
103. Miyaji A, Miyoshi T, Motokura K, Baba T (2015) Discrimination of the prochiral hydrogens at the C-2 position of *n*-alkanes by the methane/ammonia monooxygenase family proteins. Org Biomol Chem 13:8261–8270
104. Mecozzi S, Rebek J Jr (1998) The 55% solution: a formulation for molecular recognition in the liquid state. Chem Eur J 4:1016–1022
105. Miyaji A, Miyoshi T, Motokura K, Baba T (2011) The substrate binding cavity of particulate methane monooxygenase from *Methylosinus trichosporium* OB3b expresses high enantioselectivity for *n*-butane and *n*-pentane oxidation to 2-alcohol. Biotechnol Lett 33:2241–2246
106. Park S, Morley K, Horsman GP, Holmquist M, Hult K, Kazlauskas RJ (2005) Focusing mutations into the *P. fluorescens* esterase binding site increases enantioselectivity more effectively than distant mutations. Chem Biol 12:45–54
107. Shiemke AK, Cook SA, Miley T, Singleton P (1995) Detergent solubilization of membrane-bound methane monooxygenase requires plastoquinol analogs as electron donors. Arch Biochem Biophys 321:421–428
108. Smith DD, Dalton H (1989) Solubilisation of methane monooxygenase from *Methylococcus capsulatus* (Bath). Eur J Biochem 182:667–671
109. Takeguchi M, Miyakawa K, Okura I (1998) Properties of the membranes containing the particulate methane monooxygenase from *Methylosinus trichosporium* OB3b. Biometals 11:229–234
110. Miyaji A, Nitta M, Baba T (2019) Influence of copper ions removal from membrane-bound form of particulate methane monooxygenase from *Methylosinus trichosporium* OB3b on its activity for methane hydroxylation. J Biotechnol X 1:100001
111. Ribbons DW (1975) Oxidation of C1 compounds by particulate fractions from *Methylococcus capsulatus*: distribution and properties of methane-dependent reduced nicotinamide adenine dinucleotide oxidase (methane hydroxylase). J Bacteriol 122:1351–1363
112. Shiemke AK, Arp DJ, Sayavedra-Soto LA (2004) Inhibition of membrane-bound methane monooxygenase and ammonia monooxygenase by diphenyliodonium: implications for electron transfer. J Bacteriol 186:928–937
113. Choi DW, Kunz RC, Boyd ES, Semrau JD, Antholine WE, Han JI, Zahn JA, Boyd JM, de la Mora AM, DiSpirito AA (2003) The membrane-associated methane monooxygenase (pMMO) and pMMO-NADH:quinone oxidoreductase complex from *Methylococcus capsulatus*. Bath. J Bacteriol 185:5755–5764
114. Cornish A, MacDonald J, Burrows KJ, King TS, Scott D, Higgins IJ (1985) Succinate as an *in vitro* electron donor for the particulate methane mono-oxygenase of *Methylosinus trichosporium* OB3b. Biotech Lett 7:319–324
115. Tonge GM, Harrison DE, Knowles CJ, Higgins IJ (1975) Properties and partial purification of the methane-oxidising enzyme system from *Methylosinus trichosporium*. FEBS Lett 58:293–299
116. Leak DJ, Dalton H (1986) Growth yields of methanotrophs. 1. Effect of copper on the energetics of methane oxidation. Appl Microbiol Biotechnol 23:470–476

117. Jensen S, Prieme A, Bakken L (1998) Methanol improves methane uptake in starved methanotrophic microorganisms. Appl Environ Microbiol 64:1143–1146
118. Benstead J, King GM, Williams HG (1998) Methanol promotes atmospheric methane oxidation by methanotrophic cultures and soils. Appl Environ Microbiol 64:1091–1098
119. Hanczár T, Csáki R, Bodrossy L, Murrell JC, Kovács KL (2002) Detection and localization of two hydrogenases in *Methylococcus capsulatus* (Bath) and their potential role in methane metabolism. Arch Microbiol 177:167–172
120. Zahn JA, DiSpirito AA (1996) Membrane-associated methane monooxygenase from *Methylococcus capsulatus* (Bath). J Bacteriol 178:1018–1029

Chapter 3
Heterogeneous and Homogeneous Catalytic Partial Oxidations of Methane to Methanol and Its Derivatives

3.1 Introduction

As described in Chapter 2, the conversion of methane to methanol via direct oxidation of methane with molecular oxygen at ambient temperature and pressure has been established in biological systems. On the other hand, the development of an artificial, energy-efficient, one-step process for the production of methanol from methane using molecular oxygen as the oxidant is still one of the most challenging transformations. Such a process would be attractive, as it could potentially compete with the conventional production of methanol from methane via steam reforming and subsequent methanol synthesis. Therefore, great efforts have been expended to develop heterogeneous and homogeneous catalysts for the molecular oxygen-mediated oxidation of methane to methanol. This chapter presents important examples of the synthetic homogeneous and heterogeneous catalytic systems for the oxidation of methane to methanol, with emphasis on one-step catalytic systems that do not require strong oxidants. However, some systems that do not meet these conditions but have provided motivation or direction in this field are also described.

While multi-stage processes for the conversion of methane to methanol are more feasible than one-step catalytic processes, they are also more complex and less efficient, and therefore less economical. In the multi-stage processes, methanol is adsorbed on a solid surface or is derivatized once produced. In the heterogeneous systems, methoxy species are adsorbed on the solid surface. Therefore, an additional stage is required for the extraction of methanol with water. At the same time, the production from methane via methanol derivatives by using homogeneous catalysts also requires the additional stage for the conversion of methanol derivatives to methanol via processes such as hydrolysis. The methanol adsorbed on the solid surface or derivatized methanol can be considered protected or stabilized methanol, which is not oxidized further by the catalysts. Therefore, the conversion of methane to the methanol adsorbed on the solid surface or the derivatized methanol may be useful methods for obtaining methanol in high yield. Therefore, this chapter presents

T. Baba and A. Miyaji, *Catalysis and the Mechanism of Methane Conversion to Chemicals*,
https://doi.org/10.1007/978-981-15-4132-2_3

representative examples for the conversion of methane to methanol derivatives in heterogeneous and homogeneous systems.

3.2 Heterogeneous Reactions for the Production of Methanol from Methane

Many heterogeneous catalysts have been investigated for the direct and selective oxidation of methane to methanol using molecular oxygen as the oxidant. Metal ions with valences greater than 3, such as Mo, V, Ga, Fe, Co, Mn, and Pd, are active in methane oxidation reactions [1, 2]. However, in most reported reactions, carbon dioxide and carbon monoxide have been the major products, and the yield of partially oxidized methane products has been low. In addition, higher reaction temperatures have typically been required when molecular oxygen is used as the oxidant rather than N_2O [3, 4]. However, some materials based on the aforementioned metals have been found to be capable of catalytically producing methanol from methane.

3.2.1 Iron

Fe(III) species impregnated on fumed silica [5], iron-exchanged zeolite (Fe/ZSM-5) [6], mesoporous silica SBA-15-supported FeO_x [7], FeO_x/SiO_2 prepared by adsorption–precipitation [8] and sol–gel methods [9], and supported crystalline $FePO_4$ [10–15] have been reported to display various degrees of activity and selectivity in the direct oxidation of methane to formaldehyde and methanol. These studies have highlighted iron-based catalysts as promising candidates for the selective oxidation of methane to methanol and formaldehyde.

Silica-supported Fe_2O_3 catalyzes the conversion of methane under relatively mild conditions [2, 16]. A Fe/SiO_2 catalyst afforded reasonable methane conversion and methanol selectivity (13.2% and 78.8%, respectively) compared to those achieved with molybdenum and vanadium oxides (mentioned in Sects. 3.2.2 and 3.2.3) [17]. Mössbauer spectroscopic analysis of the Fe/SiO_2 catalyst revealed that 81% of the iron in the catalyst existed as supported hematite (Fe_2O_3), whereas the remaining 19% was embedded into the silica matrix as tetrahedral Fe^{3+} sites [17]. This result suggests the promising potential of suitable silica-supported iron catalysts for the partial oxidation of methane to methanol.

On the other hand, several iron-based catalysts have been investigated for the molecular oxygen-mediated oxidation of methane to methanol. However, their catalytic performances are not superior to those seen in the silica-supported Fe_2O_3 systems discussed above.

$FePO_4$ catalysts supported on SiO_2 with iron loadings of 2–16 wt% were examined in the selective oxidation of methane using molecular oxygen in a continuous-flow reactor at atmospheric pressure [12]. The major products of the reaction were formaldehyde, carbon dioxide, and carbon monoxide, and methanol were also produced as a minor product. The catalytic activity and selectivity showed a clear dependence on the $FePO_4$ loading. $FePO_4$ supported on silica gave the highest yield of formaldehyde, while alumina-supported $FePO_4$ exhibited the lowest selectivity toward formaldehyde. The highest selectivities and space–time yields for formaldehyde and methanol were observed at lower $FePO_4$ loadings. Mössbauer spectra suggested the formation of penta-coordinated Fe species when $FePO_4$ was supported on silica. X-ray photoelectron spectroscopy (XPS) results indicated that the Fe on the surface of the silica support existed in a lower oxidation state, and the catalysts also displayed a substantial enrichment of phosphorus on the surface, relative to less selective catalysts. The presence of iron that can undergo a facile reduction in a high coordination state and its isolation by the phosphate groups was proposed as the reason for the good performance of the silica-supported $FePO_4$ catalyst. The stability of the resulting formaldehyde and methanol on the support was also considered to play an important role in achieving a high yield.

The single crystalline phase $CuFe_2(P_2O_7)_2$ catalyzed the conversion of methane to formaldehyde and methanol in the presence of oxygen molecules as the oxidant at 673–898 K with high stability [18]. The performance of the catalyst could be optimized by tuning the atomic ratio of copper to iron in the pyrophosphate matrix and the calcination conditions [19]. The optimum conversion of methane to formaldehyde and methanol was achieved at a Cu/Fe atomic ratio close to 1:2. However, its methane conversion and methanol selectivity were low (<1% and <6%, respectively), while the selectivity for formaldehyde was ~95%. The methane conversion and the selectivity toward methanol could be enhanced by using N_2O as the oxidant. For example, at 773 K, methane conversion increased from 0.1 to 1%, and methanol selectivity increased from 5 to 30%. The catalyst was composed predominantly of crystalline $Fe_2(P_2O_7)_2$ with nanodomains containing copper. The interactions between these phases probably generate a disordered crystalline structure, which may play a role in the catalytic site. The pulse reaction studies demonstrated that the oxygens in the lattice react with methane molecules to produce the intermediate that converts to formaldehyde and methanol, which is assumed to be the CH_3O species on copper. The lattice oxygen could be regenerated rapidly by N_2O, resulting in the production of the oxygen radical, which did not form when molecular oxygen was used as the oxidant. The ability to generate oxygen species effectively from N_2O for mediating the conversion of methane to methanol is very similar to the process observed in iron-exchanged ZSM-5 with extra-framework Fe species, which will be discussed next.

An iron ion-exchanged zeolite was found to convert methane to methanol at room temperature using nitrous oxide or hydrogen peroxide as the oxidizing agent [20]. The active species of this reaction over iron-exchanged ZSM-5 was a mononuclear high spin tetravalent iron monoxide species ($Fe(IV) = O$), *i.e.*, the so-called α-Fe site [21]. This active species is formed by treating an iron ion (α-Fe(II)) that is constrained to the "square planar" shape in the zeolite pores with nitrous oxide (N_2O) at 523 K. The highly reactive oxygen, called α-O, is derived from N_2O. α-O extracts hydrogen from methane at 298 K, forming a methyl radical that binds with oxygen via a rebound mechanism to give a CH_3O–Fe complex. The CH_3O group can be extracted as methanol with water [22, 23]. The selectivity for methanol after aqueous extraction was shown to be 75%. This stoichiometric methanol production suggests that the catalytic oxidation of methane to methanol using molecular oxygen to methanol might be achieved over iron-exchanged ZSM-5, but the catalytic reaction has not been established yet.

3.2.2 Molybdenum

Molybdenum oxide (MoO_3) has been investigated as a catalyst for the partial oxidation of methane to methanol using molecular oxygen as the oxidant. Although the major products of MoO_3-catalyzed methane conversion are formaldehyde, carbon dioxide, and carbon monoxide [2, 24, 25], supported molybdenum oxide catalysts were recognized as the most active catalysts for the partial oxidation of methane to methanol until the beginning of this century [1, 26, 27].

Various materials have been investigated as supports for the dispersion of molybdenum oxide. SiO_2-supported MoO_3 exhibits particularly good performance; other materials, such as Al_2O_3, Al_2O_3–SiO_2, MgO, TiO_2, TiO_2–SiO_2, SnO, ZnO_2, H-ZSM-5, and US-Y, have shown quite poor performance compared to SiO_2. The Mo(V) loading has significant effects on properties of the catalysts that determine their catalytic performance, such as their structure and reducibility [28]. Generally, catalysts with low Mo(V) loadings (high dispersion) show better activity and selectivity than those with high Mo(V) loadings. It remains unclear whether the participation of lattice oxygen species (O^{2-}) or adsorbed oxygen species (O^-) plays an important role in the formation of C_1-oxygenates.

Enhancement of methane conversion and its selectivity toward methanol was observed in Mo catalysts in the presence of water vapor. In the case of the partial oxidation of methane to methanol and formaldehyde on highly dispersed MoO_3/SiO_2 catalysts at 873 K using a gas mixture consisting of oxygen, methane, and water [29], the selectivity toward the oxygenates (methanol and formaldehyde) was found to depend strongly on the water vapor fraction; the distribution of the oxygenates in the products increased with increasing content of water in the feed gas. Inversely, the distribution of CO and CO_2 in the products decreased with increasing water content. In the case of the partial oxidation of methane over silica-supported MoO_3 catalysts in the presence of excess water vapor, silicomolybdic acid ($H_4SiMo_{12}O_{40}$) was determined to be the active species [30]; the formation of this species during the reaction was partially evidenced by infrared (IR) spectroscopy measurements. Accordingly, silica-supported silicomolybdic acid catalysts were prepared and tested under the conditions described above to determine whether silicomolybdic acid could effectively catalyze the partial oxidation of methane. The results showed that silicomolybdic acid was active for the production of oxygenates at water vapor fractions of more than 50% in the feed gas. The oxygenate yield increased to 20% of methane at a water vapor fraction of around 60%, with a methane conversion of about 25% and an oxygenate selectivity of 90%. However, in the absence of water vapor, the silicomolybdic acid easily decomposed into SiO_2 and MoO_3 at 873 K.

The properties of the oxygen species generated on the surface of MoO_x by different oxidants have been investigated [31]. Barbaux et al. found that O^- was generated from O_2 and led to the formation of CO_x, while N_2O generated O^{2-} led to increased selectivity toward formaldehyde. With an emphasis on the surface species formed on the supports and their effects on reactivity and selectivity, the partial oxidation of methane to formaldehyde over MoO_3/SiO_2 was examined [32]. Smith and Ozkan found that low MoO_3 loadings led to the generation of silicomolybdic species with terminal Mo=O sites, while higher MoO_3 loadings led to the generation of polymolybdate species with Mo–O–Mo bridges; the former showed better selectivity toward formaldehyde than the latter. Interestingly, another study seemed to support the opposite view; bridging oxygens [–O–] were concluded to be responsible for selective oxidation, while terminal oxygens [=O] were associated with full oxidation [33].

Different C–H activation pathways over various terminal [=O] and bridging [–O–] active sites have been investigated by Fu et al. using density functional theory (DFT) calculations of the Mo_3O_9 model system [34]. Their theoretical calculations indicated that H abstraction is a viable pathway for the activation of C–H bonds on molybdenum oxide. The terminal sites [=O] were calculated to be more reactive in H abstraction than the bridging sites [–O–]. The H abstraction produced radical pairs, which have also been detected in electron paramagnetic resonance (EPR) experiments. These radical pairs underwent a fast rebound process resulting in the generation of stable hydroxyl and alkoxy species, which have been detected in IR experiments. These experimental observations can be explained via the mechanism below:

1. The calculated activation energy for CH_4 conversion is 188 kJ mol^{-1}, which was consistent with the values derived from experimental data (172 [35], 176 [3], and 189 [36] kJ mol^{-1}).
2. A one-electron oxidation process results in the generation of a Mo^{5+} species and an alkyl radical, which is consistent with electron paramagnetic resonance measurements [37].
3. H atom abstraction followed by fast oxygen atom insertion results in the generation of stable hydroxyl and alkoxy species, consistent with IR observations [38].

The mechanism shown in Fig. 3.1 is more favorable than the alternative (5 + 2) cycloaddition reaction mechanism. H abstraction/O rebound mechanisms are ubiquitous in various areas of chemistry, including biological reactions of metalloenzymes, heterogeneous reactions on solid surfaces, homogeneous reactions with organic metal complexes, and gas-phase and matrix-isolated species [39]. The calculation results also imply that the (2 + 2) cycloaddition reaction mechanism (Mo-O reacts with H-CH_3) is most feasible when the Mo=O bond is highly polar, while the (5 + 2) mechanism can provide another effective pathway if the two Mo=O bonds are in close proximity.

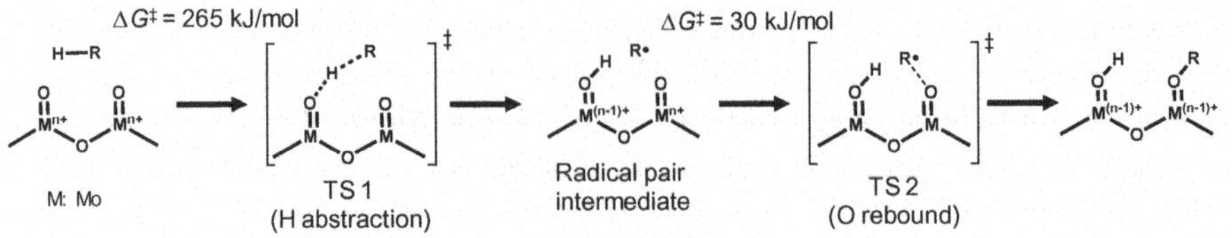

Fig. 3.1 Reaction mechanism of methane oxidation to methoxy species on Mo_3O_9 model (M: metal ion)

Evidence for the involvement of H abstraction in the oxidation of hydrocarbons by such high-valence metal-oxo compounds has been provided [39–42]. In the mechanism, the metal center and the oxygen act as the electron acceptor and the proton acceptor, respectively. Thus, the H abstraction activity should correlate not only with the reducibility of the metal center but also with the basicity of the oxygen atom. The driving force for these reactions is the formation of a strong Mo–H bond. In several biochemical processes, the involvement of one or more hydrogen atom transfer steps has been reported, such as the oxidation of methane by soluble methane monooxygenase (sMMO) as discussed in Chapter 2. An understanding of such metal-oxo based systems is vital for designing catalysts for the selective oxidation of methane with efficiency and specificity similar to those of methane monooxygenases.

3.2.3 *Vanadium*

Vanadium oxide (V_2O_5) has also been investigated as a catalyst for the partial oxidation of methane to methanol using molecular oxygen as the oxidant. Compared to molybdenum oxide catalysts, vanadium oxide catalysts show higher methane conversion activity but lower selectivity toward methanol.

As in the case of molybdenum catalysts, the dispersion of metal species and the presence of water vapor are important to the selectivity for methanol. Several supporting materials have been tested for the dispersion of vanadium species, including the mesoporous siliceous material SBA-15 [43] and the mesoporous siliceous material MCM-41 with high surface area. For VO_x/SBA-15, a clear correlation between the dispersion of vanadium oxide on the surface of silica and the selectivity toward formaldehyde was observed [44]. For V_2O_5/SiO_2 catalysts, both the dispersion of vanadium oxide and the selectivity for formaldehyde decreased when the vanadium catalyst loading was above 2 wt%. The active sites of the vanadium oxide catalyst supported on silica for the selective formation of formaldehyde were proposed to be isolated tetrahedral vanadium species containing terminal V=O groups [32, 45–48]. The catalyst VO_x/MCM-41 exhibited a high concentration of isolated active sites. This catalyst also demonstrated increased activity during the co-feeding of steam due to the generation of $VO(OH)_x(OSi)_{3-x}$ (valence of the vanadium ion is 5+) surface species or the corresponding "reduced sites" $VO(OH)_x(OSi)_{2-x}$ (valence of the vanadium ion is 4+), which were believed to be the true active sites. A V_2O_5/SiO_2 xerogel was also able to convert methane to methanol, and the vanadium species on its surface were investigated using solid-state ^{51}V nuclear magnetic resonance (NMR) and Raman spectroscopy. According to these analyses, the monomeric tetrahedral V^{5+} species on the surface were the active sites for the conversion of methane to methanol and formaldehyde, and these species, along with polymeric surface tetrahedral V^{5+} species and crystalline V_2O_5, were the active sites for the further oxidation of the primary products to carbon dioxide and carbon monoxide [49].

3.2.4 Copper

Very few studies have investigated the use of synthetic Cu-based catalysts for the selective oxidation of methane to methanol and formaldehyde, as CuO_x species, like Fe_2O_3 aggregates, favor the total oxidation of methane [26, 32, 34, 49]. Although no heterogeneous Cu materials that can catalyze the oxidation of methane to methanol have been reported, stoichiometric heterogeneous transformations involving Cu are interesting, as copper (and iron) sites are the active centers in the biologically catalyzed oxidation of methane to methanol using O_2 at ambient temperature (see Chapter 2 for a detailed discussion of the biocatalytic reaction). So far, the use of copper-exchanged zeolites such as Cu-ZSM-5 for methane oxidation has been investigated [50]. Also, the selective oxidation of methane to formaldehyde over a very low loading of CuO_x on mesoporous silica, SBA-15, has also been documented [51].

In most reactions using zeolites, the methanol must be extracted from the zeolite using water because the methoxy species are typically produced on the zeolite-based material. In addition, these materials require reactivation via calcination before the subsequent reaction cycle [50, 52]. For example, the oxidation of methane to methanol over Cu-ZSM-5 at 398 K was reported [50]. However, the zeolite was only active after pretreatment with O_2 at ≥ 573 K, and the formed methanol had to be extracted from the zeolite. As these reactions do not occur in a catalytic cycle in which all elementary reactions occur under the same conditions, they are not catalytic and require the use of multi-step processes to obtain methanol from methane.

In copper-exchanged zeolites, a mono-oxo di-nuclear copper species ($[Cu(\mu\text{-}O)Cu]^{2+}$, Fig. 3.2) formed by copper ions bonded to the oxygen atom of the framework has been observed [53, 54]. The generation of tri-oxo tri-nuclear copper

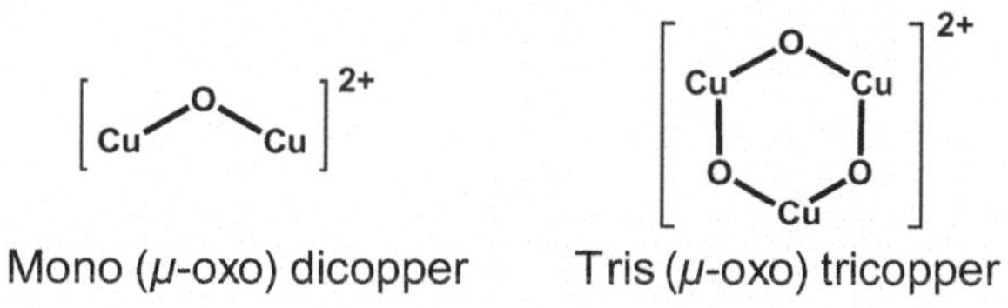

Fig. 3.2 Proposed active oxygen species in copper-exchanged zeolites

species ($[Cu_3(\mu\text{-}O)_3]^{2+}$) has also been reported [55, 56]. Although the oxygen molecules are activated by the same binuclear copper ion species as in the enzyme-catalyzed reaction, the subsequent oxo species differs from that produced in the enzyme; namely, a CH_3O-group is formed in the methane conversion reaction on the copper ion-exchanged zeolite. Methanol is then obtained by extracting this species using water. Additionally, in most copper-exchanged zeolites, expensive oxidants such as tert-butyl hydroperoxide [57], H_2O_2 [58], and N_2O [59] have been required to generate reactive oxygen species.

Interestingly, methanol can be obtained using Cu-exchanged zeolites, even in the absence of oxygen using water as the oxidizing agent [60]. A quantum chemical calculation of this process indicated that after the copper ion bonds with the carbon of methane, a zeolite lattice oxygen bonds with hydrogen, giving $Cu–CH_3$ and a Brønsted acid. The CH_3 group then reacts with the oxygen atom bridging the Cu(II) centers to form a CH_3O group. At this time, both Cu(II) centers are reduced to Cu(I). Subsequently, a water molecule reacts with the CH_3O group to dissociate methanol, a binuclear copper complex in which water molecules are crosslinked is formed, and finally, hydrogen dissociates. The reaction of water differentiates this mechanism from the formation of methanol via the reaction of CH_3O-groups formed on the surface of the zeolite with water [61]. In this reaction mechanism, the oxygen of the zeolite lattice also contributes to the reaction, in addition to the copper ions and the oxygen atoms bridging them. This is a function of the zeolite lattice, which creates an environment more similar to the reaction field of the enzyme than to a molecular catalyst involving only a metal ion and its ligands.

3.2.5 Heteropoly Compounds

Heteropoly compounds have been investigated as catalysts for the partial oxidation of methane to formaldehyde and methanol at atmospheric pressure in the temperature range 973–1023 K. $(NH_4)_6HSiMo_{11}FeO_{40}$, $(NH_4)_4PMo_{11}FeO_{39}$, and $H_4PMo_{11}VO_{40}$ showed activity toward methane oxidation when O_2 or N_2O was used as the oxidizing agent; the products of these reactions were CH_3OH, HCHO, CO, CO_2, and water [62]. The catalytic conversion of methane using solid $CsPW_{12\text{-}x}M_xO_{40}$ (M=Fe, Co, or Ni) and the corresponding reaction mechanism were also investigated [63]. According to the reported mechanism, methanol and carbon monoxide are produced stoichiometrically, while formaldehyde and carbon dioxide are produced catalytically.

3.3 Homogeneous Reactions for the Production of Methanol from Methane

Methane oxidation to methanol with a di-nuclear, bridged Ru(III) catalyst [(HSalen)$_2$Ru$_2$(μ-O)(μ-CH$_3$CO$_2$)$_2$] (salen = 2,2′-ethylenebis(nitrilomethylidene)diphenol, *N*,*N*′-ethylenebis(salicylimine)) was reported by Jasra et al. The homogeneous catalyst achieved 100% methanol selectivity in the oxidation of methane (12.5 atm) using molecular oxygen (2.5 atm) at 303 K in a 1:1 mixture of acetone and H_2O [64]. The reaction conditions of the catalyst are very close to the ideal conditions for a practical process (neutral media, ambient temperature, atmospheric pressure, etc.). In a typical experiment (CH_4: 10 atm, O_2: 5 atm), this catalyst achieved a methanol yield of ~9% based on methane with a TON of 54 and a selectivity of 96.6%. The authors suggested that the homogeneous catalyst first activated both molecular oxygen and methane in the neutral solvent. The proposed active oxygen species generated in the Ru complex is shown in Fig. 3.3. The insertion of an oxygen atom into the C–H bond of methane occurs via an anionic route. This mechanism is different from that of the di-iron site of sMMO, which reacts with the molecular oxygen to introduce oxygen into the C–H bond through a radical process, as described in Chapter 2

$$CH_4 + O_2 \xrightarrow[\text{Acetone/H}_2\text{O, 303 K}]{\text{0.5 mM Ru(III) catalyts}} CH_3OH \tag{3.1}$$

.

Metal ions with strong electron acceptor ability are expected to activate the C–H bonds of methane molecules via interaction of the d-orbitals of the metal species with the σ bond of methane. Therefore, many high-valence metal ions have been investigated for methane activation to produce methanol. Shilov reported a Pt(IV) complex that exhibited highly efficient oxidation of methane to methanol [65].

$$CH_4 + H_2O + [PtCl_6]^{2-} \xrightarrow[\text{Acetate/H}_2\text{O, \textasciitilde 373 K}]{K_2PtCl_4} CH_3OH + 2H^+ + PtCl_2 + 4Cl^- \tag{3.2}$$

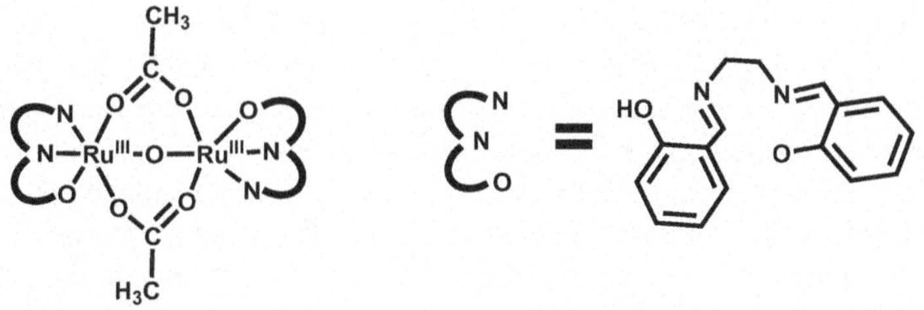

Fig. 3.3 3 Proposed active oxygen species in Ru(III) catalyst

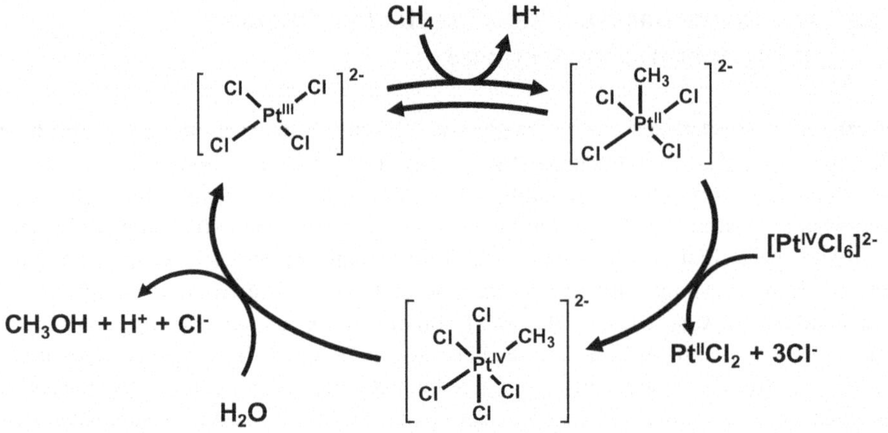

Fig. 3.4 Stochiometric methane oxidation cycle on Pt complex

The oxidation cycle is schematically shown in Fig. 3.4. By the electrophilic activation of methane by platinum, Pt-CH_3 species is produced in the reaction. The reaction requires a stoichiometric amount of $[PtCl_6]^{2-}$ for the conversion.

3.4 Homogeneous Catalysts for the Production of Methanol Derivatives from Methane

As described in the previous section, the highly selective synthesis of methanol via methane conversion by homogeneous catalysts is still difficult. This difficulty arises from the fact that the product methanol is more reactive than the reactant methane; thus, further oxidation of methanol to formaldehyde, formic acid, carbon dioxide, and carbon monoxide occurs under the conditions of the methane conversion reaction. Therefore, reactions in which methanol is obtained as its more stable ester derivatives have been devised. The obtained methanol derivatives can then be converted into methanol via hydrolysis.

In this approach, both a reaction to oxidize methane to a methanol derivative and a reaction to hydrolyze the methanol derivative are required. This is similar to the extraction of the methoxy species formed on the zeolite surface with water in the conversion of methane using metal ion-exchanged zeolites. Therefore, the conversion of methane to methanol derivatives is industrially disadvantageous. However, these reactions can proceed at temperatures of 373 K or less, have high selectivity toward the methanol derivatives, and are true catalytic reactions, unlike conversion using metal ion-exchanged zeolites. In this section, an overview of the conversion of methane to methanol derivatives by homogeneous catalysts is discussed.

3.4.1 *Transformation of Methane into Methanol Derivatives via Methyl Radicals*

The methyl radical intermediate is known to play a key role in some methane conversion reactions to methanol. Furthermore, radical processes have also been reported to accomplish the oxidation of methane to methanol derivatives under mild conditions, especially at relatively moderate temperatures (<473 K).

The $EnCl_3$-Zn-trifluoroacetic acid-catalytic system has been reported for the oxidation of methane with molecular oxygen and the subsequent methyl esterification mediated by trifluoroacetic acid (CF_3COOCH_3) [66, 67]. The oxidation by this catalytic system proceeded at 313 K, and a TON of 5.3 was achieved for the production of CF_3COOCH_3, and that for carbon dioxide was about 25 after 1 h of reaction. According to the selectivities obtained during the oxidations of several alkanes, the production of methyl radical intermediates during the oxidation of methane was proposed. Furthermore, the generation of the active oxygen species from the divalent europium and superoxide was suggested by electrochemical studies. Based on these experiments, in the catalytic system, En^{3+} is reduced to En^{2+} by Zn^0, followed by the reaction with an oxygen molecule to generate active oxygen species on europium. Then, the active oxygen species oxidize methane to methanol and trifluoroacetic acid to carbon dioxide. The TON of the catalytic reaction could be enhanced to 10 in 1 h reaction with the use of Ti(IV) oxide or Ti(IV) oxyacetylacetonate [68]. This enhancement was considered to originate from the electron transfer between Eu^{3+} and Zn^0 by Ti^{4+}, which enhances the generation of active oxygen species on europium.

In the V-containing heteropolyacid system $H_5PV_2Mo_{10}O_{40}/K_2S_2O_8$/TFAA and TFA, methane was converted into methyl trifluoroacetate in 95% yield at 353 K [69]. A possible mechanism for this reaction involves the generation of $\bullet CH_3$ through the abstraction of H from methane by V(V)=O, followed by oxidation of the methyl radical to a methyl cation and the formation of the methyl ester. However, the use of expensive $K_2S_2O_8$ and the corrosive nature of CF_3CO_2H are not acceptable for industrial applications.

3.4.2 *Electrophilic Methane Activation and Transformation to Methanol Derivatives*

Periana and colleagues found that Hg(II) in sulfuric acid could catalyze C–O bond formation via methane activation in the electrophilic substitution reaction shown in reaction (3.3) [70].

$$CH_4 + 2\,H_2SO_4 \xrightarrow[H_2SO_4,\ 453\ K]{Hg^{II}SO_4} CH_3OSO_3H + 2H_2O + SO_2 \qquad (3.3)$$

The catalytic cycle is schematically shown in Fig. 3.5. In this catalytic reaction, a nucleophilic metal–ligand ($-OSO_3H$) forms a C–O bond with CH_3 after the formation

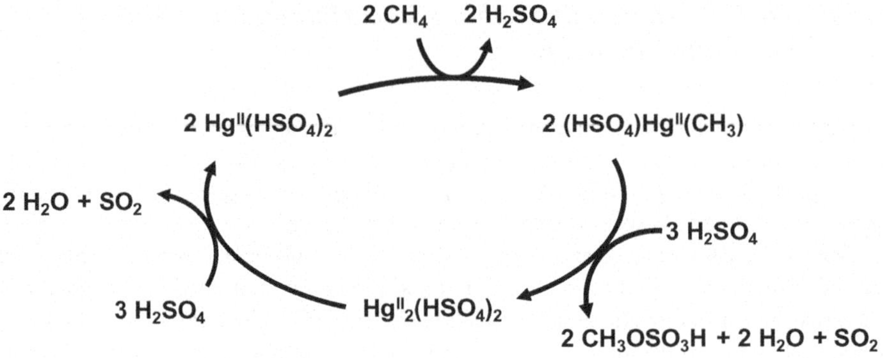

Fig. 3.5 Catalytic methane oxidation cycle on Hg complex

of the Hg–CH_3 intermediate. In other words, the CH_3 coordinated to Hg showed electrophilic character. Reaction (3.3) was found to proceed not only with mercury, but also with Ce(IV), Pd(II), and Au (III) [71, 72].

Using Pt(II) with bipyridine as the ligand and oleum (SO_3/H_2SO_4) as the solvent and oxidant at 493 K, a methyl hydrogen sulfate yield of 70% based on methane and a selectivity exceeding 95% were achieved [73]. When K_2PtCl_4 was used in 20% fuming sulfuric acid, the turnover frequency (TOF) reached approximately 24,000 h^{-1} (conversion rate 20–30%, selectivity >98%) [74]. The performance of these systems was attributed to the Pt(II)/Pt(IV)/Pt(II) catalytic cycle and the use of H_2SO_4. H_2SO_4 cannot oxidize Pt(II) catalysts or CH_3–Pt(II) species to the corresponding Pt(IV) complexes and intermediates, respectively, but is able to react with methanol to form the corresponding ester. The ester formation avoids the overoxidation of methanol. The Pt(IV) complexes could also play a role as oxidants to generate CH_3–Pt(IV) intermediates from CH_3–Pt(II) species, as in the Shilov system described above. On the other hand, when Tl(III) or Pb(IV), which are more strongly electron-withdrawing than Hg(II), were used in an aqueous trifluoroacetic acid or acetic acid solution, which are weaker than sulfuric acid, methyl trifluoroacetate or methyl acetate was produced from methane [75].

The combinations of suitable catalysts with inexpensive and effective oxidants for obtaining methanol derivatives from methane in high selectivity and yield under mild conditions have been investigated. For example, the conversion of methane (5.4 MPa) to methyl trifluoroacetate ($CH_3CO_2CH_3$) was reported by Gretz et al. [76]. In their report, >60% yield of $CH_3CO_2CH_3$ (based on Pd(II)) was achieved by using $Pd(OAc)_2$ in trifluoroacetic acid (CF_3CO_2H) at 353 K. This reaction might involve electrophilic activation of methane to form a CH_3–Pd(II) (CH_3CO_2) intermediate followed by the formation of $CF_3CO_2CH_3$ and Pd(0). A catalytic version of the methane conversion by using $Pd(OAC)_2$ has been established by adding the hydrogen peroxide/trifluoroacetic anhydride (TFAA, $(CF_3CO)_2O$) regeneration system for the generation of Pd(II) from Pd(0), which is similar to the Wacker oxidation system

[77]. The turnover number (TON) of the catalytic reaction was 5.3, and the yield was very low. However, the TON was increased by using a bridged bis-*N*-heterocyclic carbene (NHC)–Pd(II) complex as the catalyst and $K_2S_2O_8$ as the oxidant for the trifluoroacetoxylation of methane in TFAA/TFA at 363 K [78]. The TON for the methane (3 MPa) conversion reached 30 in 14 h at 363 K. According to a computational study [79], this reaction proceeds via a Pd(II)/Pd(IV)/Pd(II) catalytic cycle and involves C–H bond activation by Pd(II).

The use of environmentally friendly and inexpensive molecular oxygen (O_2) as the terminal oxidant is always desirable in oxidation reactions. However, only a few reports have been published on the partial oxidation of methane to methanol derivatives under homogeneous catalysts by using molecular oxygen as the oxidant. The poor solubility of molecular oxygen in solvents for liquid phase reaction is one of the main limitations for this type of reaction in homogeneous catalysis. For example, the Co(III)-catalyzed trifluoroacetoxylation of methane to methyl trifluoroacetate was observed with O_2 in TFA at 453 K [80]. This reaction has been improved to obtain a methyl trifluoroacetate yield of 50% (based on methane) via the oxidation of methane (1 MPa) with O_2 (0.5 MPa) and 3.8 mol% $Co(OAC)_2$ $4H_2O$ as the catalyst in TFAA/TFA (6:1) at 453 K [81]. An et al. achieved the one-pot aerobic oxidation of methane with a Pd(II) catalyst at 353 K using a *p*-benzoquinone/hydrobenzoquinone redox shuttle, NO_2/NO couple, and O_2 [82]. The TON (expressed in terms of moles of CF_3COOCH_3 per mole of Pd(II)) of the Pd catalyst was only 7, but the yield of methyl trifluoroacetate was 0.06% based on methane. A $Pd(OAc)_2$/*p*-benzoquinone/$H_5PMo_{10}V_2O_{40}$ system for the direct oxidation of methane to the methanol derivative CF_3COOCH_3 in CF_3COOH at low temperature was demonstrated. The catalysts can be regenerated by molecular oxygen [83]. Perfluorocarbons exhibit very high oxygen solubility and transport properties [84–86]. Therefore, the perfluorocarbon C_8F_{18}, which is inert in catalytic oxidation, was introduced to improve oxygen solubility and transport in CF_3COOH. In the reaction, the methanol derivative CF_3COOCH_3 was the only liquid product, and CO_2 was the only gaseous product. Higher partial pressures of O_2 led to higher yields of CF_3COOCH_3 by the promotion of $H_5PMo_{10}V_2O_{40}$ regeneration [87].

3.5 Heterogeneous Catalysts for the Production of Methanol Derivatives from Methane

A series of methane conversion reactions originating from the Hg catalyst developed by Periana and co-workers have been found to proceed at relatively low temperatures (273–493 K). The product of these reactions is the methyl ester, which is hydrolyzed to give the final target compound, methanol. These catalytic systems have inspired the development of heterogeneous catalysts for the oxidation of methane to methanol derivatives. For example, the conversion of methane via an M–CH_3 intermediate has been realized using polymeric materials with metal ligands [88] and has been developed for heterogeneous catalysis. A shared triazine-based framework (CTF) via the trimerization of 2,6-dicyanopyridine in molten $ZnCl_2$ has also been developed [89]. This solid ligand modulates $PtCl_2$ to form a solid similar to the bipyrimidine $PtCl_2$ system. The Pt–CTF system was activated by the oxidation of methane in fuming sulfuric acid to give methyl bisulfate. The catalyst required an activation period, but eventually reached a TON of greater than 250. These examples demonstrate how concepts developed in homogeneous catalysis can be useful for the development of heterogeneous catalysts. However, in the context of industrial application, this batch process showed very low space–time yield and required the post-treatment of the intermediate methyl hydrogen sulfate and re-oxidation of SO_2, which can be prohibitively expensive. Additionally, the loss of the Pt-CTF catalyst in each recycling step (about 5–10 wt%) is cost-inefficient.

3.6 Summary

This chapter discussed the direct conversion of methane to methanol over homogeneous or heterogeneous catalysts. In these catalytic reactions, the yields of the methanol obtained are low [90], mainly due to the high reactivity of methanol under the reaction conditions for the activation of methane. In other words, the current state of the art does not allow the simultaneous maximization of both methane conversion and methanol selectivity. The available reactions give high methanol selectivity only at very low conversions and are not suitable for production on industrial scales.

This chapter also reviewed the conversions of methane to methyl esters in homogeneous systems and the conditions for achieving a tradeoff between conversion and selectivity. These reactions achieved very high product selectivity even at high methane conversion [91], and also allow the production of methyl esters in a concentration range of 2 mol L^{-1} or higher [92], which is the desired product concentration necessary for achieving product separation and recovery on an industrial scale. Methanol production at this concentration has not been achieved with conventional heterogeneous catalysts, and this high yield of methanol derivative production is due to the conversion to methyl ester, which is less reactive than methanol under methane conversion reaction conditions.

Since the methyl ester obtained from methane must be hydrolyzed and converted to methanol, some reports have addressed the production of methyl esters as methanol production. However, the approaches for the production of methanol from methane via a methanol derivative and those for converting methane directly to methanol should be considered separately. The hydrolysis process is required for obtaining methanol from methanol derivatives. The methyl bisulfate or methyl trifluoroacetate hydrolysis is relatively easy on a small scale. However, on an industrial scale, several other challenging factors, such as the exothermic addition of the reaction mixture to water which needs to be performed slowly to prevent the volatilization of methanol, must be considered [90]. The process for producing methanol from methane via methanol derivatives requires hydrolysis and is hence disadvantageous. Further, in addition to the use of highly corrosive substances such as sulfuric acid and trifluoroacetic acid, the use of hydrolysis processes can be a major problem in the methanol synthesis process from methane via methanol derivatives.

On the other hand, the "methanol protection" strategy undertaken in the conversion to methanol derivatives suggests an important direction of research toward the development of practical applications for methanol synthesis from methane [90]. For example, the physical separation of methanol from a reaction mixture using a membrane can be considered for adapting to methanol production reactions. The membrane separation technique is also being studied in the production of methanol from methane using biocatalysts [93–95]. Also, other methanol protection strategies such as the use of chemical loops in the methane conversion with metal ion-exchanged zeolites and the use of multicomponent catalysts for the protection of methanol have been reported. Overall, approaches to both the catalytic reactions for the direct synthesis of methanol from methane and other approaches for methanol production from a broader perspective are essential for enabling potential breakthroughs in the development of artificial catalysts for conversion of methane to methanol.

References

1. Spencer ND (1988) Partial oxidation of methane to formaldehyde by means of molecular oxygen. J Catal 109:187–197
2. Spencer ND, Pereira CJ (1989) V_2O_5-SiO_2-catalyzed methane partial oxidation with molecular oxygen. J Catal 116:399–406
3. Liu HF, Li RS, Liew KY, Johnson RE, Lunsford JH (1984) Partial oxidation of methane by nitrous oxide over molybdenum on silica. J Am Chem Soc 106:4117–4121
4. Caceres CV, Fierro JLG, Lopez Agudo A, Blanco MN, Thomas HJ (1985) Preparation and characterization of equilibrium adsorption-prepared molybdena-alumina catalysts. J Catal 95:501–511
5. Kobayashi T, Guilhaume N, Miki J, Kitamura N, Haruta M (1996) Oxidation of methane to formaldehyde over $FeSiO_2$ and Sn-W mixed oxides. Catal Today 32:171–175
6. Michalkiewicz B (2004) Partial oxidation of methane to formaldehyde and methanol using molecular oxygen over Fe-ZSM-5. Appl Catal A: Gen 277:147–153
7. Zhang Q, Li Y, An D, Wang Y (2009) Catalytic behavior and kinetic features of FeOx/SBA-15 catalyst for selective oxidation of methane by oxygen. Appl Catal A: Gen 356:103–111
8. Arena F, Gatti G, Martra G, Coluccia S, Stievano L, Spadaro L, Famulari P, Parmaliana A (2005) Structure and reactivity in the selective oxidation of methane to formaldehyde of low-loaded FeO_x/SiO_2 catalysts. J Catal 231:365–380
9. He J, Li Y, An D, Zhang Q, Wang Y (2009) Selective oxidation of methane to formaldehyde by oxygen over silica-supported iron catalysts. J Nat Gas Chem 18:288–294
10. Otsuka K, Wang Y (2001) Direct conversion of methane into oxygenates. Appl Catal A: Gen 222:145–161
11. Wang Y (2006) Selective oxidation of hydrocarbons catalyzed by iron-containing heterogeneous catalysts. Res Chem Intermed 32:235–251
12. McCormick RL, Alptekin GO (2000) Comparison of alumina-, silica-, titania-, and zirconia-supported $FePO_4$ catalysts for selective methane oxidation. Catal Today 55:269–280

13. Alptekin GO, Herring AM, Williamson DL, Ohno TR, McCormick RL (1999) Methane partial oxidation by unsupported and silica supported iron phosphate catalysts: influence of reaction conditions and co-feeding of water on activity and sectivity. J Catal 181:104–112
14. Wang X, Wang Y, Tang Q, Guo Q, Zhang Q, Wan H (2003) MCM-41-supported iron phosphate catalyst for partial oxidation of methane to oxygenates with oxygen and nitrous oxide. J Catal 217:457–467
15. Wang Y, Wang X, Su Z, Guo Q, Tang Q, Zhang Q, Wan H (2004) SBA-15-supported iron phosphate catalyst for partial oxidation of methane to formaldehyde. Catal Today 93–95:155–161
16. Brown MJ, Parkyns ND (1991) Progress in the partial oxidation of methane to methanol and formaldehyde. Catal Today 8:305–335
17. Fajardo CAG, Niznansky D, Guyen YN, Courson C, Roger A-C (2008) Methane selective oxidation to formaldehyde with Fe-catalysts supported on silica or incorporated into the support. Catal Commun 9:864–869
18. Štolcová M, Litterscheid C, Hronec M, Glaum R (2007) Bimetallic orthophosphate and pyrophosphate catalysts for direct oxidation of methane to formaldehyde. Stud Surf Sci Catal 167:37–42
19. Polnišer R, Štolcová M, Hronec M, Mikul M (2011) Structure and reactivity of copper iron pyrophosphate catalysts for selective oxidation of methane to formaldehyde and methanol. Appl Catal A Gen 400:122–130
20. Sobolev VI, Dubkov KA, Panna OV, Panov GI (1995) Selective oxidation of methane to methanol on a FeZSM-5 surface. Catal Today 24:251–252
21. Snyder BER, Vanelderen P, Bols ML, Hallaert SD, Böttger LH, Ungur L, Pierloot K, Schoonheydt RA, Sels BF, Solomon EI (2016) The active site of low-temperature methane hydroxylation in iron-containing zeolites. Nature 536:317–321
22. Starokon EV, Parfenov MV, Pirutko LV, Abornev SI, Panov GI (2011) Room-temperature oxidation of methane by α-oxygen and extraction of products from the FeZSM-5 surface. J Phys Chem C 115:2155–2161
23. Göltl F, Michel C, Andrikopoulos PC, Love AM, Hafner J, Hermans I, Sautet P (2016) Computationally exploring confinement effects in the methane-to-methanol conversion over iron-oxo centers in zeolites. ACS Catal 6:8404–8409
24. Zhen KJ, Khan MM, Mak CH, Lewis KB, Somorjai GA (1985) Partial oxidation of methane with nitrous oxide over V_2O_5-SiO_2 catalyst. J Catal 94:501–507
25. Chen SY, Willcox D (1993) Effect of vanadium oxide loading on the selective oxidation of methane over vanadium oxide (V_2O_5)/silica. Ind Eng Chem Res 32:584–587
26. Otsuka K, Hatano M (1992) Boron, phosphorus, and alkaline earth-metal mixed oxides as active catalysts in partial oxidation of methane into formaldehyde. Chem Lett 21:2397–2400
27. Weng T, Wolf EE (1993) Partial oxidation of methane on Mo/Sn/P silica supported catalysts. Appl Catal A 96:383–396
28. Tabata K, Teng Y, Takemoto T, Suzuki E, Banares MA, Pena MA, Fierro JL (2002) Activation of methane by oxygen and nitrogen oxide. Catal Rev 44:1–58
29. Aoki K, Ohmae M, Nanba T, Takeishi K, Azuma N, Ueno A, Ohfune H, Hayashi H, Udagawa Y (1998) Direct conversion of methane into methanol over MoO_3/SiO_2 catalyst in anexcess amount of water vapor. Catal Today 45:29–33
30. Kasztelan S, Payen E, Moffat JB (1988) The formation of molybdosilicic acid on Mo/SiO_2 catalysts and its relevance to methane oxidation. J Catal 112:320–324
31. Barbaux Y, Elamrani A, Bonnelle JP (1987) Catalytic oxidation of methane on MoO_3-SiO_2: mechanism of oxidation with O_2 and N_2O studied by surface potential measurements. Catal Today 1:147–156
32. Smith MR, Ozkan US (1993) The partial oxidation of methane to formaldehyde: role of different crystal planes of MoO_3. J Catal 141:124–139
33. Grzybowska-Swierkosz B (2000) Thirty years in selective oxidation on oxides: what have we learned? Top Catal 11:23–42

34. Fu G, Xu X, Lu X, Wan H (2005) Mechanisms of methane activation and transformation on molybdenum oxide based catalysts. J Am Chem Soc 127:3989–3996
35. Kihlborg L (1963) Least squares refinement of the crystal structure of molybdenum trioxide. Ark Kemi 21:357–364
36. Spencer ND, Pereira CJ (1987) Partial oxidation of CH_4 to HCHO over a MoO_3-SiO_2 catalyst: a kinetic study. AIChE J 33:1808–1812
37. Kuba S, Heydorn PC, Grasselli RK, Gates BC, Che M, Kno¨zinger H (2001) Redox properties of tungstated zirconia catalysts: relevance to the activation of *n*-alkanes. Phys Chem Chem Phys 3:146–154
38. Busca G, Finocchio E, Lorenzelli V, Ramis G, Baldi M (1999) IR studies on the activation of C-H hydrocarbon bonds on oxidation catalysts. Catal Today 49:453–465
39. Limberg C (2003) The role of radicals in metal-assisted oxygenation reactions. Angew Chem Int Ed 42:5932–5954
40. Mayer JM (1998) Hydrogen atom abstraction by metal−oxo complexes: understanding the analogy with organic radical reactions. Acc Chem Res 31:441–450
41. Soper JD, Mayer JMJ (2003) Slow hydrogen atom self-exchange between Os(IV) anilide and Os(III) aniline complexes: relationships with electron and proton transfer self-exchange. J Am Chem Soc 125:12217–12229
42. Roberts BP (1999) Polarity-reversal catalysis of hydrogen-atom abstraction reactions: concepts and applications in organic chemistry. Chem Soc Rev 28:25–35
43. Fornés V, López C, López HH, Mart´ınez A (2003) Catalytic performance of mesoporous VO_x/SBA-15 catalysts for the partial oxidation of methane to formaldehyde. Appl Catal A: Gen 249:345–354
44. Kartheuser B, Hodnett BK (1993) Relationship between dispersion of vanadia on silica catalysts and selectivity in the conversion of methane into formaldehyde. J Chem Soc Chem Commun 1093–1094
45. Koranne MM, Goodwin JG Jr, Marcelin G (1994) Partial oxidation of methane over silica- and alumina-supported vanadia catalysts. J Catal 148:388–391
46. Irusta S, Cornaglia LM, Miró EE, Lombardo EA (1995) The role of V=O sites on the oxidation of methane to formaldehyde over V/SiO_2. J Catal 156:167–170
47. Irusta S, Marchi AJ, Lombardo EA, Miró EE (1996) Characterization of surface species on V/SiO2 and V, Na/SiO2 and their role in the partial oxidation of methane to formaldehyde. Catal Lett 40:9–16
48. Das N, Eckert H, Hu H, Wachs IE, Walzer JF, Feher FJ (1993) Bonding states of surface vanadium(V) oxide phases on silica: structural characterization by ^{51}V NMR and Raman spectroscopy. J Phys Chem 97:8240–8243
49. Wang CB, Herman RG, Shi C, Sun Q, Roberts JE (2003) V_2O_5-SiO_2 xerogels for methane oxidation to oxygenates: preparation, characterization, and catalytic properties. Appl Catal A 247:321–333
50. Groothaert MH, Smeets PJ, Sels BF, Jacobs PA, Schoonheydt RA (2005) Selective oxidation of methane by the bis(μ-oxo)dicopper core stabilized on ZSM-5 and mordenite zeolites. J Am Chem Soc 127:1394–1395
51. Li Y, An D, Zhang Q, Wang Y (2008) Copper-catalyzed selective oxidation of methane by oxygen: studies on catalytic behavior and functioning mechanism of CuO_x/SBA-15. J Phys Chem C 112:13700–13708
52. Panov GI, Sobolev VI, Dubkov KA, Parmon VN, Ovanesyan NS, Shilov AE, Shteinman AA (1997) Iron complexes in zeolites as a new model of methane monooxygenase. React Kinet Catal Lett 61:251–258
53. Smeets PJ, Groothaert MH, Schoonheydt RA (2005) Cu based zeolites: a UV-vis study of the active site in the selective methane oxidation at low temperatures. Catal Today 110:303–309
54. Woertink JS, Smeets PJ, Groothaert MH, Vance MA, Sels BF, Schoonheydt RA, Solomon EIA (2009) $[Cu_2O]^{2+}$ core in Cu-ZSM-5, the active site in the oxidation of methane to methanol. Proc Natl Acad Sci USA 106:18908–18913

55. Grundner S, Markovits MAC, Li G, Tromp M, Pidko EA, Hensen EJM, Jentys A, Sanchez-Sanchez M, Lercher JA (2015) Single-site trinuclear copper oxygen clusters in mordenite for selective conversion of methane to methanol. Nat Commun 6:1–8
56. Markovits MAC, Jentys A, Tromp M, Sanchez-Sanchez M, Lercher JA (2016) Effect of location and distribution of Al sites in ZSM-5 on the formation of Cu-oxo clusters active for direct conversion of methane to methanol. Top Catal 59:1554–1563
57. Raja R, Ratnasamy P (1997) Direct conversion of methane to methanol. Appl Catal A Gen 158:L7–L15
58. Hammond C, Forde MM, Ab Rahim MH, Thetford A, He Q, Jenkins RL, Dimitratos N, Lopez-Sanchez JA, Dummer NF, Murphy DM, Carley AF, Taylor SH, Willock DJ, Stangland EE, Kang J, Hagen H, Kiely CJ, Hutchings GJ (2012) Direct catalytic conversion of methane to methanol in an aqueous medium by using copper-promoted Fe-ZSM-5. Angew Chem Int Ed 51:5129–5133
59. Panov GI, Uriarte AK, Rodkin MA, Sobolev VI (1998) Generation of active oxygen species on solid surfaces. Opportunity for novel oxidation technologies over zeolites. Catal Today 41:365–385
60. Sushkevich VL, Palagin D, Ranocchiari M, van Bokhoven JA (2017) Selective anaerobic oxidation of methane enables direct synthesis of methanol. Science 356:523–527
61. Wang W, Hunger M (2008) Reactivity of surface alkoxy species on acidic zeolite catalysts. Acc Chem Res 41:895–904
62. Benlounes O, Mansouri S, Rabia C, Hocine S (2008) Direct oxidation of methane to oxygenates over heteropolyanions. J Nat Gas Chem 17:309–312
63. Mansouri S, Benlounes Q, Rabia C, Thouvenot R, Bettahar MM, Hocine S (2013) Partial oxidation of methane over modified Keggin-type polyoxotungstates. J Mol Catal A: Chem 379:255–262
64. Khokhar MD, Shukla RS, Jasra RV (2009) Selective oxidation of methane by molecular oxygen catalyzed by a bridged binuclear ruthenium complex at moderate pressures and ambient temperature. J Mol Catal A: Chem 299:108–116
65. Shilov AE, Shul'pin GB (1997) Activation of C–H bonds by metal complexes. Chem Rev 3:2879–2932
66. Yamanaka I, Soma M, Otsuka K (1995) Oxidation of methane to methanol with oxygen catalysed by europium trichloride at room temperature. J Chem Soc Chem Commun 2235–2236
67. Yamanaka I (2002) Reductive activation of O_2 and monooxygenation of hydrocarbons by Eu catalyst. Catal Surv Asia 6:63–72
68. Yamanaka I, Soma M, Otsuka K (1996) Enhancing effect of titanium(II) for the oxidation of methane with O2 by an EuCl3-Zn-CF3CO2H-catalytic system at 40 °C. Chem Lett 25:565–566
69. Piao D, Inoue K, Shibasaki H, Taniguchi Y, Kitamura T, Fujiwara Y (1999) An efficient partial oxidation of methane in trifluoroacetic acid using vanadium-containing heteropolyacid catalysts. J Organomet Chem 574:116–120
70. Periana RA, Taube DJ, Evitt ER, Löffler DG, Wentrcek PR, Voss G, Masuda T (1993) A mercury-catalyzed, high-yield system for the oxidation of methane to methanol. Science 259:340–343
71. Sen A, Benvenuto MA, Lin M, Hutson AC, Basickes N (1994) Activation of methane and ethane and their selective oxidation to the alcohols in protic media. J Am Chem Soc 116:998–1003
72. Jones CJ, Taube D, Ziatdinov VR, Periana RA, Nielsen RJ, Oxgaard J, Goddard WA (2004) Selective oxidation of methane to methanol catalyzed, with C-H activation, by homogeneous, cationic gold. Angew Chem Int Ed 43:4626–4629
73. Periana RA, Taube DJ, Gamble S, Taube H, Satoh T, Fujii H (1998) Platinum catalysts for the high-yield oxidation of methane to a methanol derivative. Science 280:560–564
74. Zimmermann T, Soorholtz M, Bilke M, Schueth F (2016) Selective methane oxidation catalyzed by platinum salts in oleum at turnover frequencies of large-scale industrial processes. J Am Chem Soc 138:12395–12400
75. Hashiguchi BG, Konnick MM, Bischof SM, Gustafson SJ, Devarajan D, Gunsalus N, Ess DH, Periana RA (2014) Main-group compounds selectively oxidize mixtures of methane, ethane, and propane to alcohol esters. Science 343:1232–1237

76. Gretz E, Oliver TF, Sen A (1987) Carbon-hydrogen bond activation by electrophilic transition-metal compounds. Palladium(II)-mediated oxidation of arenes and alkanes including methane. J Am Chem Soc 109:8109–8111
77. Sen A (1991) Homogeneous palladium(II) mediated oxidation of methane. Plat Metal Rev 35:126–132
78. Muehlhofer M, Strassner T, Herrmann WA (2000) New catalyst systems for the catalytic conversion of methane into methanol. Angew Chem Int Ed 41:1745–1747
79. Munz D, Myyer D, Strassner T (2013) Methane CH activation by palladium complexes with chelating bis(NHC) ligands: a DFT study. Organometal 32:3469–3480
80. Vargaftik MN, Stolarov IP, Moiseev II (1990) Highly selective partial oxidation of methane to methyl trifluoroacetate. J Chem Soc Chem Commun 1049–1050
81. Strassner T, Ahrens S, Muehlhofer M, Munz D, Zeller A (2013) Cobalt-catalyzed oxidation of methane to methyl trifluoroacetate by dioxygen. Eur J Inorg Chem 21:3659–3663
82. An Z, Pan X, Liu X, Han X, Bao X (2006) Combined redox couples for catalytic oxidation of methane by dioxygen at low temperatures. J Am Chem Soc 128:16028–16029
83. Yuan J, Wang L, Wang Y (2011) Direct oxidation of methane to a methanol derivative using molecular oxygen. Ind Eng Chem Res 50:6513–6516
84. Riess JG (2001) Oxygen carriers ("Blood substitutes")-raison d'etre, chemistry and some physiology. Chem Rev 101:2797–2920
85. Riess JG (2002) Fluorous micro- and nanophases with a biomedical perspective. Tetrahedron 58:4113–4131
86. Elibol M, Mavituna F (1999) A remedy to oxygen limitation problem in antibiotic production addition of perfluorocarbon. Biochem Eng J 3:1–7
87. Kozhevnikov IV (1997) $PMo_{12-n}V_nO^{(3+n)-}{}_{40}$ heteropolyanions as catalysts for aerobic oxidation. J Mol Catal A: Chem 117:151–158
88. Palkovits R, von Malotki C, Baumgarten M, Müllen K, Baltes C, Antonietti M, Kuhn P, Weber J, Thomas A, Schüth F (2010) Development of molecular and solid catalysts for the direct low-temperature oxidation of methane to methanol. Chemsuschem 22:277–282
89. Palkovits R, Antonietti M, Kuhn P, Thomas A, Schüth F (2009) Solid catalysts for the selective low-temperature oxidation of methane to methanol. Angew Chem Int Ed 48:6909–6912
90. Ravi M, Ranocchiari M, van Bokhoven JA (2017) The direct catalytic oxidation of methane to methanol—a critical assessment. Angew Chem Int Ed 22:16464–16483
91. Conley BL, Tenn WJ, Young KJH, Ganesh SK, Meier SK, Ziatdinov VR, Mironov O, Oxgaard J, Gonzales J (2006) Goddard III WA, Periana RA. J Mol Catal A 251:8–23
92. Periana RA, Bhalla G, Tenn WJ, Young KJH, Liu XY, Mironov O, Jones CJ, Ziatdinov VR (2004) J Mol Catal A 220:7–25
93. Furuto T, Takeguchi M, Okura I (1999) Semicontinuous methanol biosynthesis by *Methylosinus trichosporium* OB3b. J Mol Catal A: Chem 144:257–261
94. Lee SG, Goo JH, Kim HG, Oh JI, Kim YM, Kim SW (2004) Optimization of methanol biosynthesis from methane using *Methylosinus trichosporium* OB3b. Biotechnol Lett 26:947–950
95. Duan C, Luo M, Xing X (2011) High-rate conversion of methane to methanol by *Methylosinus trichosporium* OB3b. Bioresour Technol 102:7349–7353

Chapter 4
Application of Biocatalysts for the Production of Methanol from Methane

4.1 Introduction

Methane can be converted to methanol at atmospheric temperature and pressure by utilizing the isolated enzyme methane monooxygenase (MMO) or whole-cell methane-oxidizing bacteria as a biocatalyst. The utilization of MMO isolated from bacterial cells has disadvantages, such as the high cost of isolating MMO and the instability of MMO outside the bacterial cells. These disadvantages should eventually be overcome by advances in MMO protein engineering, as discussed at the end of this chapter. However, at present, methanol production using methane-oxidizing bacteria is more promising than using isolated MMO.

Methane-oxidizing archaea can also oxidize methane anaerobically. The oxidation occurs via a reverse methanogenesis pathway using several types of electron acceptors, such as sulfate, nitrate, and metal ions [1–6]. However, the use of anaerobic methane-oxidizing archaea as a biocatalyst to produce methanol from methane has not yet been reported. Therefore, in this section, only methanol production using methane-oxidizing bacteria is discussed. Representative research into the production of methanol using methane-oxidizing bacteria is summarized in Table 4.1 [7–29]. Each method is characterized by the type of biocatalyst (bacterial strain), reactor, methanol dehydrogenase inhibitor, and electron donor used.

The use of a methanol dehydrogenase inhibitor is essential to achieve the extracellular accumulation of methanol when using methane-oxidizing bacteria. After the oxidation of methane to methanol by MMO, the subsequent metabolic step is the dehydrogenation of methanol to formaldehyde by methanol dehydrogenase. Thus, inhibition of this dehydrogenation is crucial, but results in a shortage of the nicotinamide adenine dinucleotide (NADH) required as cellular energy for the synthesis of various organic compounds required for the metabolism and replication of the bacterial cells, as well as the hydroxylation of methane. The carbon assimilation pathway is also affected by methanol dehydrogenase inhibition, as the assimilation is initiated from formaldehyde. Therefore, not only methanol oxidation inhibitors,

T. Baba and A. Miyaji, *Catalysis and the Mechanism of Methane Conversion to Chemicals*,
https://doi.org/10.1007/978-981-15-4132-2_4

Table 4.1 Representative research regarding the production of methanol from methane using methane-oxidizing bacteria

Year	Author	Bacterial strains	Reactor	MDH inhibitor	Electron donor	Ref.
1987	Mehta et al.	*Methylosinus trichosporium* OB3b	Batch	Phosphate	–	7
1991	Mehta et al.	DEAE-immobilized *Methylosinus trichosporium*	Batch/Continuous	Phosphate	Formate	8
1995	Sugimori et al.	*Methylosinus trichosporium* OB3b	Batch	Cyclopropanol	Formate	9
1997	Takeguchi et al.	*Methylosinus trichosporium* OB3b	Batch	Cyclopropanol	Formate	10
1999	Furuto et al.	*Methylosinus trichosporium* OB3b	Semi-batch	Cyclopropanol	Formate	11
2004	Lee et al.	*Methylosinus trichosporium* OB3b	Batch	NaCl	Formate	12
2009	Xin et al.	*Methylosinus trichosporium* IMV3011	Batch	CO_2	–	13
2009	Markowska et al.	*Methylosinus trichosporium* OB3b	Batch	MDH inhibitor	Reductant	14
2010	Kim et al.	*Methylosinus trichosporium* OB3b	Batch/ Repeated batch/ Continuous	NaCl, EDTA	Formate	15
2011	Duan et al.	*Methylosinus trichosporium* OB3b	Membrane reactor	Phosphate	Formate	16
2013	Han et al.	Consortiumu of *M. trichosporum* OB3b, *Methylococcus capsulatus*, and *Methylosinus sporium*	Batch	NaCl		17
2014	Pen et al.	*Methylosinus trichosporium* OB3b	Membrane reactor	Phosphate, NaCl, EDTA	Formate	18
2015	Hwang et al.	*Methylosinus trichosporium* OB3b	Batch	Phosphate, EDTA	Formate	19
2015	Yoo et al.	*Methylosinus sporium*	Batch	Phosphate, NaCl, NH_4Cl or EDTA	–	20
2016	Pen et al.	*Methylosinus trichosporium* OB3b	Membrane reactor	Phosphate, NaCl, EDTA	Formate	21
2016	Patel et al.	Immobilized *Methylisinus sporium*	Batch	Phosphate	Formate	22, 23

(continued)

Table 4.1 (continued)

Year	Author	Bacterial strains	Reactor	MDH inhibitor	Electron donor	Ref.
2016	Patel et al.	*Methylocystis bryophila, Methylocella silvestris, Methyloferula stellata,* and *Methylomonas methanica*	Batch	Phosphate, $MgCl_2$	Formate	24
2016	Mardina et al.	Immobilized *Methylocella tundrae*	Batch	Phosphate	Formate	25
2016	Sheets et al.	*Methylocaldum sp.* 14B	Batch	Phosphate	Formate	26
2017	Patel et al.	Immobilized *Methylocella tundrae*	Batch	Phosphate	Formate	27, 28
2018	Patel et al.	Co-immobilized *Methylomonas methanica* and *Methylocella tundrae*	Batch	Phosphate	Formate	29

but also electron donors to supply the energy needs of the bacterial cells must be utilized in biocatalytic methanol production.

Thus, an understanding of bacterial methane metabolism is crucial to develop practical methanol production processes using whole-cell biocatalysts. Therefore, this section describes how methane is metabolized in methane-oxidizing bacteria, though the current understanding of their metabolism remains incomplete. Subsequently, the development of methanol production processes using biocatalysts is described.

4.2 Methane Metabolism in Methane-Oxidizing Bacteria

Methane-oxidizing bacterial oxidizes methane in both its dissimilation (energy acquisition steps in a biological system) and assimilation (steps involving the synthesis of various organic compounds in a bacterial cell) pathways (Fig. 4.1).

Both pathways are initiated by the MMO-catalyzed oxidation of methane to methanol. The produced methanol is dehydrogenated to formaldehyde by methanol dehydrogenase (MDH). Some of the formaldehyde is terminally converted to CO_2 for energy generation (dissimilation), while the rest is converted to various organic compounds in the carbon assimilation pathways. The conversion of formaldehyde via either pathway is rapid because of the toxic effect of formaldehyde on the methane-oxidizing bacteria [30].

(1) Dissimilation pathway

In the dissimilation pathway, the produced formaldehyde is further oxidized to formate, which is subsequently oxidized to carbon dioxide. The two oxidation reactions are catalyzed by formaldehyde dehydrogenase (FaDH) and formate dehydrogenase (FDH), respectively. In the last three steps of the methane dissimilation

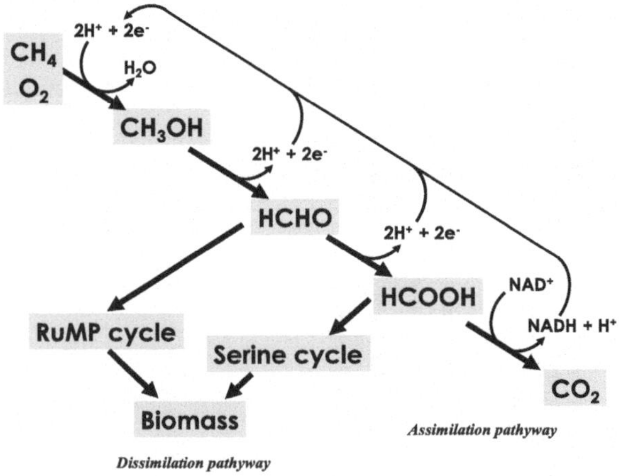

Fig. 4.1 Dissimilation and assimilation pathways of methane-oxidizing bacteria

pathway, NAD^+, the oxidized form of NADH, accepts two electrons, and NADH, the reduced form of NAD, is produced. Via the three oxidation steps, *i.e.*, the oxidation of methanol, formaldehyde, and formate, a total of six electrons (two electrons per step) is produced. Two of the electrons are provided to the methane oxidation process in the form of a reducing equivalent for simultaneous methane uptake. The remaining four electrons are provided to the electron transport chain, which produces adenosine triphosphate (ATP) through the proton motive force, as shown in Fig. 4.1. Finally, these four electrons reduce oxygen molecules as the terminal electron acceptor, forming H_2O. The overall reaction is shown as reaction 4.1.

$$CH_4 + O_2 + NADH + H^+ \rightarrow CH_3OH + H_2O + NAD^+ \quad (4.1)$$

(2) Assimilation pathway

In the assimilation pathway, formaldehyde, formate, and carbon dioxide can be used for the synthesis of cellular components through the ribulose monophosphate (RuMP) pathway and the serine pathway [30]. Furthermore, some methane-oxidizing bacteria also have a reductive pentose phosphate cycle (Calvin–Benson–Bassham (CBB) cycle) [31–37].

All the carbons incorporated into the bacterial cells via the RuMP pathway are derived from formate produced from methane [30, 38, 39]. On the other hand, in the serine pathway, only about 50% of the carbons that are incorporated into the bacterial cells are derived from formaldehyde produced from methane, while the rest are supplied from the multiple carboxylation reactions in both the serine and ethylmalonyl coenzyme A (ethylmalonyl-CoA) pathways for glyoxylate regeneration or a glyoxylate bypass [40–43].

Based on differences in their assimilation pathways, as well as in their morphology and physiological properties, methane-oxidizing bacteria have conventionally been classified into three classes (Table 4.2), namely Type I, Type II, and Type III [38, 44]. Type I was recently divided into three types (Ia, Ib, and Ic) [45–51], while Type II was divided into two types (IIa and IIb) [51, 52]. Furthermore, non-proteobacterial methane-oxidizing bacteria classified as Type III hase been found in the phylum *Verrucomicrobia* [53–55]. These bacteria use methane only as an energy source, oxidize methane to carbon dioxide, and use the CBB cycle to fix carbon dioxide [56].

Some proteobacteria such as the Type I species *Methylococcus capsulatus* and *Methylocaldum szegediense* O-12 and the Type II species *Methyloferula stellata* AR4, *Methylocella silvestris* BL2, and *Methylocapsa acidiphila* B2 also contain a complete set of genes for the CBB cycle. However, the role of the CBB cycle in the metabolism of these bacteria has yet to be experimentally validated.

Methane-oxidizing bacteria have been found in many ecosystems such as soils, peatlands, rice paddies, sediments, freshwater and marine systems, alkaline soda lakes, acidic hot springs, mud pots, cold ecosystems, and tissues of higher organisms [57]. Most methane-oxidizing bacteria utilize methane exclusively as their carbon and energy sources (obligate methanotrophs), but some methanotrophs can also utilize

Table 4.2 Classification of methane-oxidizing bacteria and their assimilation pathways

Phylogeny	Type	C1 assimilation pathway	Genus
Gammaproteobacteria	Type Ia	Mostly RuMP pathway	*Methylobacter* *Methyloglobulus* *Methylomarinum* *Methylomicrobium* *Methylomonas* *Methylofundus* *Methylosarcina* *Methylosoma* *Methylosphaera* *Methylovulum*
	Type Ib		*Methylocaldum* *Methylococcus* *Methylogaea* *Methylomagnum* *Methyloparacoccus*
	Type Ic		*Methylohalobius* *Methylomarinovum* *Methylothermus*
Alphaproteobacteria	Type IIa	Serine pathway	*Methylosinus* *Methylcystis*
	Type IIb		*Methylocapsa* *Methylocella* *Methyloferula*
Outside the phylum *Proteobacteria*	Type III	CBB cycle	*Methylacidimicrobium* *Methylacidiphilum*

other carbon and energy sources. For example, *Methylocella silvestris* BL2 shows the widest versatility currently known among methane-oxidizing bacteria [58, 59]. In addition to C1 compounds, *M. silvestris* BL2 can use various organic acids such as acetic acid, pyruvic acid, propionic acid, succinic acid, malic acid, and gluconic acid, alcohols such as ethanol and 2-propanol, and gaseous compounds such as ethane and propane [60]. Facultative methane-oxidizing bacteria with a narrower substrate range than *M. silvestris* BL2 has been found in other members of the genus *Methylocera* and *Methylocapsa aurea*. Some *Methylocystis* strains can be grown using acetic acid or ethanol, although the growth rate using these carbon sources was 3–10 times lower than when methane was used [30, 61].

4.2.1 Methane Metabolism Pathways in Methane-Oxidizing Bacteria

Details of enzymatic conversion of methane in the metabolic pathways of methane-oxidizing bacteria are described below.

(1) Methane oxidation to methanol

There are two forms of MMO: iron-containing soluble MMO (sMMO) located in the cytoplasm, and copper-containing membrane-bound MMO (pMMO) located in the cytoplasmic inner membrane [30]. The two MMOs have completely different properties as functional proteins (see Chapter 2).

All methane-oxidizing bacteria, except those of the genera *Methylocella* [62] and *Methylogerula* [36], produce pMMO. Type II methane-oxidizing bacteria also produce sMMO. On the other hand, the genera *Methylocella* and *Methylogerula* produce only sMMO.

In Type II methane-oxidizing bacteria, pMMO and sMMO are not present in the bacterial cell under the same growth conditions; their expression is controlled by the concentration of copper ions in the growth medium. pMMO is expressed at copper ion concentrations above 0.85 μmol per g-dry weight of bacterial cells, while sMMO is produced at concentrations below 0.85 μmol per g-dry weight of cells [63]. MMO expression is controlled by the bacterial gene expression regulation system, which involves copper ions, although the corresponding mechanism is still under investigation [64–69].

In the oxidation of methane to methanol, sMMO utilizes NADH generated from the dehydrogenation of formaldehyde and formate as an electron donor [70, 71]. On the other hand, the electron donor for pMMO remains controversial. As discussed in Sect. 2.3.5, the direct electron donor for pMMO is assumed to be a quinol derivative. Based on the structural similarities between pMMO and ammonia monooxygenase and the phylogenetic similarities between methane-oxidizing bacteria and ammonia-oxidizing bacteria, ubiquinol 8 (coenzyme Q_8) has been identified as a probable pMMO electron donor [72]. However, the ubiquinone reduction mechanism has not been clarified.

The other possible mechanism by which electrons may be supplied to pMMO was proposed by Higgins and co-workers, as discussed in Sect. 2.3.5. These authors presented an alternative model for the electron transport that requires reverse electron flow; in this model, electrons from methanol are provided via not only MDH and cytochrome c, but also by low potential respiratory chain components such as cytochrome b and quinone. The proteins involved in this pathway have not been yet identified biochemically. Direct electron transfer from MDH to pMMO has also been considered based on complex formation between pMMO and MDH [73, 74].

Alternatively, the NADH produced in conjunction with the dehydrogenation of formaldehyde and formate may be responsible for the ubiquinone reduction, as an enhancement in methane oxidation and methanol production was observed when exogenous formate was added [75, 76]. Hence, as shown in Chapter 2 (Fig. 2.15), pMMO is believed to use various electron sources depending on the growth conditions [75, 76].

Bacterial cells that express pMMO oxidize methane more efficiently than sMMO-containing cells because of their high affinity for methane and growth yield [76]. Therefore, the use of pMMO-expressing cells is considered to be more favorable for the production of methanol from methane. On the other hand, sMMO has a wider substrate range than pMMO. This characteristic makes bacterial cells expressing sMMO attractive for producing several chemicals for biotechnological applications.

(2) Oxidation of methanol to formaldehyde

The methanol produced in the MMO-catalyzed oxidation of methane is dehydrogenated to formaldehyde via catalysis by methanol dehydrogenase (MDH). MDH is located in the periplasm (the space between outer membrane and intracytoplasmic membrane of methane-oxidizing bacteria) [30], and contains pyrroloquinoline quinone (PQQ) as a cofactor. The reduction of PQQ is coupled with the dehydrogenation of methanol. The reduced form of PQQ ($PPQH_2$) is oxidized and transfers two electrons to the terminal oxidase by way of cytochromes c and other carriers as intermediates, or to reduce the electron donor for pMMO [77].

(3) Formaldehyde conversion

As described above, formaldehyde plays a vital role as a metabolic intermediate in the methane metabolism system. Some of the produced formaldehyde is oxidized to CO_2 for energy generation, while the rest is incorporated into the carbon assimilation pathway for cell replication.

Two systems for the dehydrogenation of formaldehyde into formate have been proposed. The first is dehydrogenation catalyzed by formaldehyde dehydrogenase (FalDH). Two types of FalDH enzymes, a PQQ-containing and cytochrome-linked type and a NAD-linked type, have been reported [78, 79]. PQQ-containing FalDH was found to be the major formaldehyde oxidizing enzyme in pMMO-expressing *Methylococcus capsulatus* cells [80]. The second system is the tetrahydromethanopterin (H_4MPT)-linked formaldehyde dehydrogenation pathway (Fig. 4.2). In the H_4MPT pathway, methylene tetrahydromethanopterin dehydrogenase (MtdB) reduces NAD^+ to produce NADH [75, 81]. In addition to these two systems, several other formaldehyde dehydrogenation systems have been found in methanol-oxidizing bacteria. These formaldehyde dehydrogenation systems may also exist in methane-oxidizing bacteria [78].

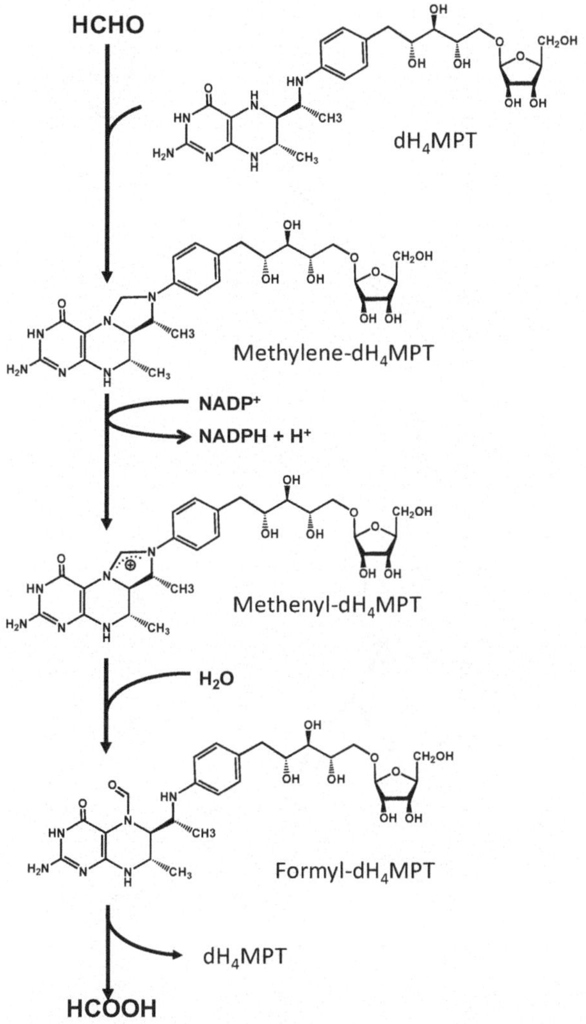

Fig. 4.2 Conversion of formaldehyde to formate

(4) Oxidation of formate to carbon dioxide

The last step in the dissimilation pathway of methane-oxidizing bacteria is the oxidation of formate to CO_2. The conversion of formate to carbon dioxide is catalyzed by the NAD-dependent enzyme FDH (EC 1.2.1.2) [82–85], which contains [2Fe–2Fe], [4Fe–4Fe], and molybdenum as cofactors for its catalytic function. This last step also provides electrons for the oxidation of methane to methanol [79, 86].

(5) Assimilation pathways

A mentioned above, there are two pathways for methane assimilation: the RuMP (Fig. 4.3) and serine (Fig. 4.4) ones. Each pathway involves different reactions and enzymes for energy production and cell replication.

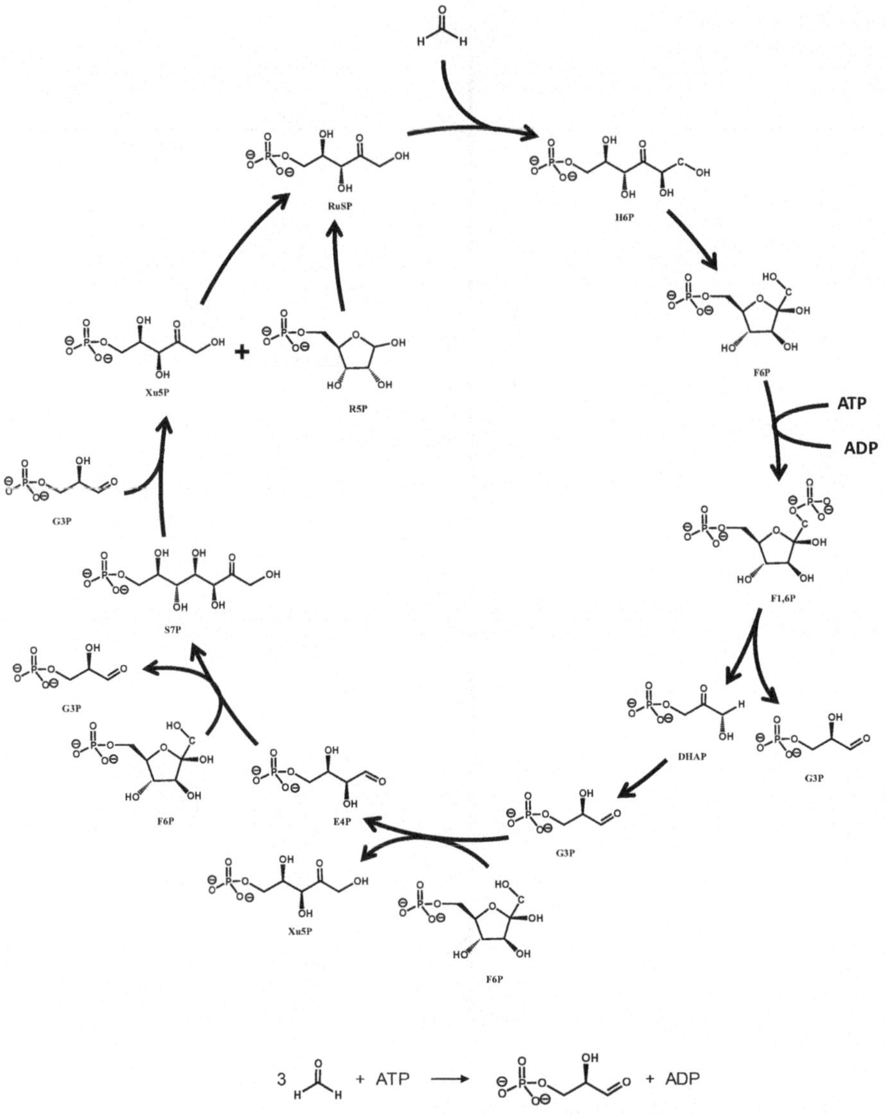

Fig. 4.3 RuMP cycle, one of the assimilation pathways of methane-oxidizing bacteria

The RuMP pathway begins with the reaction of formaldehyde and ribulose-5-phosphate to produce fructose-6-phosphate. This reaction is catalyzed by two enzymes: hexulosephosphate synthase (HPS) and hexulosephosphate isomerase (HPI), which is found only in type I methanotrophs. The fructose-6-phosphate is converted to pyruvate and glyceraldehyde-3-phosphate. Ribulose-5-phosphate is regenerated from glyceraldehyde to complete the biomass synthesis cycle, while

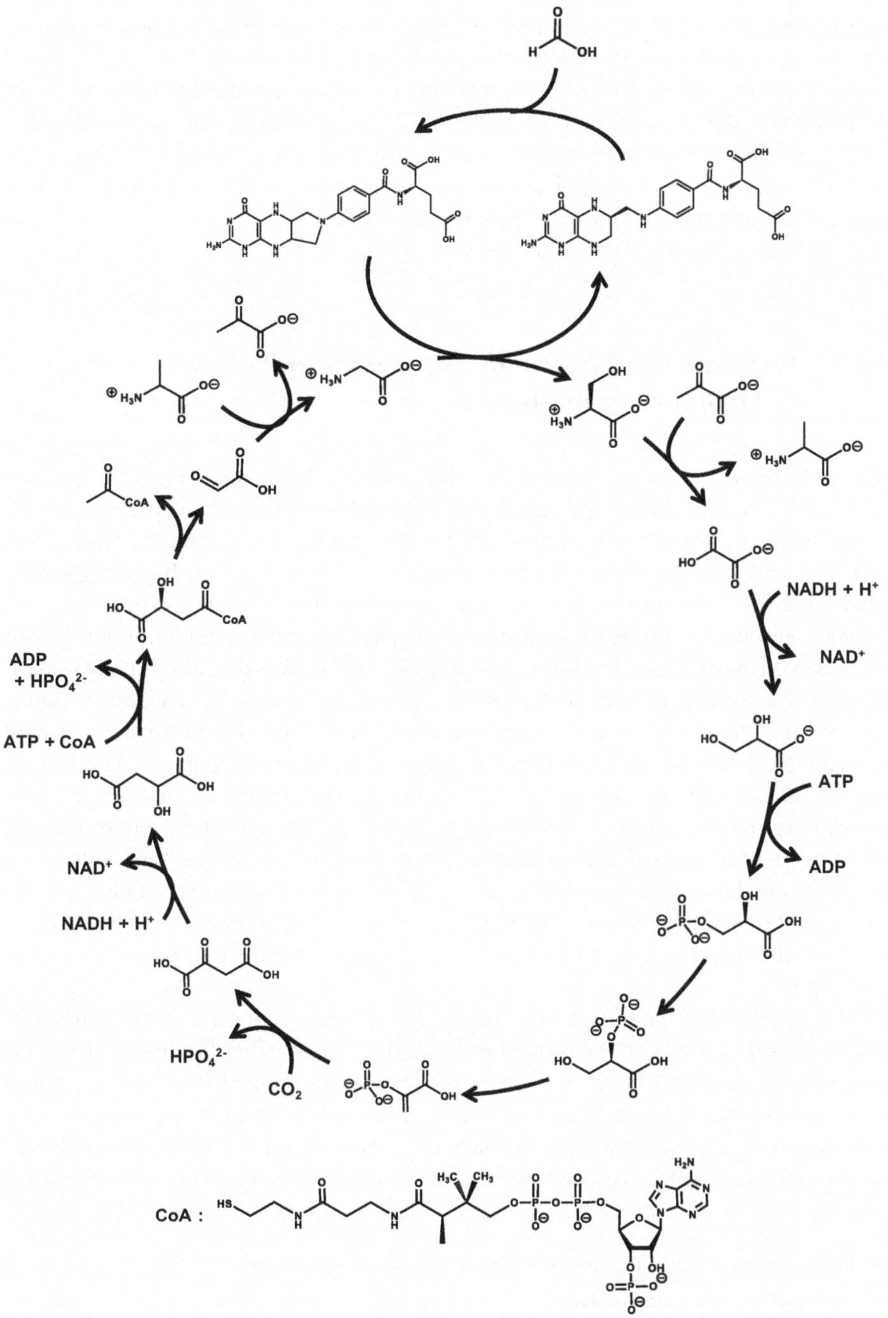

Fig. 4.4 Serine pathway, one of the assimilation pathways of methane-oxidizing bacteria

the pyruvate is used in an incomplete TCA cycle and converted to carbon dioxide. The main intermediates of the RuMP cycle are sugar phosphates [30, 75, 87–89].

In the serine cycle, formate is converted to methylene-tetrahydrofolate (H_4F) by methylene H_4F dehydrogenase (MtdA) [90]. Methylene-H_4F activates the serine cycle by mediating the reaction between formate and glycine [75, 88]. In addition, this cycle produces acetyl-CoA, which is a crucial compound in the complete TCA cycle required for energy generation and biomass synthesis [91]. The major intermediates of the serine cycle are amino acids and CoA derivatives [75].

4.2.2 Preparation of Biocatalysts for the Production of Methanol from Methane

To use methane-oxidizing bacteria as a biocatalyst, the bacteria must be cultured before the methane conversion reaction is started. Most methane-oxidizing bacteria utilize methane as their sole carbon and energy source. Therefore, methane is not only a raw material for the production of methanol, but also for the biocatalyst itself when methane-oxidizing bacteria are used in the methane conversion process.

Most commonly, nitrate mineral salts (NMS) and ammonium mineral salts (AMS) are used to grow methane-oxidizing bacteria [38]. These salts contain elements essential to the biological system, such as nitrate or/and ammonium as a nitrogen source, phosphate as the phosphorous source, and several metal ions. To optimize the yield of methane-oxidizing bacteria, the concentrations of the nitrogen source and metal ions and the salinity, pH, temperature, as well as the dissolved methane and oxygen concentrations must be adjusted. In addition, methane-oxidizing bacteria can be isolated from a variety of locations, including acidic wetlands, hot springs, mud volcanoes, and high salt lakes. Each strain requires specific growth conditions and a specific medium [92–96]. The factors affecting the growth of methane-oxidizing bacteria are not yet completely clear, and should be investigated in their natural habitats or in bioreactors.

The growth rate and cell density of the bacteria are important in the cultivation of biocatalysts. However, the growth rate and the cell density of methane-oxidizing bacterial are much lower than those of other bacteria utilized in industry, such as *Escherichia coli* and *Saccharomyces cerevisiae* (yeast). The reported growth rates for methane-oxidizing bacteria using methane in pure cultures under optimum conditions are listed in Table 4.3. Type I methanotrophs have the highest growth rates among the

Table 4.3 Growth rates in the cultivation of methane-oxidizing bacteria

Type of methane-oxidizing bacteria	Highest growth rate (hr^{-1})	Ref
Type I	~0.330	[38, 100– 102]
Type II	~0.154	[38, 103]
Type III	~0.070	[54, 104]

three types of methane-oxidizing bacteria, while the growth rates of Type III methane-oxidizing bacteria were much slower rate than those of the other two types. However, the growth rates of methanotrophs are much slower than that of industrially utilized *Escherichia coli* (~1.3 h^{-1} [97]), and slightly slower than that of *Saccharomyces cerevisiae* (~0.40 h^{-1}). One of the major reasons for the low growth rate is the agitation of culture medium that is required to overcome the low solubility (~1 mM at 30 °C and 1 atm) [98] and very low transfer rate [99] of methane gas to the water phase.

Achieving high-cell-density cultivation of methane-oxidizing bacteria is expected to require even more technical improvements than improving their growth rate. Pressurized bioreactors have been used to improve the solubility of methane and its delivery to bacteria, and high biomass densities in the range of 18–65 g L^{-1} have been achieved. However, high-pressure bioreactors consume more energy than cultivation under atmospheric pressure [105–107]. Alternatively, two-phase partitioned bioreactors using a non-aqueous phase that has a higher affinity for methane than water are being tested. For example, the addition of 10% v/v silicone oil to the culture medium led to a 3.3-fold increase in the growth rate of the strain *Methylosinus sporium* DSMZ 17706 [108]. Using the same concept, *Methylosinus trichosporium* OB3b was cultured in a medium containing 5% v/v paraffin oil and achieved a biomass density of 14 g L^{-1} after 240 h of cultivation [109]. However, much higher bacterial cell density is achieved in industry. For example, *S. cerevisiae* and *E. coli* have shown biomass densities of up to 170 g L^{-1} and 190 g L^{-1}, respectively [110, 111].

In addition to the growth rate and yield, the quality of the biocatalyst is also important. The content of MMO in methane-oxidizing bacteria determines its quality as a biocatalyst. The content of pMMO can be controlled by the copper concentration in the culture medium, with increasing copper concentration leading to increased pMMO content. However, the growth rate of methane-oxidizing bacteria decreases when the copper concentration is increased above 10–20 μM, and copper ions become toxic to bacterial cells at concentrations higher than 50 μM [112, 113]. Chan and co-workers developed a method to produce high-quality methane-oxidizing bacteria using a fermenter adapted with a hollow-fiber bioreactor. The spent media was filtered through the hollow-fiber membrane, allowing cell waste and toxic metabolites to be discarded from the reactor, while the same time the reactor was simultaneously replenished with fresh cell media. Thereby, the copper concentration of the culture solution could be controlled to avoid toxicity to the cells. Using this method, *M. capsulatus* Bath cells with high pMMO-expression (80% of the total cytoplasmic membrane) were obtained [114].

4.2.3 Production of Methanol Using Methane-Oxidizing Bacteria

Several studies have revealed various MDH inhibition strategies including, but not limited to, the addition of ethylenediaminetetraacetic acid (EDTA), cyclopropanol, or high concentrations of various salts. When MDH is inhibited, methanol accumulation stops within a few hours because of the depletion of the reducing equivalents. This problem can be resolved by supplying sodium formate as an electron source. However, the addition of expensive external electron donors such as formate limits the scalability of the process. Furthermore, the biocatalytic methanol yields have not yet reached the feasible range [18, 98, 115, 116]. In order to put the methanol production process into practical use, it will be necessary to optimize several factors (pH, temperature, cell density, and gas mixture ratio), as well as the electron donation and MDH inhibition conditions.

4.2.3.1 Bacterial Strains

As shown in Table 4.1, studies have been performed on pure cultures such as *Methylosinus trichosporium* [7, 8, 10–13, 15, 16, 18, 87, 117–119], *Methylosinus sporium* [19, 22], *Methylocella tundrae* [25], *Methylocystis bryophila* [24], and strain 14B. In addition, a consortium of strains (*Methylosinus trichosporium* OB3b, *Methylococcus capsulatus*, and *Methylosinus sporium*) and a combination of *Methylocella silvestris*, *Methyloferula stellate*, and *Methylomonas methanica* were also tested for methanol production [23]. All the studied strains except strain 14B are classified as type II methane-oxidizing bacteria.

However, Type I strains have higher growth rates as shown in Table 4.2, exhibit higher methane affinity, and require less energy for growth, making them more promising for methanol production [75]. In addition, Type I methane-oxidizing bacteria are more energy efficient. The RuMP pathway in Type I methane-oxidizing bacteria requires only one mole of ATP to assimilate three moles of formaldehyde as shown in Fig. 4.3. Compared to the RuMP pathway, the serine pathway in Type II methane-oxidizing bacteria requires more energy; three moles of ATP and two moles of NADH are needed to assimilate two moles of formaldehyde and one mole of carbon dioxide, as shown in Fig. 4.4 [88].

However, from another perspective, Type II methane-oxidizing bacteria show an advantage for the production of methanol from methane. Up to 50% of the carbon dioxide produced by Type II methane-oxidizing bacteria can be incorporated into cell synthesis, compared to only 15% in Type I methane-oxidizing bacteria. Type II methane-oxidizing bacteria were found to be suitable when biogas generated from an anaerobic digester composed of about 40% carbon dioxide was used for methanol production [75].

4.2.3.2 MDH Inhibition

The dehydrogenation of methanol to formaldehyde is catalyzed by PQQ-dependent MDH. The electrons generated in this reaction are donated to the terminal oxidase of methane-oxidizing bacteria via cytochrome c_L [30, 78, 79]. Since this dehydrogenation reaction is faster than the oxidation reaction of methane to methanol, only trace concentrations of methanol are observed in the reaction mixture unless methanol dehydrogenation is suppressed [17, 87, 115].

The most widely used strategies for the inhibition of MDH rely on the use of (i) methane-oxidizing bacteria whose MDH is inhibited in advance by chemical modification followed by its use for methane oxidation reaction and (ii) methane-oxidizing bacteria for methane oxidation reactions in the presence of MDH inhibitors. Cyclopropanol [10, 11] is used in the former strategy, and phosphate [7, 8, 16, 19, 22, 24–26], NaCl [12, 15, 17–19], EDTA [15, 19], $MgCl_2$ [7, 8, 16, 17, 19, 22, 24, 25], and NH_4Cl [17, 19] are used in the latter. These inhibitors can be used alone or in combination.

(1) Cyclopropanol

Cyclopropanol combines with a coenzyme of MDH, 2,7,9-tricarboxy-1H-pyrrolo-[2,3-f]quinolone-4,5-dione (PQQ), resulting in irreversible deactivation. Cyclopropanol has been reported to be the most efficient MDH inhibitor. It is considered an irreversible inhibitor because it produces stable compounds from interaction with free and MDH-bound PQQ [15, 17, 109]. The optimum pyrrole concentration is 0.67 μM; MDH activity is reduced by 50% at this concentration, whereas MMO activity decreases by only 5.2% [10, 11, 15, 17, 116]. However, cyclopropanol is not used universally because of its difficult preparation and instability under aerobic conditions.

(2) Salt

Salts such as NaCl, NH_4Cl, and phosphate disrupt electron transport, which in turn decreases the activity of MDH [115] by weakening the ionic interaction between MDH and cytochromes c_L. Phosphate is the most widely used inhibitor of methanol oxidation, and is typically used in the concentration range 40–100 mM [7, 8]. This form of inhibition is considered non-competitive and reversible [7, 8]. NaCl is advantageous as an inhibitor for methanol production because of its stability, abundance, and low cost. According to some studies, the optimum NaCl concentration for methanol production is approximately 200 mM [12], although NaCl concentrations exceeding 100 mM distort the cell morphology. NH_4Cl is an effective MDH inhibitor, and high conversion efficiency from methane to methanol is observed when 40 mM NH_4Cl is used. However, the methanol yield is low [15, 19, 22, 24], because NH_4Cl also inhibits the oxidation of methane to methanol by MMO [15, 19, 22, 24]. Therefore, NH_4Cl is not a suitable MDH inhibitor for the methanol synthesis reaction.

(3) Carbon dioxide

High concentrations of carbon dioxide partially inhibit MDH activity. However, the maximum productivity of methanol under carbon dioxide-inhibited conditions is very low (3 μmol L^{-1} h^{-1}) compared with other methanol synthesis methods using methane-oxidizing bacteria [13]. However, the use of carbon dioxide as an inhibitor has the advantage of not requiring chemical modification of methane-oxidizing bacteria or the addition of expensive external electron donors or MDH inhibitors to the reaction mixture.

(4) Combinations of inhibitors

EDTA has not been used alone as an MDH inhibitor. EDTA binds to the cytochrome-binding area of MDH, blocking electron transfer from MDH to cytochrome c. In fact, the oxidation of methanol by MDH can be inhibited by 0.05–1 mM of EDTA. However, EDTA also reduces MMO activity, probably because of the removal of metal ions from MMO via the chelation effect of EDTA [25, 26]. The combination of EDTA and other inhibitors such as NaCl has been tested as a method to decrease the concentration of the other inhibitor. The methanol production observed for a combined EDTA/NaCl system exceeded that of the NaCl-only system approximately two-fold [15, 18].

$MgCl_2$ also has not been used as a sole inhibitor, but has been added along with either phosphate or NaCl [15, 17, 19]. The mechanism by which $MgCl_2$ suppresses the oxidation of methanol is still unclear [109], but various studies have reported that methanol production increases significantly when 5–20 mM $MgCl_2$ is added to phosphate buffer [7, 16, 22, 24, 25]. Mg ions promote sMMO activity and the growth of methane-oxidizing bacteria [16]. The use of $MgCl_2$ in conjunction with phosphate resulted in greater productivity than the use of EDTA with phosphate, but EDTA showed a greater inhibitory effect on MDH activity than $MgCl_2$ [22, 25]. The use of 400 mM phosphate and 10 mM $MgCl_2$ resulted in a cell density of 17.3 g L^{-1} and the highest methanol concentration of 1.13 g L^{-1}, although phosphate concentrations above 100 mM have been reported to inhibit MMO activity [10, 22, 25]. These results demonstrate that higher methanol production can be achieved by increasing the inhibitor concentration as well as the cell density [16].

It has been suggested that 50% of the methanol oxidation activity of MDH should be retained to achieve the highest methanol production [10]. This hypothesis is based on the assumption that MDH supplies electrons to MMO. 100% MDH inhibition would thus result in no carbon assimilation and no electron production. Thus, the methane-to-methanol conversion ratio should be considered along with the methane uptake rate. For example, NH_4Cl and NaCl have high methane-to-methanol conversion rates during the initial period of the reaction, but a significant decrease in the methane conversion rate is observed as the reaction proceeds [17, 19]. This decrease was attributed to the decreased MMO activity and resulting lack of electrons from MDH.

4.2.3.3 Electron Donor

Theoretically, two electrons (one mole of NADH) are required to produce one mole of methanol. Formaldehyde or formate can be as used as the electron donor for these reactions, as two electrons are generated by the dehydrogenation of formaldehyde and formic acid in the cell. However, formaldehyde is toxic and has an inhibitory effect on cell metabolism. Therefore, most studies have added formate as the reducing agent for methanol production, as shown in Fig. 4.5 and Table 4.1.

The optimum concentration of formate has been reported for different methanol production processes; for example, concentrations of 14.3, 20, and 40 mM for *Methylosinus trichosporium* [8, 10, 11, 12, 15, 16, 18, 19], 40 mM for both *Methylosinus sporium* and strain 14B [22, 26], and 100 mM for *M. tundrae* and *M. bryophila* [24, 25]. The optimum concentration of formate depended not only on the bacterial strain used, but also on other parameters of the reactor.

The use of formate is not economical on the industrial scale. Therefore, alternative electron supply strategies should be investigated, such as the replacement of formate with other metabolic compounds. Xin et al. suggested that poly-β-hydroxybutyrate (PHB) could be used as the electron donor. In the conversion of carbon dioxide to methanol, methane-oxidizing bacteria containing 38.6% (w/w) PHB produced 94 μmol L^{-1} methanol without losing their vitality [118]. Despite low methanol productivity, the use of PHB as an electron donor has been shown to be feasible. The use of hydrogen as an electron donor may also be possible. Two distinct hydrogenases have been found in *Methylococcus capsulatus* Bath: a soluble hydrogenase and a membrane-bound hydrogenase. Both these enzymes are expressed constitutively under normal growth conditions. The soluble hydrogenase is capable of reducing NAD^+ using molecular hydrogen. Alternatively, the use of facultative methane-oxidizing bacteria could be considered, as these bacteria can use carbon compounds other than methane as their energy sources.

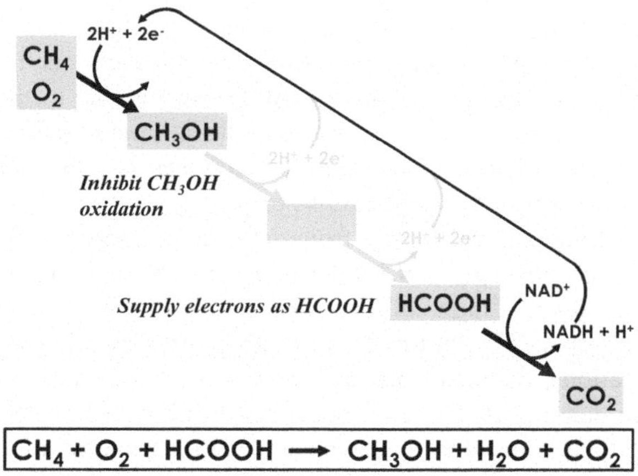

Fig. 4.5 Supply of electrons using FDH and formate

4.2.3.4 pH and Temperature

All reported biocatalytic oxidations of methane to methanol were performed at pH and temperatures of 6.40–7 and 25–37 °C, respectively. Outside these ranges, notable decreases in methanol production were observed, except in the case of *M. bryophila*, which has the ability to produce methanol at a higher pH of 8.5 [24].

4.2.3.5 Copper and Iron Ions

As mentioned above, copper ions are crucial in determining the expression of pMMO and sMMO. The addition of copper ions to the reaction mixture is expected to increase methanol productivity by increasing the enzyme activity of pMMO, which is known to show higher methane conversion than sMMO. In fact, at copper ion concentrations between 1 and 5 μM, the amount of methanol produced increases with the copper concentration. However, when the copper ion concentration exceeds 10–20 μM, both methanol production and bacterial growth are suppressed. Additionally, enhanced MMO activity and increased methanol production were observed in a reaction mixture containing 10 μM of iron ions [10, 11, 19, 21, 22, 24–26]. The addition of both 10 μM of iron ions and 5 μM of copper ions to the reaction mixture has been reported to double the methanol production of *M. sporium* [22].

4.2.3.6 Cell Density

Theoretically, increased cell density should increase the methanol productivity of a reactor, which is expressed as the amount of methanol (mol or g) per unit volume of the reactor (L) per unit reaction time (h). However, according to studies so far, the methanol productivity of a reactor does not increase linearly with the cell density, as shown in Fig. 4.6a. The productivity increases linearly until the cell density reaches approximately 1 mg-dry weight cells per liter, and then remains almost constant at cell densities above 1 mg-dry weight cells per L. As shown in Fig. 4.6b, the low productivity above this cell density is due to the lower productivity per unit biocatalyst, which is expressed as the amount of methanol (g) per unit of biocatalyst (g-dry weight cells) per unit reaction time (h); this value decreases at cell densities above 1 mg-dry weight cells per liter.

Methanol production from methane in reactors with cell densities above 10 g-dry weight cells per liter has been reported by only one research group. They recently developed a high-cell-density reactor by increasing the concentration of the MDH inhibitor [16]. Furthermore, another study mentioned that at high methane concentrations, methanol production was 3.2 times higher than in cultures with higher biomass [18]. These results suggest that increasing cell density requires increasing the concentrations of methane and MDH inhibitors and overcoming mass transfer limitations. Duan et al. added 5% paraffin oil to the reaction solution to increase the mass transfer of methane to the reaction solution, and examined the methanol

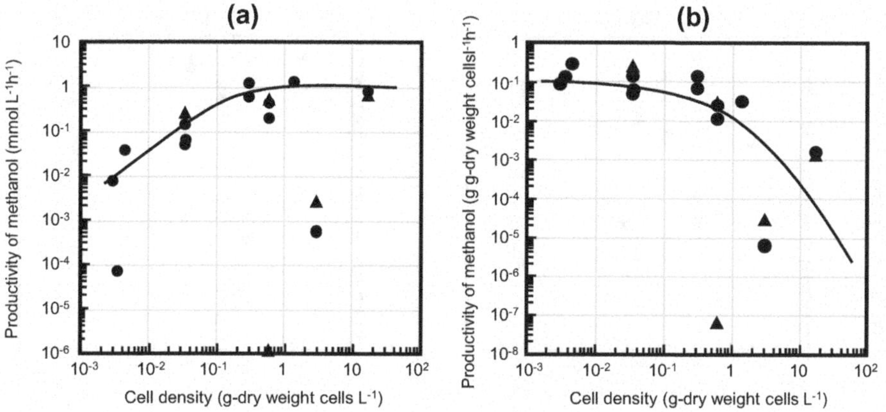

Fig. 4.6 Influence of the cell density on the methanol productivity per volume (**a**) and per bacterial cell (**b**)

accumulation at higher concentrations of MDH inhibitors in order to inhibit further increases in methanol concentration [16]. The methanol production of high-cell-density processes is expected to improve in the near feature via increases in the productivity of methanol per bacterial cell at high cell density.

4.2.3.7 Mass Transfer of Methane and Oxygen

Theoretically, one mole of methanol is produced from one mole of methane and one mole of oxygen, as shown in Eq. 4.1. However, this ratio is different from the optimum methane:oxygen ratio in the reactor headspace. This is due not only to the further oxidation of methanol, but also to the limitations of oxygen and methane mass transfer. The low gas–liquid mass transfer of methane into the culture medium has been demonstrated to be a growth-limiting factor for methane-oxidizing bacteria [30]. The low mass transfer is also responsible, in part, for the low methanol productivity.

To increase the mass transfer of methane and oxygen, several parameters of the reactor design have been investigated, such as the ratio of liquid phase to the gas phase, the pressure within the reactor, the composition of the gas phase, the speed of agitation of the reaction mixture, and the use of non-aqueous phases with higher affinities for methane than water.

For example, the liquid-to-gas ratios used in the previous studies ranged from 1:1.5 to 1:9 [7, 8, 10–13, 15–17, 19, 22, 118, 119]. One study reported a significant increase in methanol production when the ratio was increased from 1:1.67 to 1:9, especially at higher cell densities [18]. On the other hand, another study indicated that 1:5 was the optimum ratio, with the methanol production at 1:7 and 1:9 being slightly lower [22]. In previous studies, methanol formation reactions have been carried out using gas-phase methane contents ranging from 20 to 50% [7, 8, 10–13, 15–17, 19, 22, 118, 119]. As shown in Fig. 4.7, the methane content apparently has

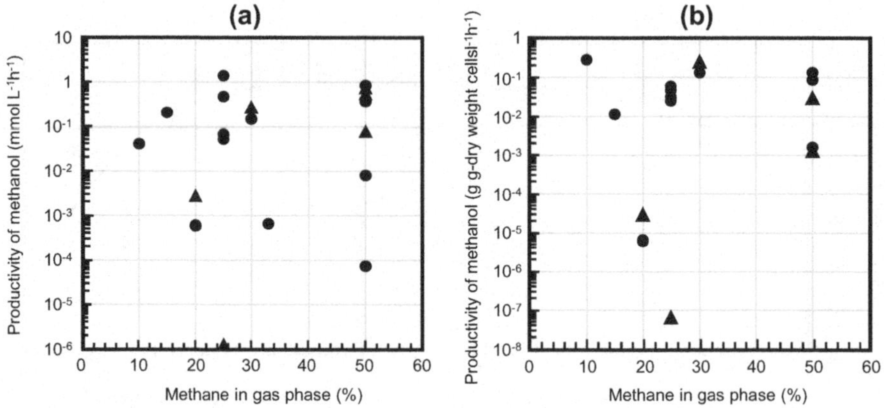

Fig. 4.7 Influence of the methane composition content in the gas phase on the methanol productivity per volume of the reaction mixture (**a**) and per unit weight of bacterial cells (**b**)

little correlation with the methanol productivity. The optimum conditions for these parameters depend on other factors such as the size of the reactor, agitation speed, and cell density. Thus, these parameters should be determined for each process.

To increase the mass transfer of methane, researchers have recently focused on the development of membrane reactors for the production of methanol from methane using methane-oxidizing bacteria. In one such membrane reaction, two porous membranes are used, which allows methane and oxygen (air) to be fed into the reactor separately, avoiding the risk of explosion (methane in air is explosive between 5% v/v and 15% v/v). The gases are delivered through membrane contactors. The large surface area avoids bubble formation and increases the gas–liquid mass transfer. The membrane reactor is stirred using a dense silicon tube. This reactor achieved improved methane mass transfer to the reaction mixture during the bioconversion of methane, and conversion of methane could be achieved at high cell densities of up to 17.3 g-dry cell g per liter. However, the methanol productivity was insufficient, as shown in Fig. 4.6. A novel recirculating macroporous membrane bioreactor was also developed. This reactor enables double the mass transfer of methane observed in a batch reactor under similar conditions, resulting in a methanol production of 0.15 mmol L^{-1} h^{-1} [75]. The productivity is not drastically higher than those reported previously, as shown in Fig. 4.6. These results indicate that membrane reactors cannot yet take advantage of the full potential of methane-oxidizing bacteria, and further improvement of membrane reactors can be expected in the near future.

As an alternative strategy for increasing the methane mass transfer, the addition of paraffin oil to the culture medium was tested. Using an oil concentration of 5% (v/v), an *M. trichosporium* OB3b cell density of 14 g-dry weight cells per liter was achieved [30]. However, the cell density did not increase further with increasing paraffin concentration, suggesting that methane mass transfer is not the only limiting factor. Although the addition of paraffin to the liquid medium has been shown to enhance methanotroph growth, its effect on methanol synthesis has not been reported.

4.3 Comparison of the Methanol Productivity of Biocatalysts and Heterogeneous Catalysts

In order to apply the biocatalytic synthesis of methanol in a practical context, biocatalytic processes that can meet industrial performance demands must be developed. One important performance parameter in industry is productivity. In catalytic reactions, productivity is defined as the amount of product produced per unit reaction time and unit of the catalyst. For example, the productivity of methanol production is given as g-methanol per g-catalyst per hour.

The productivities of biocatalytic processes for methanol production are summarized in Fig. 4.8. As shown in the figure, the productivity of biocatalysts is still insufficient for industrial methanol production, while the selectivity for methanol

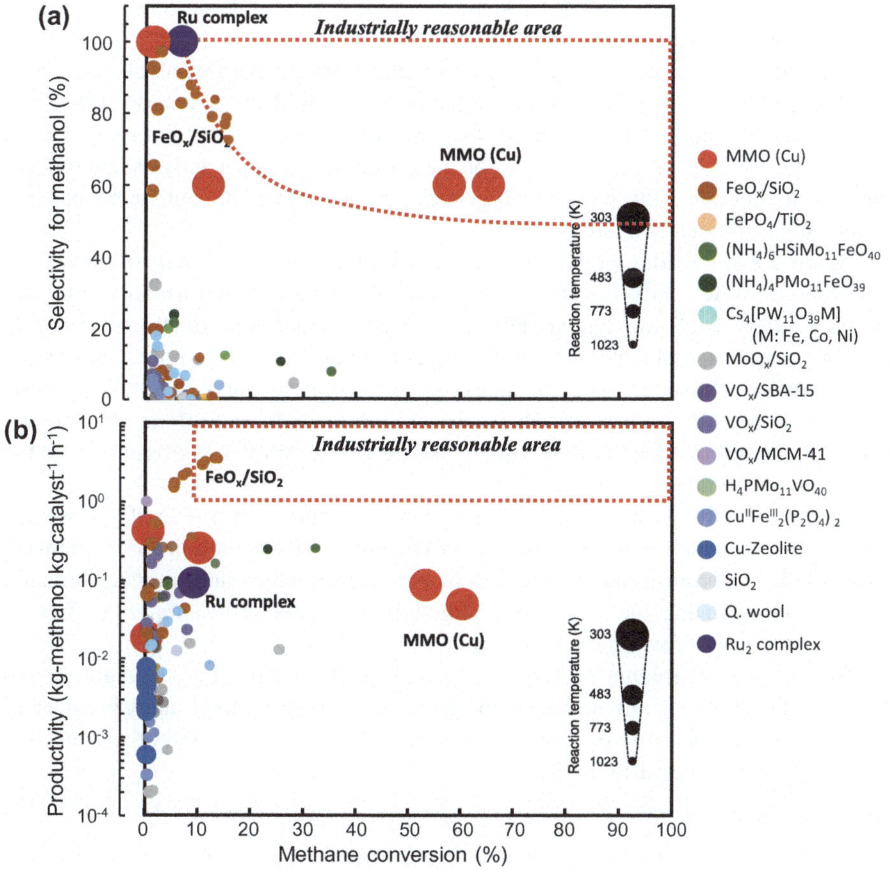

Fig. 4.8 Selectivity (**a**) and productivity (**b**) of methanol from methane and molecular oxygen molecule using homogeneous, heterogeneous (data in ref. [120], and [121]) and biological catalysts

and the reaction temperature of the biocatalysts are reasonable. To develop such bioprocesses, the productivity would need to be increased more than ten times.

There are three potential strategies to increase the productivity of biocatalytic processes. The first is to increase the amount of methanol allowed to accumulate. However, methanol production does not increase further at concentrations above ~10 mM because of the inhibition of MMO by methanol. This limitation can be overcome by increasing the cell density at the expense of a decrease in productivity. The second strategy is to decrease the amount of bacterial cells used for methanol production. The strategy was investigated by Takeguchi et al. [10], who achieved a maximum productivity of 0.05 g-methanol per g-dry cell per h. Under these conditions, the highest amount of methanol was obtained after 100 h. The final strategy is to increase the rate of methanol production per bacterial cell. For this strategy, the bacterial culture conditions must be optimized, along with the use of additives to enhance the growth rate of the bacterial culture and the MMO reaction. For this strategy, a combination of MMO system with other electron donation systems such as photosynthetic system may be useful [122, 123].

In addition to biocatalysts, heterogeneous and homogeneous catalysts have been developed with the aim of producing methanol in the liquid and gas phase [120, 121]. Among them, many catalysts require stronger oxidants than molecular oxygen such as N_2O, O_3, SO_3, $K_2S_2O_8$, and H_2O_2 or their mixtures. These catalysts can oxidize methane to methanol via an active oxygen species on a metal site, but are not capable of molecular oxygen activation.

Figure 4.8 shows the methanol productivity of various (bio)catalysts, revealing that although few catalysts achieve commercially attractive performance, biocatalysts are relatively promising. Notably, biocatalytic methanol production proceeds at ambient temperature and pressure in the aqueous phase and using molecular oxygen as the oxidant. Thus, the use of biocatalysts is a low-energy and sustainable process. Taking into account both the productivity and the reaction conditions, the production of methanol using methane-oxidizing bacteria is a promising alternative for the production of methanol from methane.

The use of not only pure methane, but also its mixtures with several gases, such as hydrogen and carbon dioxide, in biocatalytic methanol production has been investigated [118] in order to use natural gas or biogas as the methane source. Methanol can be obtained using such processes; although the productivities are not high, they are expected to improve in the near future.

The use of recombinant bacteria containing MMO systems may also allow high methanol production from methane to be achieved. However, this strategy is currently hampered by the inability to express functional MMO in *E. coli* and other cells that can be applied in industrial processes.

References

1. Beal EJ, House CH, Orphan VJ (2009) Manganese- and iron-dependent marine methane oxidation. Science 325:184–187
2. Boetius A, Ravenschlag K, Schubert CJ, Rickert D, Widdel F, Gieseke A, Amann R, Jørgensen BB, Witte U, Pfannkuche O (2000) A marine microbial consortium apparently mediating anaerobic oxidation of methane. Nature 407:623–626
3. Cui M, Ma A, Qi H, Zhuang X, Zhuang G (2015) Anaerobic oxidation of methane: an "active" microbial process. MicrobiologyOpen 4:1–11
4. Fei Q, Guarnieri MT, Tao L, Laurens LM, Dowe N, Pienkos PT (2014) Bioconversion of natural gas to liquid fuel: opportunities and challenges. Biotechnol Adv 32:596–614
5. Haroon MF, Hu S, Shi Y, Imelfort M, Keller J, Hugenholtz P, Yuan Z, Tyson GW (2013) Anaerobic oxidation of methane coupled to nitrate reduction in a novel archaeal lineage. Nature 500:567–570
6. Scheller S, Goenrich M, Boecher R, Thauer RK, Jaun B (2010) The key nickel enzyme of methanogenesis catalyses the anaerobic oxidation of methane. Nature 465:606–608
7. Mehta PK, Mishra S, Ghose TK (1987) Methanol accumulation by resting cells of *Methylosinus trichosporium* (I). J Gen Appl Microbiol 33:221–229
8. Mehta PK, Ghose TK, Mishra S (1991) Methanol biosynthesis by covalently immobilized cells of *Methylosinus trichosporium*: batch and continuous studies. Biotechnol Bioeng 37:551–556
9. Sugimori D, Takeguchi M, Okura I (1995) Biocatalytic methanol production from methane with *Methylosinus trichosporium* OB3b: an approach to improve methanol accumulation. Biotechnol Lett 17:783–784
10. Takeguchi M, Furuto T, Sugimori D, Okura I (1997) Optimization of methanol biosynthesis by *Methylosinus trichosporium* OB3b: an approach to improve methanol accumulation. Appl Biochem Biotechnol 68:143–152
11. Furuto T, Takeguchi M, Okura I (1999) Semicontinuous methanol biosynthesis by *Methylosinus trichosporium* OB3b. J Mol Catal A 144:257–261
12. Lee SG, Goo JH, Kim HG, Oh JI, Kim YM, Kim SW (2004) Optimization of methanol biosynthesis from methane using *Methylosinus trichosporium* OB3b. Biotechnol Lett 26:947–950
13. Xin J, Cui J, Niu J, Hua S, Xia C, Li S, Zhu L (2004) Biosynthesis of methanol from CO_2 and CH_4 by methanotrophic bacteria. Biotechnol 3:67–71
14. Markowska A, Michalkiewicz B (2009) Biosynthesis of methanol from methane by *Methylosinus trichosporium* OB3b. Chem Papers 63:105–110
15. Kim H, Han GGH, Kim SW (2010) Optimization of lab scale methanol production by *Methylosinus trichosporium* OB3b. Biotechnol Bioproc Eng 15:476–480
16. Duan C, Luo M, Xing X (2011) High-rate conversion of methane to methanol by *Methylosinus trichosporium* OB3b. Bioresour Technol 102:7349–7353
17. Han JS, Ahn CM, Mahanty B, Kim CG (2013) Partial oxidative conversion of methane to methanol through selective inhibition of methanol dehydrogenase in methanotrophic consortium from landfill cover soil. Appl Biochem Biotechnol 171:1487–1499
18. Pen N, Soussan L, Belleville MP, Sanchez J, Charmette C, Paolucci-Jeanjean D (2014) An innovative membrane bioreactor for methane bio-hydroxylation. Bioresour Technol 174:42–52
19. Hwang IY, Hur DH, Lee JH, Park CH, Chang IS, Lee JW, Lee EY (2015) Batch conversion of methane to methanol using *Methylosinus trichosporium* OB3b as biocatalyst. J Microbiol Biotechnol 25:375–380
20. Yoo YS, Han JS, Ahn CM, Kim CG (2015) Comparative enzyme inhibitive methanol production by *Methylosinus sporium* from simulated biogas. Environ Technol 36:983–991
21. Pen N, Soussan L, Belleville MP, Sanchez J, Paolucci-Jeanjean D (2016) Methane hydroxylation by *Methylosinus trichosporium* OB3b: monitoring the biocatalyst activity for methanol production optimization in an innovative membrane bioreactor. Biotechnol Bioprocess Eng 21:283–293

22. Patel SKS, Jeong JH, Mehariya S, Otari SV, Madan B, Haw JR, Lee JK, Zhang L, Kim IW (2016) Production of methanol from methane by encapsulated *Methylosinus sporium*. J Microbiol Biotechnol 26:2098–2105
23. Patel SKS, Mardina P, Kim D, Kim SY, Kalia VC, Kim IW, Lee JK (2016) Improvement in methanol production by regulating the composition of synthetic gas mixture and raw biogas. Bioresour Technol 218:202–208
24. Patel SKS, Mardina P, Kim SY, Lee JK, Kim IW (2016) Biological methanol production by a Type II methanotroph *Methylocystis bryophila*. J Microbiol Biotechnol 26:717–724
25. Mardina P, Li J, Patel SK, Kim IW, Lee JK, Selvaraj C (2016) Potential of immobilized whole-cell *Methylocella tundrae* as a biocatalyst for methanol production from methane. J Microbiol Biotechnol 26:1234–1241
26. Sheets P, Ge X, Li YF, Yu Z, Li Y (2016) Biological conversion of biogas to methanol using methanotrophs isolated from solid-state anaerobic digestate. Bioresour Technol 201:50–57
27. Patel SKS, Singh RK, Kumar A, Jeong JH, Jeong SH, Kalia VC, Kim IW, Lee JK (2017) Biological methanol production by immobilized *Methylocella tundrae* using simulated biohythane as a feed. Bioresour Technol 241:922–927
28. Patel SKS, Jeon MS, Gupta RK, Jeon Y, Kalia VC, Kim SC, Cho BK, Kim DR, Lee JK (2019) Hierarchical macroporous particles for efficient whole-cell immobilization: application in bioconversion of greenhouse gases to methanol. ACS Appl Mater Interfaces 11:18968–18977
29. Patel SKS, Kumar V, Mardina P, Li J, Lestari R, Kalia VC, Lee JK (2018) Methanol production from simulated biogas mixtures by co-immobilized *Methylomonas methanica* and *Methylocella tundrae*. Bioresour Technol 263:25–32
30. Hanson RS, Hanson TE (1996) Methanotrophic bacteria. Microbiol Rev 60:439–471
31. Rasigraf O, Kool DM, Jetten MSM, Sinninghe Damsté JS, Ettwig KF (2014) Autotrophic carbon dioxide fixation via the Calvin-Benson-Bassham cycle by the denitrifying methanotroph "*Candidatus Methylomirabilis oxyfera*". Appl Environ Microbiol 80:2451–2460
32. Baxter NJ, Hirt RP, Bodrossy L, Kovacs KL, Embley MT, Prosser JI, Murrell CJ (2002) The ribulose-1,5-bisphosphate carboxylase/oxygenase gene cluster of *Methylococcus capsulatus* (Bath). Arch Microbiol 177:279–289
33. Taylor SC, Dalton H, Dow CS (1981) Ribulose-1,5-bisphosphate carboxylase/oxygenase and carbon assimilation in *Methylococcus capsulatus* (Bath). J Gen Mcrobiol 122:89–94
34. Chen Y, Crombie A, Rahman MT, Dedysh SN, Liesack W, Stott MB, Alam M, Theisen AR, Murrell JC, Dunfield PF (2010) Complete genome sequence of the aerobic facultative methanotroph *Methylocella silvestris* BL2. J Bacteriol 192:3840–3841
35. Dedysh SN, Khmelenina VN, Suzina NE, Trotsenko YA, Semrau JD, Liesack W, Tiedje JM (2002) *Methylocapsa acidiphila* gen. nov., sp. nov., a novel methane-oxidizing and dinitrogen-fixing acidophilic bacterium from Sphagnum bog. Int J Syst Evol Microbiol 52:251–261
36. Vorobev AV, Baani M, Doronina NV, Brady AL, Liesack W, Dunfield PF, Dedysh SN (2011) *Methyloferula stellata* gen. nov., sp. nov., an acidophilic, obligately methanotrophic bacterium that possesses only a soluble methane monooxygenase. Int J Syst Evol Microbiol 61:2456–2463
37. Ward N, Larsen Ø, Sakwa J, Bruseth L, Khouri H, Durkin AS, Dimitrov G, Jiang L, Scanlan D, Kang KH, Lewis M, Nelson KE, Methé B, Wu M, Heidelberg JF, Paulsen IT, Fouts D, Ravel J, Tettelin H, Ren Q, Read T, DeBoy RT, Seshadri R, Salzberg SL, Jensen HB, Birkeland NK, Nelson WC, Dodson RJ, Grindhaug SH, Holt I, Eidhammer I, Jonasen I, Vanaken S, Utterback T, Feldblyum TV, Fraser CM, Lillehaug JR, Eisen JA (2004) Genomic insights into methanotrophy: the complete genome sequence of *Methylococcus capsulatus* (Bath). PLoS Biol 2:e303
38. Whittenbury R, Phillips KC, Wilkinson JF (1970) Enrichment, isolation and some properties of methane-utilizing bacteria. J Gen Microbiol 61:205–218
39. Anthony C (1986) Bacterial oxidation of methane and methanol. Adv Microb Physiol 27:113–210
40. Peyraud R, Kiefer P, Christen P, Massou S, Portais JC, Vorholt JA (2009) Demonstration of the ethylmalonyl-CoA pathway by using ^{13}C metabolomics. Proc Natl Acad Sci USA 106:4846–4851

41. Matsen JB, Yang S, Stein LY, Beck D, Kalyuzhnaya MG (2013) Global molecular analyses of methane metabolism in methanotrophic alphaproteobacterium, *Methylosinus trichosporium* OB3b. Part I. Transcriptomic study. Front Microbiol Chem 4:1–16
42. Yang S, Matsen JB, Konopka M, Green-Saxena A, Clubb J, Sadilek M, Orphan VJ, Beck D, Kalyuzhnaya MG (2013) Global molecular analyses of methane metabolism in methanotrophic alphaproteobacterium, *Methylosinus trichosporium* OB3b. Part II. Metabolomics and ^{13}C-labeling study. Front Microbiol 4:70
43. Vekeman B, Kerckhof FM, Cremers G, de Vos P, Vandamme P, Boon N, Op den Camp HJM, Heylen K (2016) New *Methyloceanibacter* diversity from North Sea sediments includes methanotroph containing solely the soluble methane monooxygenase. Environ Microbiol 18:4523–4536
44. Knief C (2015) Diversity and habitat preferences of cultivated and uncultivated aerobic methanotrophic bacteria evaluated based on pmoA as molecular marker. Front Microbiol 6:1346
45. Krause S, van Bodegom PM, Cornwell WK, Bodelier PL (2014) Weak phylogenetic signal in physiological traits of methane-oxidizing bacteria. J Evol Biol 27:1240–1247
46. Deutzmann JS, Wörner S, Schink B (2011) Activity and diversity of methanotrophic bacteria at methane seeps in eastern Lake Constance sediments. Appl Environ Microbiol 77:2573–2581
47. Dumont MG, Pommerenke B, Casper P, Conrad R (2011) DNA-, rRNA- and mRNA-based stable isotope probing of aerobic methanotrophs in lake sediment. Environ Microbiol 13:1153–1167
48. Siljanen HM, Saari A, Krause S, Lensu A, Abell GC, Bodrossy L, Bodelier PL, Martikainen PJ (2011) Hydrology is reflected in the functioning and community composition of methanotrophs in the littoral wetland of a boreal lake. FEMS Microbiol Ecol 75:430–445
49. Lüke C, Frenzel P (2011) Potential of pmoA amplicon pyrosequencing for methanotroph diversity studies. Appl Environ Microbiol 77:6305–6309
50. Henneberger R, Lüke C, Mosberger L, Schroth MH (2012) Structure and function of methanotrophic communities in a landfill-cover soil. FEMS Microbiol Ecol 81:52–65; Dumont MG, Lüke C, Deng YC, Frenzel P (2014) Classification of pmoA amplicon pyrosequences using BLAST and the lowest common ancestor method in MEGAN. Front Microbiol 5:e34
51. Deng Y, Cui X, Lüke C, Dumont MG (2013) Aerobic methanotroph diversity in Riganqiao peatlands on the Qinghai-Tibetan Plateau. Environ Microbiol Rep 5:566–574
52. Dunfield PF, Yuryev A, Senin P, Smirnova AV, Hou S, Ly B, Saw JH, Zhou Z, Ren Y, Wang J, Mountain BW, Crowe MA, Weatherby TM, Bodelier PL, Liesack W, Wang L, Alam M (2007) Methane oxidation by an extremely acidophilic bacterium of the phylum *Verrucomicrobia*. Nature 450:879–882
53. Islam T, Jensen S, Reigstad LJ (2008) Methane oxidation at 55°C and pH 2 by a thermoacidophilic bacterium belong-ing to the *Verrucomicrobia* phylum. Proc Natl Acad Sci USA 105:300–304
54. van Teeseling MC, Pol A, Harhangi HR, van der Zwart S, Jetten MS, Op den Camp HJ, van Niftrik L (2014) Expanding the verrucomicrobial methanotrophic world: description of three novel species of *Methylacidimicrobium* gen. nov. Appl Environ Microbiol 80:6782–6791
55. Khadem AF, Pol A, Wieczorek A, Mohammadi SS, Francoijs KJ, Stunnenberg HG, Jetten MS, Op den Camp HJ (2011) Autotrophic methanotrophy in Verrucomicrobia: *Methylacidiphilum fumariolicum* SolV uses the Calvin-Benson-Bassham cycle for carbon dioxide fixation. J Bacteriol 193:4438–4446
56. McDonald IR, Bodrossy L, Chen Y, Murrell JC (2008) Molecular ecology techniques for the study of aerobic methanotrophs. Appl Environ Microbiol 74:1305–1315
57. Dedysh SN, Knief C, Dunfield PF (2005) *Methylocella* species are facultatively methanotrophic. J Bacteriol 187(13, July):4665–4670
58. Theisen AR, Ali MH, Radajewski S, Dumont MG, Dunfield PF, McDonald IR, Dedysh SN, Miguez CB, Murrell JC (2005) Regulation of methane oxidation in the facultative methanotroph *Methylocella silvestris* BL2. Mol Microbiol 58:682–692

59. Crombie AT, Murrell JC (2014) Trace-gas metabolic versatility of the facultative methanotroph *Methylocella silvestris*. Nature 510:148–151
60. Belova SE, Baani M, Suzina NE, Bodelier PL, Liesack W, Dedysh SN (2011) Acetate utilization as a survival strategy of peat-inhabiting *Methylocystis* spp. Environ Microbiol Rep 3:36–46
61. Im J, Lee SW, Yoon S, Dispirito AA, Semrau JD (2011) Characterization of a novel facultative *Methylocystis* species capable of growth on methane, acetate and ethanol. Environ Microbiol Rep 3:174–181
62. Dedysh SN, Liesack W, Khmelenina VN, Suzina NE, Trotsenko YA, Semrau JD, Bares AM, Panikov NS, Tiedje JM (2000) *Methylocella palustris* gen. nov., sp. nov., a new methane-oxidizing acidophilic bacterium from peat bogs, representing a novel subtype of serine-pathway methanotrophs. Int J Syst Evol Microbiol 3:955–969
63. Stanley SH, Prior SD, Leak DJ, Dalton H (1983) Copper stress underlies the fundamental change in intracellular location of methane monooxygenase in methane-oxidising organisms: studies in batch and continuous cultures. Biotechnol Lett 5:487–492
64. Murrell JC, Gilbert B, McDonald IR (2000) Molecular biology and regulation of methane monooxygenase. Arch Microbiol 173:325–332
65. Csáki R, Bodrossy L, Klem J, Murrell JC, Kovács KL (2003) Genes involved in the copper-dependent regulation of soluble methane monooxygenase of Methylococcus capsulatus (Bath): cloning, sequencing and mutational analysis. Microbiology 149:1785–1795
66. Scanlan J, Dumont MG, Murrell JC (2009) Involvement of MmoR and MmoG in the transcriptional activation of soluble methane monooxygenase genes in *Methylosinus trichosporium* OB3b. FEMS Microbiol Lett 301:181–187
67. Semrau JD, Jagadevan S, DiSpirito AA, Khalifa A, Scanlan J, Bergman BH, Freemeier BC, Baral BS, Bandow NL, Vorobev A, Haft DH, Vuilleumier S, Murrell JC (2013) Methanobactin and MmoD work in concert to act as the 'copper-switch' in methanotrophs. Environ Microbiol 15:3077–3086
68. Kenney GE, Sadek M, Rosenzweig AC (2016) Copper-responsive gene expression in the methanotroph *Methylosinus trichosporium* OB3b. Metallomics 8:931–940
69. Dispirito AA, Semrau J, Murrell JC, Vuileumier S (2016) Methanobactin and the link between copper and bacterial methane oxidation. Microbiol Mol Biol Rev 80:387–409
70. Patel RN, Horse DS (1971) Physiological studies of methane and methanol-oxidising bacteria: oxidation of C-1 compounds by *Methylococcus capsulatus*. J Bacteriol 107:187–192
71. Stiring DI, Dalton H (1978) Purification and properties of an NAD(P)+ -linked formaldehyde dehydrogenase from *Methylococcus capsulatus* (Bath). JGen Microbiol 107:19–29
72. Semrau JD, DiSpirito AA, Yoon S (2010) Methanotrophs and copper. FEMS Microbiol Rev 34:496–531
73. Myronova N, Kitmitto A, Collins RF, Miyaji A, Dalton H (2006) Three-dimensional structure determination of a protein supercomplex that oxidizes methane to formaldehyde in *Methylococcus capsulatus* (Bath). Biochemistry 45:11905–11914
74. Culpepper MA, Rosenzweig AC (2014) Structure and protein-protein interactions of methanol dehydrogenase from *Methylococcus capsulatus* (Bath). Biochemistry 53:6211–6219
75. Kalyuzhnaya MG, Puri AW, Lidstrom ME (2015) Metabolic engineering in methanotrophic bacteria. Metab Eng 29:142–152
76. Trotsenko YA, Murrell JC (2008) Metabolic aspects of aerobic obligate methanotrophy. Adv Appl Microbiol 63:183–229
77. Tonge GM, Harrison DE, Knowles CJ, Higgins IJ (1975) Properties and partial purification of the methane-oxidising enzyme system from *Methylosinus trichosporium*. FEBS Lett 58:293–299
78. Chistoserdova L, Lidstrom ME (2013) Aerobic methylotrophic prokaryotes. In: Rosenberg E, DeLong EF, Thompson F, Lory S, Stackebrandt E (eds) The prokaryotes, 4th edn. Springer, New York, pp 267–285
79. Smith TJ, Trotsenko YA, Murrell JC (2010) Physiology and biochemistry of the aerobic methane oxidizing bacteria. In: Timmis KN (eds) Handbook of Hydrocarbon and Lipid Microbiology. Springer, Berlin, Heidelberg, pp 765–779

80. Zahn JA, Bergmann DJ, Boyd JM, Kunz RC, DiSpirito AA (2001) Membrane-associated quinoprotein formaldehyde dehydrogenase from *Methylococcus capsulatus* Bath. J Bacteriol 183:6832–6840
81. Chistoserdova L, Kalyuzhnaya MG, Lidstrom ME (2009) The expanding world of methylotrophic metabolism. Annu Rev Microbiol 63:477–499
82. Jollie DR, Lipscomb JD (1991) Formate dehydrogenase from *Methylosinus tricho-sporium* OB3b. Purification and spectroscopic characterization of the cofactors. J Biol Chem 266:21853–21863
83. Ferry JG (1990) Formate dehydrogenase. FEMS Microbiol Rev 7:377–382
84. Jollie DR, Lipscomb JD (1990) Formate dehydrogenase from *Methylosinus trichosporium* OB3b. Methods Enzymol 188:331–334
85. Yoch DC, Chen YP, Hardin MG (1990) Formate dehydrogenase from the methane oxidizer *Methylosinus trichosporium* OB3b. J Bacteriol 72:4456–4463
86. Quayle JR (1972) The metabolism of one-carbon compounds by micro-organisms. Adv Microbiol Physiol 7:119–203
87. Hwang IY, Lee SH, Choi YS, Park SJ, Na JG, Chang IS, Kim C, Kim HC, Kim YH, Lee JW, Lee EY (2014) Biocatalytic conversion of methane to methanol as a key step for development of methane-based biorefineries. J Microbiol Biotechnol 24:1597–1605
88. Karthikeyan OP, Chidambarampadmavathy K, Nadarajan S, Lee PK, Heimann K (2015) Effect of CH_4/O_2 ratio on fatty acid profile and polyhydroxybutyrate content in a heterotrophic-methanotrophic consortium. Chemosphere 141:235–242
89. Khmelenina VN, Rozova N, But CY, Mustakhimov II, Reshetnikov AS, Beschastnyi AP, Trotsenko YA (2015) Biosynthesis of secondary metabolites in methanotrophs: biochemical and genetic aspects. Prikl Biokhim Mikrobiol 51:140–150
90. Crowther GJ, Kosály G, Lidstrom ME (2008) Formate as the Main Branch Point for Methylotrophic Metabolism in Methylobacterium extorquens AM1. J Bacteriol 190(14):5057–5062
91. Chidambarampadmavathy K, Karthik O, Heimann K (2015) Role of copper and iron in methane oxidation and bacterial biopolymer accumulation. Eng Life Sci 15:387–399
92. Dedysh SN, Panikov NS, Tiedje JM (1998) Acidophilic methanotrophic communities from Sphagnum peat bogs. Appl Env Microbiol 64:922–929
93. Pol A, Heijmans K, Harhangi HR, Tedesco D, Jetten MSM, den Camp HJMO (2007) Methanotrophy below pH1 by a new Verrucomicrobia species. Nature 450:874–U817
94. Bowman JP, McCammon SA, Skerratt JH (1997) *Methylosphaera hansonii* gen. nov., sp. nov., a psychrophilic, group I methanotroph from Antarctic marine-salinity, meromictic lakes. Microbiol 143:1451–1459
95. Heyer J, Berger U, Hardt M, Dunfield PF (2005) *Methylohalobius crimeensis* gen. nov., sp. nov., a moderately halophilic, methanotrophic bacterium isolated from hypersaline lakes of Crimea. Int J System Evol Microbiol 55:1817–1826
96. Kaluzhnaya M, Khmelenina V, Eshinimaev B, Suzina N, Nikitin D, Solonin A, Lin JL, McDonald I, Murrell JC, Trotsenko Y (2001) Taxonomic characterization of new alkaliphilic and alkalitolerant methanotrophs from soda lakes of the Southeastern Transbaikal region and description of *Methylomicrobium buryatense* sp.nov. Systematic and Appl Microbiol 24:166–176
97. Andersen KB, von Meyenburg K (1980) Are growth rates of *Escherichia coli* in batch cultures limited by respiration? J Baceriol 144:114–123
98. López JC, Quijano G, Pérez R, Muñoz R (2014) Assessing the influence of CH_4 concentration during culture enrichment on the biodegradation kinetics and population structure. J Environ Manage 146:116–123
99. Fennel DE, Underhill SE, Jewell WJ (1992) Methanotrophic attached-film reactor development and biofilm characteristics. Biotechnol Bioeng 40:1218–1232
100. Hirayama H, Abe M, Miyazaki M, Nunoura T, Furushima Y, Yamamoto H, Takai K (2014) *Methylomarinovum caldicuralii* gen. nov., sp. nov., a moderately thermophilic methanotroph isolated from a shallow submarine hydrothermal system, and proposal of the family *Methylothermaceae fam.* nov. Int J Syst Evol Microbiol 64:989–999

101. Hirayama H, Suzuki Y, Abe M, Miyazaki M, Makita H, Inagaki F, Uematsu K, Takai K (2011) *Methylothermus subterraneus* sp. nov., a moderately thermophilic methanotroph isolated from a terrestrial subsurface hot aquifer. Int J Syst Evol Microbiol 61:2646–2653
102. Wise MG, McArthur JV, Shimkets LJ (2001) *Methylosarcina fibrata* gen. nov., sp. nov. and *Methylosarcina quisquiliarum* sp.nov., novel type I methanotrophs. Int J Syst Evol Microbiol 51:611–621
103. Dedysh SN, Belova SE, Bodelier PLE, Smirnova KV, Khmelenina VN, Chidthaisong A, Trotsenko YA, Liesack W, Dunfield PF (2007) *Methylocystis heyeri* sp. nov., a novel type II methanotrophic bacterium possessing "signature" fatty acids of type I methanotrophs. Int J Syst Evol Microbiol 57:472–479
104. Op den Camp HJM, Islam T, Stott MB, Harhangi HR, Hynes A, Schouten S, Jetten MSM, Birkeland NK, Pol A, Dunfield PF (2009) Environmental, genomic and taxonomic perspectives on methanotrophic *Verrucomicrobia*. Environ Microbiol Rep 1:293–306
105. Helm J, Wendlandt KD, Jechorek M, Stottmeister U (2008) Potassium deficiency results in accumulation of ultra-high molecular weight poly-β-hydroxybutyrate in a methane-utilizing mixed culture. J Appl Microbiol 105:1054–1061
106. Wendlandt KD, Geyer W, Mirschel G, Hemidi FAH (2005) Possibilities for controlling a PHB accumulation process using various analytical methods. J Biotechnol 117:119–129
107. Wendlandt KD, Jechorek M, Helm J, Stottmeister U (2001) Producing poly-3-hydroxybutyrate with a high molecular mass from methane. J Biotechnol Tailored Biopolymers 86:127–133
108. Ordaz A, López JC, Figueroa-González I, Muñoz R, Quijano G (2014) Assessment of methane biodegradation kinetics in two-phase partitioning bioreactors by pulse respirometry. Water Res 67:46–54
109. Han B, Su T, Wu H, Gou Z, Xing XH, Jiang H, Chen Y, Li X, Murrell JC (2009) Paraffin oil as a "methane vector" for rapid and high cell density cultivation of *Methylosinus trichosporium* OB3b. Appl Microbiol Biotechnol 83:669–677
110. Chang HN, Yoo IK, Kim BS (1994) High density cell culture by membrane-based cell recycle. Biotechnol Adv 12:467–487
111. Shiloach J, Fass R (2005) Growing *E. coli* to high cell density—a historical perspective on method development. Biotechnol Adv 23:345–357
112. Nguyen HHT, Shiemke AK, Jacobs SJ, Hales BJ, Lidstrom ME, Chan SI (1994) The nature of the copper ions in the membranes containing the particulate methane monooxygenase from *Methylococcus capsulatus* (Bath). J Biol Chem 269:14995–15005
113. Nguyen HHT, Elliott SJ, Yip JH, Chan SI (1998) The particulate methane monooxygenase from *Methylococcus capsulatus* (Bath) is a novel copper-containing three-subunit enzyme. Isolation and characterization. J Biol Chem 273:7957–7966
114. Yu SS, Chen KH, Tseng MY, Wang YS, Tseng CF, Chen YJ, Huang DS, Chan SI (2003) Production of high-quality particulate methane monooxygenase in high yields from *Methylococcus capsulatus* (Bath) with a hollow-fiber membrane bioreactor. J Bacteriol 185:5915–5924
115. Ge X, Yang L, Sheets JP, Yu Z, Li Y (2014) Biological conversion of methane to liquid fuels: status and opportunities. Biotechnol Adv 32:1460–1475
116. Park D, Lee J (2013) Biological conversion of methane to methanol. Korean J Chem Eng 30:977–987
117. Tabata K, Okura I (2008) Hydrogen and methanol formation utilizing bioproesses. J Jpn Petrol Inst 51:255–263
118. Xin JY, Zhang YX, Zhang S, Li SB (2007) Methanol production from CO_2 by resting cells of the methanotrophic bacterium *Methylosinus trichosporium* IMV 3011. J Basic Microbiol 47:426–435
119. Xin JY, Cui JR, Niu JZ, Hua SF, Xia CG, Li SB, Zhu LM (2004) Production of methanol from methane by methanotrophic bacteria. Biocatal Biotransform 22:225–229
120. Kondratenko EV, Peppel T, Seeburg D, Kondratenko VA, Kalevaru N, Martina A, Wohlrab S (2017) Methane conversion into different hydrocarbons or oxygenates: current status and future perspectives in catalyst development and reactor operation. Catal Sci Technol 7:366–381

121. Khokhar MD, Shukla RS, Jasra RV (2009) Selective oxidation of methane by molecular oxygen catalyzed by a bridged binuclear ruthenium complex at moderate pressures and ambient temperature. J Mol Catal A 299:108–116
122. Ito H, Mori F, Tabata K, Okura I, Kamachi T (2014) Methane hydroxylation using light energy by the combination of thylakoid and methane monooxygenase. RSC Adv 4:8645–8648
123. Ito H, Kondo R, Yoshimori K, Kamachi T (2018) Methane hydroxylation with water as an electron donor under light irradiation in the presence of reconstituted membranes containing both photosystem II and a methane monooxygenase. ChemBioChem 19:2152–2155

Chapter 5
C–C Bond Formation via the Condensation of Methane in the Presence or Absence of Oxygen

5.1 Introduction

As discussed in Chapter 1, the formation of C–C bonds to produce higher hydrocarbons (carbon number > 2) via the condensation of methane is difficult to achieve using direct conversion methods, whether in the presence or absence of oxygen. In contrast, higher hydrocarbons are produced industrially in the indirect conversion method by Fischer-Tropsch synthesis, as shown in Fig. 1.2. In this chapter, the condensation of methane, which is conventionally referred to as the oxidative coupling of methane or the dehydrogenative coupling of methane, is discussed. Hydrocarbons with carbon numbers of greater than two are referred to as C_2^+ hydrocarbons in this chapter.

The oxidative coupling of methane, which is shown in reaction (5.1), is more thermodynamically favorable than the dehydrogenative coupling of methane in reaction (5.2).

$$2\,CH_4 + 1/2\,O_2 \rightleftharpoons C_2H_6 + H_2O \qquad \Delta G = -12.6\ \text{kJ mol}^{-1}\ \text{at } 1000\ \text{K} \tag{5.1}$$

$$2\,CH_4 \rightleftharpoons C_2H_6 + H_2 \qquad \Delta G = 71.0\ \text{kJ mol}^{-1}\ \text{at } 1000\ \text{K} \tag{5.2}$$

The oxidative coupling of methane (OCM) involves the catalytic conversion of methane and oxygen to ethane (C_2H_6) and H_2O. The ethane is converted to ethylene (C_2H_4) in situ; further reactions also produce higher hydrocarbons, such as benzene. Unfortunately, the presence of oxygen leads irreversibly to oxidation, resulting in the thermodynamically stable end-product, CO_2 together with CO.

Although the dehydrogenative coupling of methane presents a thermodynamic disadvantage because of the large positive change in free energy, over-oxidation does not occur, and CO and CO_2 are not formed. The dehydrogenative coupling of

T. Baba and A. Miyaji, *Catalysis and the Mechanism of Methane Conversion to Chemicals*,
https://doi.org/10.1007/978-981-15-4132-2_5

methane is also often referred to as the non-oxidative coupling of methane. In this chapter, the oxidative and dehydrogenative methane coupling reactions are discussed. The discussion focuses mainly on the active sites of the catalysts, their reaction mechanisms, and their catalytic performance under various reaction conditions. In particular, the OCM reaction network, including the role of the methyl radical (•CH_3), the catalytically active sites for its formation, and the following coupling reactions of •CH_3 in the gas phase are discussed, while for the dehydrogenative coupling of methane, the catalytic performance and catalytic properties are emphasized.

5.2 Oxidative Coupling of Methane

The oxidative coupling of methane (OCM) can theoretically produce C_2^+ hydrocarbons. However, the products are generally limited to C_2H_6 and C_2H_4, even though C_2H_2 is thermodynamically more favorable than C_2H_6 and C_2H_4. The formation of C_2H_6 and C_2H_4 together with other higher hydrocarbons was first reported in a US patent [1], while OCM was reported by Fang et al. in 1981 [2]. Furthermore, Keller et al. reported the conversion of methane to C_2^+ hydrocarbons using various metal oxides [3]. They examined 26 metal oxides supported on alumina at 773–1273 K at atmospheric pressure and explored both the simultaneous and sequential feeding of CH_4, air, and nitrogen onto a fixed bed of the catalyst.

Much of the research into the energetics of the OCM up to the early 1990s has been reviewed in refs. [4] and [5]. The typical reaction temperatures in these studies ranged from nearly 950 to 1200 K. Hundreds of materials have been prepared and tested as catalysts for this reaction, principally during the 1990s and in recent years. Zavyalova et al. published a statistical analysis of the catalytic data published before 2011 [6], while Kondratenko et al. also reviewed the literature before 2015 [7].

The extensive studies of the OCM reaction have improved its conversion of methane and the selectivity for C_2^+ hydrocarbons. A summary of the important features of this reaction is provided below.

(1) At relatively low methane conversion, the selectivity towards C_2^+ hydrocarbons increases with increasing reaction temperature.
(2) The C_2H_4/C_2H_6 ratio increases with increasing methane conversion. Thus, C_2H_6 is the primary product of the OCM reaction.
(3) C_2^+ hydrocarbons can be formed by supplying methane over a pre-oxidized catalyst in the absence of oxygen in the gas phase, although the rate of methane conversion decreases with increasing time on stream.
(4) The •CH_3 radical is the intermediate of the OCM reaction, which occurs via a combination of heterogeneous and homogeneous steps [8–11]. Thus, the initial abstraction of a hydrogen atom from CH_4 (cleavage of a C–H bond) takes place on the surface of the catalyst to produce the •CH_3 radical, which is subsequently desorbed from the catalyst surface and undergoes recombination to produce C_2H_6 in the gas phase, not on the catalyst surface.

Details of the processes described in points (1)–(4) are given in the subsequent sections of this chapter.

5.3 Classification of Catalysts for the Oxidative Coupling of Methane

OCM, which involves methyl radical dimerization to produce C_2H_6, is usually carried out at high temperature. Methyl radicals ($\bullet CH_3$) produced on the catalyst surface diffuse to the gas phase and then engage in coupling reactions to afford C_2H_6 and C_2^+ hydrocarbons, i.e., it is rather difficult to control the selectivity of these reactions by using a suitable catalyst. Moreover, $\bullet CH_3$ are preferentially oxidized to carboxylates and ultimately to CO_2 rather than undergo homocoupling in the gas phase and/or on the catalyst surface [12]. Thus, OCM catalysts should not only initiate the formation of methyl radicals but also suppress the non-selective surface oxidation of methane and C_2^+ hydrocarbons to CO_2. Therefore, the yields (selectivities) of C_2H_6 and C_2^+ hydrocarbons do not exceed ~30%, as mentioned in Sect. 5.4.6.

Various systems for the classification of OCM catalysts have been developed. Zavyalova et al. classified OCM catalysts into four groups based on the type of catalyst material, as follows [6]:

(1) Non-reducible metal oxides,
(2) Reducible metal oxides
(3) Halogen-containing oxide materials
(4) Solid electrolytes.

Amenomiya et al. classified the catalysts into three major groups based on their location on the periodic table [5]:

Group 1: Alkali and alkaline earth metal compounds
Group 2: Lanthanide and actinide compounds
Group 3: Other metal compounds.

Group 3 can be further divided into transition metal and post-transition metal compounds (Group 11–15 elements). This classification scheme was used to classify both one- and two/multi-component catalysts.

Furthermore, Lunsford classified the catalysts more concretely into the following five groups [13]:

(1) Highly basic pure oxides, of which the early members of the lanthanide oxide series excluding CeO_2 are the best catalysts
(2) Group IA or IIA ions supported on basic oxides, such as Li/MgO and Sr/La_2O_3
(3) Monophasic oxides
(4) Transition metal oxides that contain Group IA ions
(5) Any of these materials that are promoted with chloride ions, such as $LiCa_2B_3O_4Cl_6$

Dubois et al. concluded that most, if not all, effective catalysts are strongly basic oxides that act as p-type semiconductors at elevated temperature and exhibit oxygen anion mobility [14]. Moreover, these catalysts contained free reducible surface ions, such as Na_2WO_4/SiO_2. The Na_2WO_4/SiO_2 catalyst was reported for the first time by Yan et al. in 1992 [15], and is discussed in detail in Sect. 5.4.6.

5.4 Role of Oxygen Species in Methyl Radical Formation Over Various Metal Oxide Catalysts

5.4.1 Evidence of Methyl Radical Formation During OCM

The formation of the $\bullet CH_3$ radical in the OCM reaction, which was long considered to be a well-established fact, was experimentally confirmed by Driscol et al. [8, 16]. Evidence for this heterogeneous–homogeneous reaction process was also provided by the detection of the $\bullet CH_3$ radical by using multiphoton ionization mass spectroscopy [17, 18]. Luo et al. also confirmed the formation of the $\bullet CH_3$ radical over Li/MgO in the OCM process using synchrotron vacuum ultraviolet photoionization spectroscopy [9].

The methyl radical has also been detected in the gas phase using a matrix isolation electron spin resonance (MIESR) system when methane was brought into contact with Na/CaO [19], Na/CeO_2 [20], Sr/La_2O_3 [21], Li/NiO_2 [22], Li/ZnO [23], La_2O_3 [24], other lanthanide oxides [25], and Li–MgO–Cl [26].

5.4.2 O^- Oxygen Species for Methyl Radical Formation

In addition to the $\bullet CH_3$ radical, $[Li^+O^-]$ centers were also detected in lithium-doped MgO (Li/MgO) via electron spin resonance (ESR) spectroscopy when Li/MgO was heated to 773 K or higher in air and quenched in liquid nitrogen [27]. Here, $[Li^+O^-]$ is a trapped hole at one of the six O^- ion sites surrounding a Li^+ ion. In Li/MgO, the Li^+ ions are located at Mg^{2+} cation sites and trap holes. They have a linear configuration that can be expressed as $Mg^{2+}–O^{2-}–Li^+–O^-–Mg^{2+}$ [28, 29]. Such $[Li^+O^-]$ centers were found to be formed in the alkali metal-doped alkaline earth oxides MgO, CaO, and SrO [30].

Driscoll et al. also detected $[Li^+O^-]$ centers in Li/MgO via ESR spectroscopy. The amount of the $\bullet CH_3$ radical that was generated from CH_4 over Li/MgO varied according to the amount of $[Li^+O^-]$ centers, which, in turn, was a function of the amount of lithium doped into MgO [8]. Ito et al. reported that the amount of $[Li^+O^-]$ centers (the intensity of ESR spectra due to $[Li^+O^-]$ centers) and the total amount of converted CH_4 both showed a very similar dependence on the oxygen pressure [27]. On the basis of these results, the $\bullet CH_3$ is formed on $[Li^+O^-]$ centers via the hemolytic abstraction of a hydrogen atom from CH_4, as shown in reaction (5.3).

$$[Li^+O^-] + CH_4 \longrightarrow [Li^+OH^-] + \bullet CH_3 \quad (5.3)$$

This indicates that the O^- oxygen species abstracts a hydrogen atom from CH_4 to generate a $\bullet CH_3$ radical in the initial stage of the OCM reaction. Thus, the O^- oxygen species are responsible for the formation of the $\bullet CH_3$ radical. The regeneration of the $[Li^+O^-]$ centers was suggested in ref. [27].

Furthermore, the formation of $\bullet CH_3$ radicals has been observed on not only Li/MgO, but also Li/CaO [31], Na/CaO [32], and Li/ZnO [23]. The alkali metals (M) Li and Na generate $[M^+O^-]$ centers that abstract a hydrogen atom to produce a $\bullet CH_3$ radical as a reaction intermediate in the OCM reaction. However, in ref. [32], no $[K^+O^-]$ centers were detected on K/CaO, nor were $[M^+O^-]$ centers observed on Na/MgO and K/MgO. The catalytic activities of these systems for the OCM reaction were poor.

5.4.3 Peroxide O_2^{2-} Ions: Metal Peroxides as Catalysts

Simple peroxides, such as Na_2O_2 [33, 34], BaO_2 [33, 35], and SrO_2 [33] have been found to be capable of stoichiometric OCM reaction. This provides experimental evidence that an O_2^{2-} species abstracts a hydrogen atom from methane to produce a $\bullet CH_3$ radical. Otsuka et al. proposed that the oxygen species was likely diatomic, and on the basis of kinetic studies, concluded that a diatomic oxygen species was indirectly, if not directly, responsible for the generation of $\bullet CH_3$ radicals from CH_4 on various OCM catalysts [36].

Kharas et al. observed an O_2^{2-} species on the surface of the metal oxide complex $BaPbO_5$ via X-ray photoelectron spectroscopy (XPS) [37]. Lunsford et al. also revealed the presence of peroxide (O_2^{2-} species) on the catalyst Ba/MgO via in situ Raman spectroscopy [38]; a strong correlation between the catalytic activity for the OCM and the near-surface concentration of O_2^{2-} species on Ba/MgO was also observed in the same system [39].

Moreover, Mestl et al. characterized surface peroxide (O_2^{2-}) structures on La_2O_3 and Na- and Sr-modified La_2O_3 via laser Raman spectroscopy [40]. Au et al. suggested that the interaction between O_2^{2-} ions and CH_4 might generate carbene radicals (CH_2:), and that these carbene radicals could account for the highly selective production of C_2H_4 by $BaCO_3$/LaOBr catalysts [41].

5.4.4 *Superoxide Ion (O_2^-) Species*

Superoxide (O_2^-) ions have been observed on the surface of La_2O_3 via ESR spectroscopy [42–44]. However, O_2^- species were demonstrated to be unstable on La_2O_3 under OCM reaction conditions, and suggested that their instability resulted from the low partial pressures of oxygen commonly used in the OCM reaction [43, 44]. Iwamoto et al. showed that O_2^- ions were not active toward simple alkanes such as CH_4 on MgO at temperatures up to 448 K; this temperature is lower than those typically used in the OCM reaction (~1000 K) [45]. Otsuka et al. also demonstrated that the O_2^- oxygen species contained in simple metal peroxides, such as Na_2O_2, were not active in the OCM reaction [33].

5.4.5 *Various Oxygen Species for the Formation of Methyl Radicals on the Catalyst Surface*

Sections 5.4.2, 5.4.3, and 5.4.4 discuss the abstraction of a hydrogen atom from CH_4 to produce a •CH_3 radical by O^- (oxygen anion radical), O_2^{2-} (peroxide), and/or O_2^- (superoxide) oxygen species on the catalyst surface. These oxygen species may be generated on various metal-doped oxides, including multi-component oxides such as Li/MgO or single-component metal oxides such as MgO. Furthermore, the oxidation states of the metal ions do not change. However, oxygen vacancies exist in these catalysts even at low oxygen pressures [46]. These vacancies ($V_o^{\cdot\cdot}$), which lack two electrons and have an effective double positive charge ($^{\cdot\cdot}$), react with O_2 to produce not only O_2^{2-} peroxide oxygen species but also positively charged holes ($h^{\cdot}$), as expressed in reaction (5.4).

$$V_o^{\cdot\cdot} + O_2 \longrightarrow O_2^{2-} + 2\,h^{\cdot} \tag{5.4}$$

The O_2^{2-} and 2 $h^{\cdot}$ formed in the above reaction further generate various oxygen species by the following reactions.

$$O_2^{2-} + 2\,h^{\cdot} \longrightarrow 2\,O^- + 2\,h^{\cdot} \tag{5.5}$$

$$2\,O^- + 2\,h^{\cdot} \longrightarrow 2\,O^{2-} + 4\,h^{\cdot} \tag{5.6}$$

Furthermore, it has been reported that O_3^- ions are generated by the reaction of O_2 with O^- on the surface of MgO [47, 48] and $SrTiO_{(3\pm\delta)}$ [49].

As discussed in the preceding sections, the catalytic activities of various catalysts, including Li/MgO, are influenced by the oxygen pressure, which possibly indicates the reversible formation of various oxygen species on the catalyst surface.

It was proposed that these oxygen species exist in the following equilibrium, with the equilibrium position depending on the nature of the catalysts [50, 51].

$$O_2\,(g) \rightleftharpoons O_2\,(ads) \xrightleftharpoons{+e} O_2^- \xrightleftharpoons{+e} O_2^{2-} \rightleftharpoons 2\,O^- \quad (5.7)$$

Therefore, various types of oxygen species could possibly abstract a hydrogen atom from CH_4 to generate a $\bullet CH_3$ radical.

5.4.6 Oxygen Ion Species in Catalysts

The OCM reaction proceeds over not only non-reducible metal oxides, such as Li/MgO, but also over reducible metal oxides. For example, Sofranko et al. reported that various transition metal oxides, such as manganese oxide supported on silica (Mn/SiO_2), catalyze the OCM reaction to produce C_2^+ hydrocarbons at 973 K [52, 53]. They also prepared sodium-promoted Mn/SiO_2 and Mn/MgO and alumina-supported manganese oxide [54].

To further increase the stability of the catalysts and the maximum yield of C_2^+ hydrocarbons, tungsten-manganese catalysts have been developed, as first reported by Fang et al. in 1992 [55–57]. At 1073 K, a methane conversion of 36.8% and a C_2 hydrocarbon ($C_2H_6 + C_2H_4$) yield of 23.9% were achieved using Mn (1.9 wt%) and Na_2WO_4 (5 wt%) supported on silica ($Mn/Na_2WO_4/SiO_2$) as the catalyst [55]. The $Mn/Na_2WO_4/SiO_2$ catalyst showed high stability for 30 h [56], which represented a significant improvement over non-reducible catalysts such as Li/MgO. The authors also reported that the activity of the catalyst was related to the W–O–Si species that originated from the interaction between Na_2WO_4 and SiO_2, while the presence of small amounts of Mn increased the concentration of the oxygen in the surface lattice and the rate of methane conversion. Furthermore, the X-ray photoelectron spectroscopy (XPS) spectra showed that the manganese species existed as Mn_2O_3 crystals [57].

Jiang et al. reported the observation of a structure containing a W= O and three W–O–Si surface bonds that were generated by the reconstruction of surface WO_4 tetrahedral units in MnO_x–Na_2WO_4/SiO_2, and speculated that these species were relevant to the OCM reaction [58]. They also reported that W^{6+} was transformed into W^{5+} because of the interaction between Na_2WO_4 and the SiO_2 support. On the other hand, Ji et al. asserted that Na–O–Mn and Na–O–W species played a role in catalysis of the OCM reaction based on laser Raman spectra and XPS data of Na–W–Mn/SiO_2 catalysts with different contents of sodium, tungsten, and manganese [59].

Yildz et al. investigated the effect of the support materials on the catalytic performance of Mn/Na_2WO_4 catalysts, and demonstrated that SiO_2 was indeed the best support material [60]. In $Mn/Na_2WO_4/SiO_2$ catalyst systems, Mn–O–Si and W–O–Si species, as well as W=O, were observed, and the oxidation states of both W (W^{6+}/W^{5+}) and Mn (Mn^{3+}/Mn^{2+}) changed under the OCM reaction conditions [61–68]. These results demonstrated that the oxygen species Mn–O–Si, W–O–Si, W=O, Na–O–Mn, and Na–O–W may abstract a hydrogen atom from CH_4 to generate a $\bullet CH_3$ radical during the initial stage of the OCM reaction. However, thus far, it has not been possible to identify the site of the active oxygen species on $Mn–Na_2WO_4/SiO_2$ catalysts.

5.5 Reaction Network in the Oxidative Coupling of Methane

As discussed in Sects. 5.2 and 5.3, the OCM reaction comprises heterogeneous catalytic and homogeneous non-catalytic processes and converts methane mainly into ethane and ethylene. These processes depend on the reaction conditions, especially the reaction temperature. In addition to the formation of these hydrocarbons, non-selective oxidation to CO and CO_2 also occurs. CO and CO_2 are collectively referred to as CO_x in this section. A general scheme of the reactions leading to the formation of C_2H_6 and C_2H_4 and the total oxidation products CO_x is shown in Scheme 5.1 [4]. Both CO and CO_2 are produced during the OCM reaction, but CO is converted into CO_2 over most OCM catalysts. As shown in Scheme 5.1 C_2H_4 is formed stepwise through **Path 1** and **Path 4**, rather than being formed directly from CH_4. Furthermore, CO_x is derived from all three hydrocarbons. The reaction network shown in Scheme 5.1 is explained in more detail in the following sections.

5.5.1 C_2H_6 and CO_x as Primary Products

Ekstrom et al. performed an isotopic labeling experiment to determine the origin of CO_x by adding $^{13}C_2H_4$ or $^{13}C_2H_6$ to CH_4 and measuring the fraction of $^{13}CO_x$ in the products [69]. The results showed that CO_x was formed via both **Path 3** and **Path**

Scheme 5.1 OCM reaction network

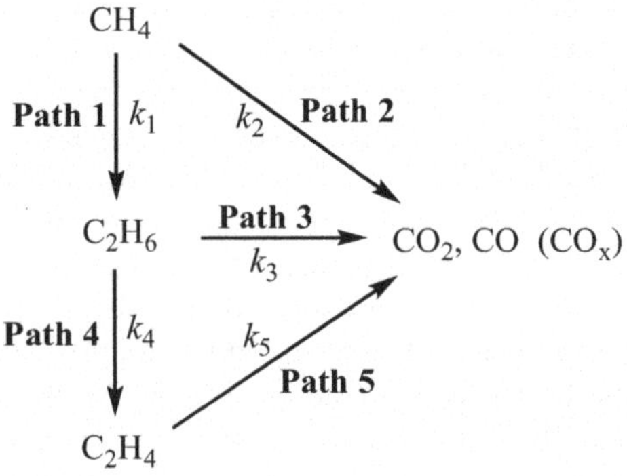

Table 5.1 Initial formation rates of the products in the oxidative coupling of methane

Compounds	Initial formation rate/mol s^{-1}	
	Li/CaO[a]	Li/MgO[b]
C_2H_6	1.67×10^{-5}	1.54×10^{-5}
C_2H_4	0	0
CO	1.04×10^{-4}	0
CO_2	1.94×10^{-5}	5.00×10^{-6}

(Reported in refs. [69] and [70])
Reaction temperature: 983 K, O_2: 7 kPa, CH_4: 67 kPa, He: 26 kPa
[a]Reported in ref. [69]
[b]Reported in ref. [70]

5 in Scheme 5.1. However, it was not clear whether CH_4 was directly converted to CO_x via **Path 2**. Korf et al. showed that C_2H_6 and CO_x were primary products by measuring the initial reaction rates of the products [70]. The results obtained using Ba/CaO and Li/MgO as catalysts at 983 K are summarized in Table 5.1. The results in refs. [69 and 70] indicate that methane is converted directly to CO_x via **Path 2**, as well as to C_2H_6 via **Path 1**.

To further investigate the reaction network of the OCM reaction, Otsuka et al. carried out the reaction using different reactants (CH_4, C_2H_6, and C_2H_4) and examined the products in the effluent gas using Sm_2O_3 as the catalyst at a temperature of 883 K and constant oxygen pressure of 0.92 kPa [71]. The results of this experiment are summarized in Table 5.2. Since the conversions of the reactants such as CH_4 were sufficiently low under these reaction conditions, the partial pressure of each product in the outlet gas of the reactor was equal to the formation rate of each product.

In Run 1 of Table 5.2, methane was converted into all the reaction products (C_2H_6, C_2H_4, CO, and CO_2) shown in Scheme 5.1. In Run 2, the partial pressure of C_2H_6 was adjusted to be nearly the same as that of C_2H_6 in the products of Run 1, while the oxygen pressure was maintained at 0.92 kPa. While the partial pressure of C_2H_4 in the products was very similar in Run 1 and Run 2, the formation of CO_x (CO and CO_2) decreased significantly in Run 2, indicating that the rate of the direct conversion

Table 5.2 Relationship between the partial pressures of the reactants and products in the Sm_2O_3-catalyzed OCM reaction

Run no.	Reactant pressure/kPa				Product pressure/kPa			
	O_2	CH_4	C_2H_6	C_2H_4	C_2H_6	C_2H_4	CO	CO_2
1	9.2×10^{-1}	7.7	–	–	1.2×10^{-2}	1.6×10^{-4}	9.2×10^{-3}	1.1×10^{-2}
2	9.2×10^{-1}	–	1.3×10^{-2}	–	–	2.0×10^{-4}	2.4×10^{-4}	2.0×10^{-4}
3	9.2×10^{-1}	–	–	2.0×10^{-4}	0	–	–	6.4×10^{-6}

(Reprinted from ref. [71], Copyright 2019, with permission from Elsevier)

of CH_4 to CO_x (**Path 2** in Scheme 5.1) was faster than those of C_2H_6 and C_2H_4 to CO_x (**Path 3** and **Path 5** in Scheme 5.1). In Run 3, the reactant pressure of C_2H_4 was adjusted to be nearly the same as its product pressure in Run 1 and Run 2. The rates of CO_x formation in Run 3 were much slower than in Run 1 and Run 2. Therefore, CO_x (CO and CO_2) were concluded to be produced mainly from CH_4 at low methane conversion using Sm_2O_3 as the catalyst. Ito et al. also reported that CO_x was formed directly from methane rather than via the further oxidation of C_2H_6 when either CH_4 or C_2H_6 was added to the reactant feed over Li/MgO catalysts [27].

5.5.2 Estimation of the First-Order Rate Constant of Each Reaction Path

In ref. [4], Lee et al. estimated the magnitude of the first-order rate constant of each reaction path in Scheme 5.1 based on the results reported using Sm_2O_3 as the catalyst in ref. [71]. The relative magnitudes of the constants calculated by these authors are listed in Table 5.3. According to these results, ethylene is more reactive via **Path 5** than methane in **Path 1** and **Path 2**. Furthermore, ethane is more easily oxidized via **Path 3** than ethylene via **Path 5**. Thus, the conversion of ethane to CO_x proceeds more readily that of ethylene.

In Table 5.3, the ratio of k_5 (the rate constant of **Path 5**) to k_2 (that of **Path 2**), which is referred to as R, is larger than 1. A greater than unity R value was also reported when other metal oxides were used as the catalyst [72, 73]. For example, the R values were 7.7 at 1073 K over CeO_2 and 4.7 at 973 K over Li/MgO.

In the OCM reaction, the coupling of $\bullet CH_3$ radicals in the gas phase to produce C_2H_6 is independent of the catalyst used. As shown in Table 5.3, at low methane conversion, the production of CO_x via **Path 2** is slower than that of C_2H_6 via **Path 1**, even when the oxygen pressure is about 8 times lower than that of CH_4. However, as the methane conversion increased, the contribution of the over-oxidation of C_2H_6 and C_2H_4 (**Path 3** and **Path 5**) increases these reaction rates, indicating that the production of CO_x (CO_2 and CO) via **Path 3** and **Path 5** occurs faster than the conversion of methane to C_2H_6 via **Path 1**. Therefore, the selectivity towards C_2^+ hydrocarbons decreases with increasing methane conversion, meaning that the yield of C_2^+ hydrocarbons does not increase at higher methane conversion.

Table 5.3 Relative first-order rate constants for each of the reaction paths in Scheme 5.1 using Sm_2O_3 as the catalyst

Reaction temperature/K	Relative rate constant				
	Path 1	Path 2	Path 3	Path 4	Path 5
883	1.3	1.0	18.4	6.3	6.4
1023	4.7	1.0	31.0	46.4	10.8

(Reported in ref. [71])

As discussed above, the coupling of the methyl radical mainly proceeds in the gas phase to produce C_2H_6. This reaction is rather different from the partial oxidation of CH_4 to CH_3OH, in which the CH_3OH is produced on the surface of a solid catalyst or biological catalyst, even though the $\bullet CH_3$ radical is an intermediate in both cases. Further details of the partial oxidation reaction are given in Chapters 2–4.

5.6 Relationship Between the Conversion of Methane and the Selectivity Towards C_2^+ Hydrocarbons

Figure 5.1 shows the relationship between the conversion of methane and the selectivity towards C_2^+ hydrocarbons in the OCM reaction. The data were obtained using various catalysts, reaction temperatures between 843 and 1273 K, and under classical conditions, i.e., single-pass methane conversion with O_2 in a continuous-flow reactor. The composition of the catalysts and the corresponding literature sources are summarized in Table 5.4 [74–113]. Some data points in Fig. 5.1 and Table 5.4 were also taken from membrane reactor experiments.

The dashed line in Fig. 5.1 represents a 30% yield of C_2^+ hydrocarbons. As illustrated by the figure, higher C_2^+ selectivity was obtained at lower methane conversion, resulting in a maximum C_2^+ hydrocarbon yield of about 30%. The 33.2% yield from ref. [75] was achieved over a NaCl–Mn–Na_2WO_4/SiO_2 catalyst at 1023 K. The high

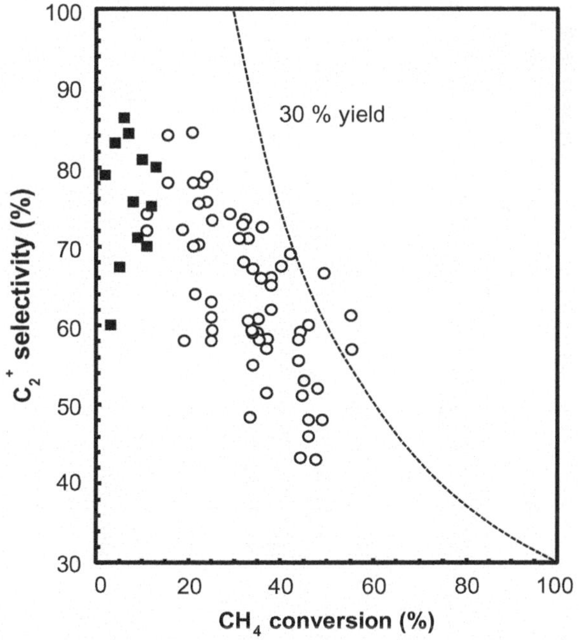

Fig. 5.1 Relationship between the conversion of methane and the selectivity for C_2^+ hydrocarbons based on the results listed in Table 5.4

Table 5.4 Composition and catalytic performance of various catalysts in the oxidative coupling of methane, and the corresponding reaction conditions

No.	Catalysts	Reaction temp./K	Ratio of CH_4/O_2	CH_4 conv./%	O_2 conv./%	C_2^+ H. C. selec./%	C_2^+ H. C. yield/%	Ref.
1	Sn–Ba–Ti Perovskite	1048	1.0	49.4	–	66.6	32.6	[74]
2		1073	2.0	49.0	–	58.0	28.4	[74]
3		1048	7.5	20.8	–	84.4	17.5	[74]
4	NaCl–Mn–Na_2WO_4/SiO_2	1023	1.6	55.0	80.7	61.2	33.2	[75]
5	NaCl–Mn/SiO_2	1023	1.6	55.2	52.2	56.9	29.7	[75]
6	KCl–Mn–Na_2WO_4/SiO_2	1023	1.6	35.9	45.5	72.4	26.0	[75]
7	Na–W–Mn/SiO_2	1069	0.3	42	–	69	29	[76]
8	Sr–Ba–Ti	1023	3.0	40.1	–	67.5	27.1	[77]
9	Chloride–Mn–Na_2WO_4/SiO_2	973	1.6	46	–	60	27	[78]
10	Mn–Na_2WO_4/SiO_2	973	1.6	23	–	78	18	[78]
11	$CaZrLiO_3$	1048	2.0	44.2	–	59.1	26.1	[79]
12	$BaTiGaO_3$	1048	2.0	43.9	–	55.5	24.4	[79]
13	Na–W–Mn/SiO_2	1175	2.0	45	97	53	24	[80]
14	K–W–Mn/SiO_2	1112	2.1	35	70	59	21	[80]
15	Nano-($2Li_2O$/MgO)/Al_2O_3	1023	1.6	38	–	66	25.1	[81]
16	Mn–Na–WO_X/SiO_2	1123	2.0	48	–	52	25	[82]
17		1123	4.0	32	–	68	22	[82]
18	Li/W/Mn//SiO_2	1178	2.0	38	85	65	25	[83]
19	Li/W/Mn/SiO_2	1127	2.1	45	92	53	24	[83]
20	Na/W/Mn/SiO_2	1153	1.9	46	99	48	22	[83]
21	Na–W–Mn–Ba–Ti/Ce–SiO_2	1033	4.0	32.5	–	73.4	23.9	[84]

(continued)

Table 5.4 (continued)

No.	Catalysts	Reaction temp./K	Ratio of CH_4/O_2	CH_4 conv./%	O_2 conv./%	C_2^+ H. C. selec./%	C_2^+ H. C. yield/%	Ref.
22	Na–W–Mn–Ba–Ti/SiO_2	1033	4.0	31.8	–	72.7	23.1	[84]
23	Na–W–Mn–Zr–S–P/SiO_2	1023	3.0	44.6	–	51.1	23.8	[85]
24	Na_2WO_4/Mn/SiO_2	948	3.5	35.8	–	65.9	23.6	[86]
25		948	3.5	34.0	–	67.1	22.8	[86]
26	Na_2WO_4/Mn/SiO_2	1103	2.0	49	–	48	23	[87]
27	Na_2WO_4/Mn/SiO_2	1073	5.0	33	–	71	23	[88]
28	Na_2WO_4/Ce/SiO_2	1073	5.0	31	–	71	22	[88]
29	Na_2WO_4/SiO_2	1073	5.0	21	–	78	16	[88]
30	$SrTi_{0.9}Mg_{0.1}O_3$	1173	2.0	37.1	–	58.2	21.6	[89]
31		1123	4.0	23.9	–	75.6	18.1	[89]
32	Na_2WO_4/Mn/SiO_2	1048	3.0	29	–	74	22	[90]
33	Li/MgO	998	2.0	35.1	–	60.8	21.4	[91]
34	Li/MgO	1073	2.0	46	–	46	21	[92]
35	Mn/Na_2WO_4/SiO_2	1123	3.0	37	–	57	21	[93]
36	Mn–Na_2WO_4/SiO_2	1098	2.5	35.4	–	58.1	20.6	[94]
37	Ba–Ti-Ce	1048	2.0	47.5	98.2	43.0	20.4	[95]
38	Ba–Ti-Sn	1048	2.0	44.2	99.5	43.2	19.1	[95]
39	Ce/Na_2WO_4/Mn/SiO_2	1113	3.0	34.1	98.4	59.0	20.1	[96]
40		1113	5.0	22.2	99.6	70.2	15.6	[96]
41	Na_2WO_4/Mn/SiO_2	1073	3.0	33.8	–	59.3	20.0	[97]
42	Na_2SO_4–WO_3–Mn/SiO_2	1098	5.0	33.0	–	60.5	20.0	[98]

(continued)

Table 5.4 (continued)

No.	Catalysts	Reaction temp./K	Ratio of CH_4/O_2	CH_4 conv./%	O_2 conv./%	C_2^+ H. C. selec./%	C_2^+ H. C. yield/%	Ref.
43	SrO	973	3.0	37.0	–	51.4	19.0	[99]
44	$Ca–La_2O_3$	973	3.0	33.5	–	48.4	16.2	[99]
45	$KCl–DyCl_3$	1023	0.5	34	–	55	19	[100]
46	$KCl–YbCl_3$	1023	0.5	25	–	61	15	[100]
47	$KCl–SmCl_3$	1023	0.5	21	–	70	15	[100]
48	Na–W–Zr–Mn–S–P	1069	0.3	23.9	–	78.8	18.8	[101]
49	$Mn/Na_2WO_4/SiO_2$	1073	4.0	25.2	–	73.3	18.5	[102]
50	Sr/Al_2O_3	1083	4.9	38	–	62	18	[103]
51	Sr/La_2O_3	1083	4.9	25	–	63	16	[103]
52	$Na_2WO_4/Mn/SiO_2$	1073	32.5	43.8	100	58.1	17.8	[104]
53	$Na_2WO_4/Mn/SiO_2$	973	6.0	22.1	97.7	75.4	16.7	[105]
54	$Na_2WO_4–MnO_x/SiO_2$	873	7.5	15.5	–	78.0	15.6	[106]
55	$Mn_xO_y–Na_2WO_4SiO_2$	1023	4.0	25.2	–	59.4	15.0	[107]
56	$La_2O_3–CeO_2$	843	5.0	25	–	58	15	[108]
57	Li/Ce/MgO	1073	5.0	21.5	–	64.0	13.8	[109]
58	$La_2O_3–CaO$	1073	8.0	18.6	–	72.1	13.4	[110]
59	$Mn/Na_2WO_{43-\delta}$	1073	9.0	15.5	–	84	13	[111]
60	$Na_2WO_4–Mn_xO_y/SiO_2$	1048	4.0	10.9	–	74.1	8.1	[112]
61	$La–Ce–SiO_2$	1115	3.0	11	14	72	8	[113]

performance of this system was attributed to the presence of chlorine, which, unfortunately, was removed from the catalysts with increasing time on stream, resulting in decreasing activity. The selectivity and yield of C_2^+ hydrocarbons achieved using a membrane reactor were more formidable than those obtained using a continuous-flow reactor.

5.7 Membrane Reactor for OCM

As discussed above, fluidized bed reactors come close to achieving the desired isothermal performance, although the thus obtained C_2^+ hydrocarbon yields do not exceed 30% because of poor selectivity. Catalytic membrane reactors, which properly implement the oxygen dosing policy, offer higher C_2^+ hydrocarbon selectivity than fluidized bed reactors [114–119]. For example, Othman et al. developed a catalytic hollow fiber membrane micro-reactor using micro-structured $La_{0.6}Sr_{0.4}Co_{0.8}Fe_{0.2}O_{3-\delta}$ hollow fiber as an oxygen separation membrane, as well as a functional substrate for in situ deposition of a $Bi_{1.5}Y_{0.3}Sm_{0.2}O_{3-\delta}$ catalyst by a sol-gel method [114]. A maximum C_2^+ yield of 39% and a selectivity of 79% were achieved at 1173 K.

The further objectives for designing an efficient OCM membrane reactor are (1) establishment of fine oxygen distribution to obtain a high level of methane conversion with high selectivity toward ethylene, (2) minimization of the contribution of the undesired gas-phase reaction, and (3) prevention of the formation of hot spots, which decrease reaction performance.

5.8 Non-oxidative Coupling of Methane

Direct methane conversion via methane coupling can be accomplished not only in the presence of oxygen (OCM reaction), but also in the absence of oxygen (non-oxidative coupling of methane reaction, non-OCM reaction) [120]. Non-oxidative methane coupling reactions are classified as low-temperature two-step methane coupling reactions or one-step high-temperature coupling reactions.

5.8.1 Two-Step Methane Coupling

As mentioned in Sect. 5.1, the non-oxidative coupling (dehydrogenative coupling) of methane is thermodynamically unfavorable. Belgued et al. attempted to circumvent this thermodynamic limitation for the conversion of methane into C_2^+ hydrocarbons [121]. In this early investigation into the low-temperature activation of methane, a 6 wt% Pt/SiO_2 catalyst (EUROPT) was first exposed to methane at 523 K and then flushed with H_2, resulting in the formation of C2 to C7 saturated hydrocarbon products, mainly C_2H_6. Thus, the overall reaction was carried out via periodic two-step operation in the absence of oxygen [122–125].

5.8.2 One-Step Non-oxidative Coupling of Methane

At high temperature, the overall reaction for the thermal coupling of methane is generally believed to involve stepwise dehydrogenation [126].

$$2\,CH_4 \xrightarrow{H_2} C_2H_6 \xrightarrow{H_2} C_2H_4 \xrightarrow{H_2} C_2H_2 \xrightarrow{H_2} 2\,C \qquad (5.8)$$

As shown in reaction (5.8), the non-oxidative coupling of methane (non-OCM reaction) gives hydrogen (H_2) as a reaction product because no oxygen or oxygen-containing species are involved. Theoretically, the selectivity towards C_2H_6 and C_2H_4 could be increased. However, the direct conversion equilibrium of C_2H_6 or C_2H_4 is thermodynamically limited, and these products are only formed at high temperature. In reaction (5.8), the formation of graphic carbon (C) is much more favorable than that of C2 hydrocarbons (C_2H_6 and C_2H_4). At 773 K, the equilibrium conversions of C_2H_6 and C_2H_4 are only about 0.6% and 1%, respectively.

Based on the Gibbs free energy change, the direct conversion of methane to C_2H_6 and C_2H_4 should be practical at temperatures above ~1000 K, with a significantly greater amount of C_2H_4 being formed than C_2H_6. The direct formation of C_2H_6 shown in reaction (5.9) is always significantly endothermic with very little temperature dependence.

$$2\,CH_4 \rightleftarrows C_2H_6 + H_2 \qquad (5.9)$$

At 1173 K, the conversion of CH_4 to C_2H_6 is only 5%, while the formation of aromatic hydrocarbons, especially benzene and naphthalene, is more favorable than that of C_2H_4. Details of the formation of aromatic hydrocarbons via dehydroaromatization are discussed in Chapter 6.

5.8.3 Catalysts for the Non-oxidative Coupling of Methane and the Activation of Methane on the Surface of Metal Catalysts

In the non-OCM reaction, mono-metallic and bi-metallic catalysts consisting of metals such as Co [127], Ru [127, 128], Pt [129], Rh [130], Re [131], Pd [132], Ni [133], and Au [134] supported on Al_2O_3, SiO_2, or zeolites (molecular sieves) [120] are typically used. Unfortunately, these catalysts suffer from the accumulation of coke on the catalyst surface.

Koerts et al. reported that CH_4 was dissociatively adsorbed on transition metal catalysts, such as Ru/SiO_2, resulting in the formation of carbonaceous species (CH_x species) on the catalyst surface and hydrogen [135]. In a discussion of the high catalytic performance of Co–Pt/Na–Y and Co–Pt/Al_2O_3, Guczi et al. also referred to adsorbed CH_x, discussing several forms of carbon species such as CH_2 and CH. These CH_x species weakly bonded to catalyst surfaces participated in C–C bond formation over the bi-metallic catalysts [136, 137].

5.8.4 High Catalytic Performance of Catalysts with Single Iron Sites in the Non-OCM Reaction

As discussed above, high reaction temperatures are essential for non-OCM reactions. In 2014, Guo et al. reported that a single iron site embedded in a silica matrix enabled the non-oxidative conversion of methane exclusively to ethylene and aromatic hydrocarbons such as benzene [138]. The maximum conversion of methane (48.1%) was achieved at 1363 K using 0.5 wt% Fe/SiO_2 as the catalyst. Only ethylene, benzene, and naphthalene were produced; neither coke nor CO_2 was detected despite the relatively high reaction temperature. The selectivities varied from 40.9 to 52.1% for ethylene, from 21.0 to 29.1% for benzene, and from 23.6 to 38.2% for naphthalene in the temperature range 1223–1363 K. In comparison, a blank test (empty reactor with no catalyst) under the same conditions gave a CH_4 conversion of 2.5% with 95% of the product being coke; a test using SiO_2 alone as the catalyst yielded virtually the same result.

5.9 Summary of OCM Features

OCM is one of the processes for the direct conversion of methane into hydrocarbons. This reaction requires high temperatures (~1000 K) and affords products limited to C_2H_6 and C_2H_4 (C_2 hydrocarbons). The corresponding mechanism is not simple, as OCM proceeds both heterogeneously and homogeneously. Once the formation of $\bullet CH_3$ is initiated by the catalyst, the gas-phase self-coupling of $\bullet CH_3$ together with the coupling of $\bullet CH_3$ with other alkyl radicals proceeds extensively. Furthermore, the facile oxidation of $\bullet CH_3$ on the catalyst surface and in the gas phase affords CO_2 and CO. Therefore, the formation of C_2 hydrocarbons is controlled mainly by the homogeneous process, and the yield of these hydrocarbons does not exceed 30%.

In the non-oxidative coupling of methane (non-OCM), methane is mainly activated on the catalyst surface to produce $\bullet CH_3$, and the formation of higher hydrocarbons such as benzene proceeds homogeneously.

In view of the above, OCM is expected to become one of the main models for refining the approach to describe complex multistep and multipathway reactions and develop methods for determining and estimating the kinetic parameters of separate steps.

References

1. Mitchell LL, Waghorne RH (1980) Catalyst forb the conversion of relatively low molecular weight hydrocarbons to higher molecular weight hydrocarbons and the regeneration of the catalysts. US Patent 4205194, 1980
2. Fang T, Yeh CT (1981) Interaction of methane with ThO_2/SiO_2 surface at 1073 K. J Catal 69:227–229
3. Keller GF, Bhasin MM (1982) Synthesis of ethylene *via* oxidative coupling of methane: determination of active catalysts. J Catal 73:9–19
4. Lee JS, Oyama ST (1988) Oxidative coupling of methane to higher hydrocarbons. Catal Rev -Sci Eng 30:249–280
5. Amenomiya Y, Briss VI, Goledzinouski M, Galuszka J, Sanger AR (1990) Conversion of methane by oxidative coupling. Catal Rev -Sci Eng 32:163–227
6. Zavyalova U, Holena M, Schlög R, Baerns M (2011) Statistical analysis of past catalytic data on oxidative methane coupling for new insight into the composition of high-performance catalysts. ChemCatChem 3:1935–1947
7. Kondratenko EV, Peppel T, Seeburg D, Kondratenko VA, Kalevaru N, Martin A, Wohlrab S (2017) Methane conversion into different hydrocarbons or oxygenates: current status and future prospective in catalyst development and reactor operation. Catal Sci Technol 7:366–381
8. Driscoll DJ, Martir W, Wang JX, Lunsford JH (1985) Formation of gas-phase methyl radical over MgO. J Am Chem Soc 107:58–63
9. Luo L, Tang X, Wang W, Wang Y, Sun S, Qi F, Huang W (2013) Methyl radical in oxidative coupling of methane directly confirmed by Synchrotron VUV Photoionization mass spectroscopy. Scientific Reports https://doi.org/10.1038/serp01625, 3, 1625
10. Wolf EE (2014) Methane to light hydrocarbons *via* oxidative methane coupling: lessons from the past to search for a selective heterogeneous catalyst. J Phys Chem Lett 5:986–988
11. Parishan S, Nowicka E, Fleischer V, Schulz C, Colmenares MG, Rosowski F, Schomäcker R (2018) Investing into consecutive reaction of ethane and ethylene under the OCM reaction conditions over $Mn_xO_y–Na_2WO_4/SiO_2$. Catal Lett 148:1659–1675
12. Tang Y, Lunsford JH (1991) Mechanistic and kinetic studies of the reactions of gas-phase methyl radical with metal oxides. J Am Chem Soc 113:4741–4746
13. Lunsford JH (1995) The catalytic oxidative coupling of methane. Angew Chem Int Engl 34:970–980
14. Dubois JL, Cameron CJ (1990) Common features of oxidative coupling of methane co-feed catalysts. Appl Catal 67:49–71
15. Yan QJ, Wang Y, Jin YS, Chen Y (1992) Methane oxidative coupling over Na_2WO_4/SiO_2. Catal Lett 13:221–228
16. Driscoll DJ, Lunsford JH (1985) Gas-phase radical formation during the reaction of methane, ethane, ethylene and propylene over selected oxide catalysts. J Phys Chem 89:4415–4418
17. Feng Y, Niiranen J, Gutman D (1991) Kinetics studies of the catalytic oxidation of methane, 1. Methyl radical production on 1% Sr/La_2O_3. J Phys Chem 95:6558–6563
18. Feng Y, Niiranen J, Gutman D (1991) Kinetics studies of the catalytic oxidation of methane, 2. Methyl radical recombination and ethane formation over 1% Sr/La_2O_3. J Phys Chem 95:6564–6568
19. Campbell KD, Lunsford JH (1988) Contribution of gas-phase radical coupling in the catalytic oxidation of methane. J Phys Chem 92:5792–5796

20. Tong Y, Rosynek MP, Lunsford JH (1989) Secondary reactions of methyl radicals with lanthanide oxides: their role in the selective oxidation of methane. J Phys Chem 93:2896–2898
21. Xu M, Lunsford JH (1991) Effect of temperature on methyl radical generation over Sr/La_2O_3 catalysts. Catal Lett 11:295–300
22. Tong Y, Lunsford JH (1990) Gas-phase coupling of methyl radicals during the partial oxidation of methane over transition metal oxide catalysts. J Chem Soc Chem Commun 11:792–793
23. Zhang HS, Wang JX, Driscoll DJ, Lunsford JH (1988) Activation and oxidative dimerization of methane over lithium-promoted zinc oxide. J Catal 112:366–374
24. Lin CH, Campbell KD, Wang JX, Lunsford JH (1986) Oxidative dimerization of methane over lanthanum oxide. J Phys Chem 90:534–537
25. Campbell D, Zhang H, Lunsford JH (1988) Methane activation by the lanthanide oxides. J Phys Chem 92:750–753
26. Campbell KD, Morales E, Lunsford JH (1987) Gas-phase coupling of methyl radical during the catalytic partial oxidation of methane. J Am Chem Soc 109:7900–7901
27. Ito T, Wang JX, Lin CH, Lunsford JH (1985) Oxidative dimerization of methane over lithium-promoted magnesium oxide catalyst. J Am Chem Soc 107:5062–5068
28. Chen Y, Tohver HT, Narayan J, Abraham MM (1977) High-temperature and ionization-induced effect in lithium-doped MgO single crystals. Phys Rev B 16:5535–5542
29. Olson DN, Orera VM, Chen Y, Abraham MM (1980) Thermally generated $[Li]^0$ center in CaO. Phys Rev B 21:1258–1263
30. Schirmer OF (1971) Trapped-hole center containing lithium in MgO, CaO and SrO. J Phys Chem Solid 32:499–509
31. Lin TH, Ito T, Wang JX, Lunsford JH (1987) Oxidative dimerization of methane over magnesium and calcium oxide catalysts promoted with Group IA ions: the role of M^+O^- centers. J Am Chem Soc 109:4808–4810
32. Lin TH, Wang JX, Lunsford JH (1988) Oxidative dimerization of methane over sodium-promoted calcium oxide. J Catal 111:302–316
33. Otsuka K, Said AA, Jinno K, Komatsu T (1987) Peroxide anions as possible active species in oxidative coupling of methane. Chem Lett 16:77–80
34. Otsuka K, Murakami Y, Wada Y, Said AA, Morikawa A (1990) Oxidative coupling of methane, ethane, and propane with sodium peroxide at low temperatures. J Catal 121:122–130
35. Sinev MY, Korchak VN, Krylov OV (1986) Highly selective formation of ethane in the reduction barium peroxide by methane. Kinet Katal 27:1274
36. Otsuka K, Jinno K (1986) Kinetic studies on partial oxidation of methane over samarium oxides. Inorg Chim Acta 121:237–241
37. Kharas KCC, Lunsford JH (1989) Catalytic partial oxidation of methane over barium metaplumbate $BaPBO_3$: possible involvement of peroxide ion. J Am Chem Soc 111:2336–2337
38. Lunsford JH, Yang X, Haller K, Laane MJG, Knözinger H (1993) *In situ* Raman spectroscopy of peroxide ions on barium/magnesium oxide catalysts. J Phys Chem 97:13810–13813
39. Dissanayake D, Lunsford JH, Rosynek MP (1993) Oxidative coupling of methane over oxide-supported barium catalysts. J Catal 143:286–298
40. Mestl G, Knözinger H, Lunsford JH (1993) High temperature *in situ* Raman spectroscopy of working oxidative coupling of catalysts. Ber Bunsen-Ges Phys Chem 97:319–321
41. Au CT, He H, Lai SY, Ng CF (1996) The oxidative coupling of methane over $BaCO_3$/LaOBr-catalysts of high ethylene yields. J Catal 163:399–408
42. Wang JX, Lunsford JH (1986) Evidence for the thermal generation of superoxide ions on La_2O_3. J Phys Chem 90:3890–3891
43. Louis C, Chang TL, Kermarec M, Van TL, Tatibouët J, Che M (1992) EPR study of the stability and the role of the O_2^- species on La_2O_3 in the oxidative coupling of methane. Catal Today 13:283–289
44. Yang T, Feng L, Shen S (1994) Oxygen species on the surface of La_2O_3/CaO and its role in the oxidative coupling of methane. J Catal 145:384–389
45. Iwamoto M, Lunsford JH (1980) Surface reactions of oxygen ions. 5. Oxidation of alkanes and alkenes by O_2^- on magnesium oxide. J Phys Chem 84:3079–3084

46. Voskrenskaya EN, Roguleva VG, Anshits AG (1995) Oxidant activation over structural defects of oxide catalyst in oxidation of methane coupling. Catal Rev -Sci Eng 37:101–143
47. Takita Y, Lunsford JH (1979) Surface reactions of oxygen ions. 3. Oxidation of alkanes by O_3^- on magnesium oxide. J Phys Chem 83:683–688
48. Takita Y, Iwamoto M, Lunsford JH (1980) Surface reactions of oxygen ions. 4. Oxidation of alkanes by O_3^- on magnesium oxide. J Phys Chem 84:1710–1712
49. Oliva C, Bonoldi L, Cappelli S, Fabbrini L, Rossetti I, Forai L (2005) Effect of preparation parameters on $SrTiO(3 \pm \delta)$ catalyst for the flameless combustion of methane. J Mol Catal A: Chemical 226:33–40
50. Shvets VA, Vorotyntsev VM, Kazanskii VB (1969) ESR spectra of surface complexes formed during the adsorption of oxygen on vanadium pentoxide deposited silica gel. Kinet Katal 10:356–363
51. Kazanskii VB (1977) Possible mechanism of heterogeneous chain generation with the preparation of O^- surface radical during catalytic oxidation on oxide. Kinet Katal 18:43–54
52. Sofranko JA, Leonard JJ, Jones CA (1987) The oxidative conversion of methane to higher hydrocarbons. J Catal 103:302–310
53. Jones CA, Leonard JJ, Sofranko JA (1987) The oxidative conversion of methane to higher hydrocarbons over alkali-promoted Mn/SiO_2. J Catal 103:311–319
54. Sofranko JA, Leonard JJ, Jones CA, Gaffney AM, Withers HP (1988) Catalytic oxidative coupling of methane over sodium-promoted Mn/SiO_2 and Mn/MgO. Catal Today 3:127–135
55. Fang X, Li S, Lin J, Chu Y (1992) Oxidative coupling of methane on tungsten-manganese catalysts. Fenzi Cuihua 6:427–433
56. Fang X, Li S, Lin J, Gu J, Yan D (1992) Preparation and characterization of tungsten-manganese catalysts for oxidative coupling of methane. Fenzi Cuihua 6:255–262
57. Jiang Z, Fang X, Yu C, Li S, He J, Luo X (1992) Manganese (III) oxide-sodium tungstate (Na_2WO_4)/silicon dioxide catalyst for oxidative coupling of methane. I. Surface dispel son state and oxide/support interaction. Fenzi Cuihua 6:477–480
58. Jiang ZC, Yu CJ, Fang XP, Li SB, Wang HL (1993) Oxide/support interaction and surface reconstruction in the Na_2WO_4/SiO_2 system. J Phys Chem 97:12870–12875
59. Ji S, Xiao TC, Li SB, Xu CZ, Hou RL, Coleman KS, Green MLH (2002) The relation ship between the structure and the performance of Na-W-Mn/SiO_2 catalyst for the oxidative coupling of methane. Appl Catal A: General 225:271–284
60. Yildiz M, Simm U, Otemba T, Aksu Y, Kailasam K, Thomas A, Schomäcker R, Arndt S (2014) Support material variation for the MnO-Na_2WO_4/SiO_2 catalyst. Catal Today 228:5–14
61. Sinev M, Ponomareva E, Sinev I, Lomonosov V, Gordienko Y, Fattakhova Z, Shaskin D (2019) Oxygen pathways in oxidative coupling of methane and related processes case study: NaWMn/SiO_2 catalyst. Catal Tody 333:36–46
62. Arndt S, Otremba T, Simon U, Yildiz M, Schubert H, Schomäcker R (2012) Mn-Na_2WO_4/SiO_2 as catalysts for the oxidative coupling of methane. What is really known? Appl Catal A: General 425–426:53–61
63. Fleisher V, Simon U, Parishan S, Colmenares MG, Görke O, Gurlo A, Riedel W, Thum L, Schmidt J, Risse T, Dinse KP, Schomäcker R (2018) $Na_2WO_4/Mn/SiO_2$ catalyst composition in the oxidative coupling of methane by chemical looping experiments. J Catal 360:102–117
64. Lomonosov VI, Gordienko YA, Sinev MY, Rogov VA, Sadykov VA (2018) Thermochemical properties of the lattice oxygen in W, Mn-containing mixed oxide catalysts for the oxidative coupling of methane. Russ J Phys Chem A 92:430–437
65. Gu S, Oh HS, Choi JW, Jae J, Ghoi J, Ha JM (2018) Effects of metal or metal oxide additives on oxidative coupling of methane using Na_2WO_4/SiO_2 catalysts: reducibility of metal additives to manipulate the catalytic activity. Appl Catal A: General 562:114–119
66. Riedel W, Thum L, Möser J, Fleische V, Simon U, Siemensmeyer K, Schnegg A, Schomäcker R, Risse T, Dinse KP (2018) Magnetic properties of reduced and reoxidized Mn--Na_2WO_4/SiO_2: a catalyst for oxidative coupling of methane (OCM). J Phys Chem C 112:22605–22614

67. Ismagilov IZ, Matus EV, Kuznetsov VV, Kerzhentsev MA, Yashnik SA, Larina TV, Prosvirin IP, Navarro RM, Fierro JLG, Gerritsen G, Abbennuis HCL, Ismagilov ZR (2016) Effect of preparation mode on the properties of Mn-Na-W/SiO_2 catalyst for oxidative coupling of methane: conventional methods *vs.* post nanotechnology. Eurasian Chem-Technol J 18(2):93–110
68. Shubin A, Zilberberg I, Ismagilov I, Matus E, Kerzhentsev M, Ismagilov Z (2018) Hydrogen abstraction from methane on cristobalite supported W and Mn oxo complexes: a DFT study. J Mol Catal 445:307–315
69. Ekstrom A, Lapszewicz JA, Campbell I (1989) Origin of the low limits in the higher hydrocarbons yields in the oxidative coupling reaction methane. Appl Catal 56:L29–L34
70. Korf SJ, Roos JA, Derksen JWHC, Vreeman JA, Van Ommen JG, Ross JRH (1990) Oxidative coupling of methane over Ba/CaO catalysts. A comparison with Li/MgO. Appl Catal 59:291–309
71. Otsuka K, Jinno K, Morikawa A (1986) Active and selective catalysts for synthesis of C_2H_6 and C_2H_4 *via* oxidative coupling of methane. J Catal 100:353–359
72. Shi C, Rosynek MP, Lunsford JH (1994) Origin of carbon oxides during the oxidative coupling of methane. J Phys Chem 98:8371–8376
73. Nelson PF, Cant NW (1990) Oxidation of C_2 hydrocarbon products during the oxidative coupling of methane over a Li/MgO catalyst. J Phys Chem 94:3756–3761
74. Yaghobi N, Ghoreisky MHR (2008) Oxidative coupling of methane in a fixed bed reactor over perovskite catalysts: a simulation study using experimental kinetic model. J Nat Gas Chem 17:8–16
75. Hiyoshi N, Ikeda T (2015) Oxidative coupling of methane over alkali chloride-Mn-Na_2WO_4/SiO_2 catalysts: promoting effect of molten alkali chloride. Fuel Proc Technol 133:29–34
76. Chua YT, Mohamed AR, Bhatia S (2008) Oxidative coupling of methane for production of ethylene over sodium-tungsten-manganese-supported-silica catalyst. Appl Catal A: General 343:142–148
77. Hiyoshi N, Sato K (2016) Oxidative coupling of methane over Mn-Na_2WO_4/SiO_2. Fuel Proc Technol 151:148–154
78. Yaghobi N, Ghoreisky MHR (2008) Comparative study on the kinetic modeling if the oxidative coupling of methane in laboratory and bench scales. Chem Prod Proc Model 3:16/1–16/26. https://doi.org/10.2202/1934-2659.1133
79. Hamidzadeh M, Sahebdelfar S, Jafari N, Mohammad R (2010) Method for preparing ceramic catalyst for converting methane to higher hydrocarbons by oxidative condensation (or coupling) of methane, Brit. UK Pat Appl (2010) GB 2469877A
80. Dedov AG, Nipan GD, Loktev AS, Tyunyaev AA, Ketsko AV, Pakhomenko KV, Miseev II (2011) Oxidative coupling of methane: influence of the phase composition of silica-based catalysts. App Catal A: General 406:1–21
81. Fallah B, Falamaki C (2010) A new nano-(2Li2O/MgO) catalyst/porous α-alumina composite for the oxidative coupling of methane reaction. AIChE J 56:717–728
82. Albrecht M, Rodemerck U, Kondratenko E (2014) Higher hydrocarbon production through oxidative coupling of methane combined with hydrogenation of carbon oxide. Chem Ind Tech 86:1894–1900
83. Nipan GD, Artukh VA, Yusupov VS, Loktev AS, Spesivtsev NA, Dedov AG, Moiseev II (2014) Pressure effect on the formation of active components of a catalyst for methane oxidative coupling. Doklady Phys Chem 455:60–63
84. Wu J, Zghang M, Feng J, Ke L, Zhao Q, Xue W (2014) Preparation and application of supported catalysts for methane oxidation coupling for preparing lower olefins, Patent CN 103657640 A 20140326
85. Zheng W, Cheng D, Chen F, Zhan X (2010) Characterization and catalytic behavior of Na-W-Mn-Zr-S-Na_2WO_4/SiO_2 prepared by different methods in oxidative coupling of methane. J Nat Gas Chem 19:515–521

86. Ahari JS, Sadeghi MT (2011) Optimization of OCM reaction conditions over Na-W-Mn/SiO_2 catalyst as elevated pressure. J Taiwan Inst Chem Eng 42:751–759
87. Sadjadi S, Simon U, Godini HR, Görke O, Schomäcker R, Wonzny G (2015) Reactor material and gas dilution effect on the performance of miniplant-scale fluidized-bed reactors for oxidative coupling of methane. Chem Eng J 281:678–687
88. Gholipour Z, Malekzadeh A, Hatami R, Mortazavi Y, Khodadadi A (2010) Oxidative coupling of methane over (Na_2WO_4 + Mn or Ce)/SiO_2 catalysts: in situ measured of electrical conductivity. J Nat Gas Chem 19:35–42
89. Ivanov DV, Isupova LA, Gerasinov EY, Isupova LA, Gerasimov EY, Dovlitova LS, Glazneva TS (2014) Oxidative methane coupling over Mg, AL, Ca, Ba, Pb-promoted $SrTiO_3$ and $SrTiO_4$: influence of surface composition and micro structure. App Catal A: General 485:10–19
90. Fleischer V, Littlewood P, Parishan Schomäcker SR (2016) Chemical looping as reactor concept for the oxidative coupling of methane over Na_2WO_4/Mn/SiO_2. Chem Eng J 306:646–654
91. Ahari JS, Zarrinpashne S, Sadeghi MT (2013) Micro-kinetic modeling of OCM reactions over Mn/Na_2WO_4/SiO_2 catalyst. Fuel Proc Technol 115:79–87
92. Raouf F, Taghizadeh M, Yousefi M (2013) Activity enhancement of Li/MgO catalysts by lithium chloride as lithium precursor for the oxidative coupling of methane. Reac Kinet Mech Catal 110:373–385
93. Vatani A, Jabbari E, Askarieh M, Torangi MA (2014) Kinetic modeling of oxidative coupling of methane over Li/MgO catalyst by genetic algorithm. J Nat Gas Sci Eng 20:347–356
94. Godini HR, Gili A, Görke O, Arndt S, Simon U, Thomas A, Schomäcker R (2014) Sol-gel method for synthesis of Mn-Na_2WO_4/SiO_2 catalyst for methane oxidative coupling. Catal Today 236:12–22
95. Khodadadian M, Taghizadeh M, Hamidzadeh M (2011) Effect of various barium precursors and promotes on catalytic activity of Ba-Ti perovskite catalysts for oxidative coupling of methane. Fuel Proc Technol 92(6):1164–1168
96. Shahri SMK, Pour AN (2010) Ce-promoted Mn/Na_2WO_4/SiO_2 catalyst for oxidative coupling of methane at atmosphere. J Nat Gas Chem 19:47–53
97. Wang W, Ji S, Pan D, Li C (2011) A novel particle/monolithic two-stage catalysts bed reactor and their catalytic performance for oxidative coupling of methane. Fuel Proc Technol 92:541–546
98. Ghiasi M, Malekzadeh A, Hoseini H, Mortazavi Y, Khodadadi A, Talebizadeh T (2011) Kinetic study of oxidative coupling of methane over Mn and/or W promoted Na_2WO_4/SiO_2 catalyst. J Nat Gas Chem 20:428–434
99. Song J, Sun Y, Ba R, Huang S, Zhao Y, Zhang J, Sun Y, Zhu Y (2015) Monodisperse Sr-La_2O_3 hybrid nanofibers for oxidative coupling of methane of synthesize C_2 hydrocarbons. Nanoscale 7:2260–2264
100. Huang K, Chen FQ, Lü DW (2001) Artificial neural network-aided design of a multi-component catalyst for methane oxidative coupling. Appl Catal A: General 219:61–68
101. Branco JB, Ferreira AC, do Rego AMB, Ferreira AM, Lopes G, Gasche TA (2014) Oxidative coupling of methane over KCl-$LaCl_3$ eutectic molten salt catalyst. J Mol Liquids 191:100–106
102. Serres T, Aquino C, Minodatos C, Schuurman Y (2015) Influence of the composition texture of Mn-Na-W catalysts on the oxidative coupling of methane. App Catal A: General 504:509–518
103. Baiya T, van Vegten N, Verel R, Jiang Y, Yulokov M, Kohn T, Jeschke G, Baiker A (2011) Sr-Al_2O_3 mixed oxide: a promising class of catalyst for oxidative coupling of methane. J Catal 281:241–253
104. Ren S, Qin S, Zhu J, Peng X, Hu C (2010) New Co-La/SiO_2 catalyst for the simultaneous production of C_2H_4 and syngas with Na_2WO_4/Mn/SiO_2. Ind Eng Chem Res 49:2078–2083
105. Ahari JS, Sadeghi MT, Zarrinpashne S (2011) Effects of operating parameters on oxidative coupling of methane over Na-W-Mn/SiO_2 catalyst at elevated pressure. J Nat Gas Chem 20:204–213

106. Malekzadeh A, Dalai AK, Khodadadi A, Mortzovi Y (2008) Structural features of Na_2WO_4-MO_X/SiO_2 catalyst in oxidative coupling of methane. Catal Commun 9:960–965
107. Yildiz M, Aksu Y, Simom U, Othemba T, Kailasam K, Görke C, Rosowski F, Thomas A, Schomäcker R, Arndt S (2016) Silica material variation for Mn_XO_Y-Na_2WO_4/SiO_2. Appl Catal A: General 525:168–179
108. Noon D, Zohour B, Senkan S (2014) Oxidative coupling of methane with La_2O_3-CeO_2 nanofiber fabrics: a reaction engineering study. J Nat Gas Sci Eng 18:406–411
109. Gonçalves RLP, Muniz FC, Passos FB, Schmal M (2010) Promoting effect of Ce on the oxidative coupling of methane catalysts. Catal Lett 135:26–32
110. Rane VH, Chaudhari ST, Choudhary VR (2010) Oxidative coupling of methane over La-promoted CaO Catalysts: influence of precursors ands catalyst preparation method. J Nat Gas Chem 19:25–30
111. Daneshpayeh M, Khodadadi A, Mostoufi N, Mortazavi Y, Satudeh-Gharebagh R (2009) Kinetic modeling of oxidative coupling of methane over $M/Na_2WO_4/SiO_2$ catalyst. Fuel Proc Technol 90:403–410
112. Colmenres MG, Simon U, Yildiz M, Arndt S, Schomaecker R, Thomas T, Rosowski F, Gurlo A, Goerke O (2016) Oxidative coupling of methane on the Na_2WO_4-Mn_xO_y catalyst: COK-12 as an inexpensive alternative to SBA-15. Catal Commun 85:75–78
113. Dedov AG, Loktev AS, Tel'pukhoskaya NO, Parkhiomenko KV, Moiseev II (2008) Mesoporous amorphous rare earth silicates new catalysts of methane oxidative coupling. Doklady Phys Chem 422:253–255
114. Othman NH, Wu Z, Li K (2015) An oxygen permeable membrane micro reactor with an in-situ deposited Bi1.5Y0.3Sm0.2O3-δ catalyst for oxidative coupling of methane. J Membr Sci 488:182–193
115. Bhatia S, Thien CY, Mohamed AR (2009) Oxidative coupling of methane (OCM) in a catalytic membrane reactor and comparison of its performance with other catalytic reactor. Chem Eng J 148:525–532
116. Godini HR, Xiao S, Kim M, Holst N, Jaso S, Görke O, Steinbach J, Wozny G (2014) Experimental and model-based analysis of membrane reactor performance for methane oxidative coupling: effect of radial heat and mass transfer. J Ind Eng Chem 20:1993–2002
117. Godini HR, Gili A, Görke O, Simon U, Hou K, Wozny G (2014) Performance analysis of a porous packed bed membrane reactor for oxidative coupling of methane: structural and operational characteristics. Energy Fuels 28:877–890
118. Olivier L, Haag S, Mirocatos C, van Veen AC (2009) Oxidative coupling of methane using catalyst modified dense perovskite membrane reactor. Catal Today 142:34–41
119. Othman NH, Wu Z, Li K (2014) A micro-structured La0.6Sr0.4Co0.8Fe0.2O3-δ hollow fiber membrane reactor for oxidative coupling of methane. J Membr Sci 468:31–41
120. Choudhary TV, Asyolu E, Goodman W (2003) Non oxidative activation of methane. Catal Rev 45:151–203
121. Belgued M, Paréja P, Amariglio A, Amariglio H (1991) Conversion of methane into higher hydrocarbons on platinum. Nature 352:789–790
122. Amariglio H, Paréja P, Amariglio A (1995) Periodic operation of catalysts s a means of overcoming a thermodynamic constraint. The case of the methane homologation on metals. Catal Today 25:113–125
123. Belgued M, Amariglio A, Paréja P, Amariglio H (1996) Oxygen-free conversion of methane to higher alkanes through an isothermal two-step reaction on platinum (EUROPT-1). I. Chemisorption of methane. J Catal 159:441–448
124. Belgued M, Amariglio A, Paréja P, Amariglio H (1996) Oxygen-free conversion of methane to higher alkanes through an isothermal two-step reaction on platinum (EUROPT-1). II. Hydrogenation of the adspecies resulting from the chemisorption of methane. J Catal 159:449–457
125. Belgued M, Amariglio A, Lefort L, Paréja P, Amariglio H (1996) Oxygen-free conversion of methane to higher alkanes through an isothermal two-step reaction on ruthenium. J Catal 161:282–291

126. Holmen A, Olsvik O, Rokstad OA (1995) Pyrolysis of natural gas: Chemistry and process concepts. Fuel Proc Technol 42:249–267
127. Guczi L, Sarma KV, Borlcó L (1997) Low-temperature methane activation under nonoxidative conditions over supported ruthenium-cobalt bimetallic catalysts. J Catal 167:495–502
128. Choudhary TV, Goodman DW (2002) Methane activation on ruthenium: the nature of the surface intermediates. Top Catal 20:35–42
129. Gerecker D, Motagamwala AH, Rivera-Dones KR, Miller JB, Huber GW, Marikakis M, Dumesic J (2017) Methane conversion to ethylene and aromatics on PtSn catalysts. ACS Catal 7:2088–2100
130. Solmosi F, Csrényi J (1995) Enhanced formation of ethane in the conversion of methane over Cu-Rh/SiO_2. Catal Lett 34:343–350
131. Bazin D, Borkó L, Koppány Z, Stefler G, Sajó LI, Schay Z, Guczi L (2002) Re-Co/NaY and Re-Co/Al_2O_3 bimetallic catalysts: *in situ* EXAFS study and catalytic activity. Catal Lett 84:169–182
132. Moya SF, Martins RL, Ota A, Kunkes EL, Behrens M, Schmal M (2012) Nanostructured supported palladium catalysts—non-oxidative methane coupling. Appl Catal A: General 411–412:105–113
133. Moya SF, Martins RL, Schmal M (2011) Monodispersed and nanostructured Ni/SiO_2 catalyst and its activity for non oxidative methane activation. Appl Catal A: General 396:159–169
134. Lang SM, Bernhardt TM, Barnett RN, Landman U (2010) Methane activation and catalytic ethylene formation on free Au_2^+. Angew Chem Int Ed 49:980–983
135. Koerts T, Deelene MJAG, van Santen RA (1992) Hydrocarbon formation from methane by a low-temperature two-step reaction sequence. J Catal 138:101–114
136. Guczi L, Sarma KV, Borkó L (1996) Non-oxidative methane coupling over Co-Pt/NaY bimetallic catalysts. Catal Lett 39:43–47
137. Guczi L, Sarma KV, Borkó L (1999) Studies on deuterium exchange in surface CH_X species formed from methane over platinum. Ruthenium and cobalt under non-oxidative conditions. React Kinet Catal Lett 68:95–104
138. Guo X, Fang G, Li G, Ma H, Fan H, Yu L, Ma C, Wu X, Deng D, Wei M, Tan D, Si R, Zhang S, Li J, Sun L, Tang Z, Pan X, Bao X (2014) Direct non oxidative conversion of methane to ethylene, aromatics and hydrogen. Science 344:616–619

Chapter 6
Conversion of Methane to Aromatic Hydrocarbons

6.1 Introduction

Aromatic hydrocarbons such as benzene are important for the production of polymers and raw materials for chemical manufacturing. Benzene is usually produced via the catalytic reforming of naphtha, which yields a mixture of benzene, toluene, and the xylenes (BTX). Benzene is also a volatile byproduct of the conversion of coal to coke. Benzene from coke over distillate is available only to the extent that coke is required by the steel industry.

As discussed in Chap. 5, the oxidative coupling of methane, i.e., the direct conversion of methane with the assistance of oxidants such as O_2, is more thermodynamically favorable than the non-oxidative conversion of methane to hydrocarbons. Unfortunately, the presence of oxygen can cause over-oxidation of the hydrocarbons to CO_2 and CO.

However, aromatic hydrocarbons such as benzene can be produced by the direct conversion of methane in the absence of oxygen, as mentioned in Chap. 5. For example, at high reaction temperatures (~1300 K), Fe (0.5 wt%)/SiO_2 catalysts selectively produce ethylene and aromatic hydrocarbons such as benzene without the production of CO_2 and CO under non-oxidative reaction conditions. The maximum methane conversion was 48.1% at 1363 K, with a hydrocarbon yield of 48% [1]. This reaction is often known as "dehydrocyclization", "aromatization", or "dehydroaromatization". For the purpose of this chapter, the conversion of methane to aromatic hydrocarbons will be referred to as the "methane dehydroaromatization" (MDA) reaction.

The conversion of methane to benzene (C_6H_6) is stoichiometrically expressed in reaction (6.1); as shown, nine moles of hydrogen (H_2) are produced along with each mole of benzene.

$$6\,CH_4 \longrightarrow C_6H_6 + 9\,H_2 \tag{6.1}$$

T. Baba and A. Miyaji, *Catalysis and the Mechanism of Methane Conversion to Chemicals*,
https://doi.org/10.1007/978-981-15-4132-2_6

In this chapter, the catalytic properties of various zeolite catalysts, the roles of the catalysts, including their active sites, and the reaction mechanisms of the formation of benzene in the MDA reaction will be the main focus.

6.2 Initial Reports of the Methane Dehydroaromatization Reaction

The methane dehydroaromatization (MDA) reaction was first carried out by Bragin et al. in 1982 using zeolite catalysts [2]. They reported that the high-silica zeolite TsVM (Si/Al ratio of 33.3), which was modified with 0.1–1.0 wt% of group VIII metal. The MDA reaction was carried out at 873 K in a pulsed micro-reactor. TsVM gave C_6–C_{10} hydrocarbons, and a benzene yield of 6.4% was achieved.

Shortly after, in 1983, zeolite catalysts such as Fe_2O_3-modified H^+-exchanged ZSM-5 (H-ZSM-5) were found to produce aromatic hydrocarbons using N_2O as the oxidant [3]. The aromatic hydrocarbon yields of these reactions were very low. Anderson et al. reported that when the reaction was carried out at a CH_4/N_2O ratio of 98:2 at 734 K using H-ZSM-5 as the catalyst, the methane conversion was 2.7%, the aromatic selectivity was 6.7%, and the major product was CO_2 (87.1%) [4].

In 1993, Wang et al. reported that Mo- or Zn-impregnated H-ZSM-5 zeolites catalyzed the conversion of methane to aromatic hydrocarbons (mainly benzene) without an oxidant [5]. When the methane conversion was carried out at 973 K, using Mo (2 wt%)/H-ZSM-5 or Zn (2 wt%)/H-ZSM-5 in a flow reactor, the methane conversions were 7.2 and 2.3%, respectively. In both cases, the selectivity towards benzene was 100%, although the material balance including coke formation was not reported. In a temperature-programmed experiment, the starting temperature of the MDA reaction was found to be about 50 K lower for Mo/H-ZSM-5 than for Zn/H-ZSM-5, with ethane and H_2 being detected at the same time that benzene began to evolve.

Since then, many researchers have invested great effort into the investigation of the MDA process, and promising catalysts have been developed [6–11]. In addition to the development of improved catalysts, optimization of the reactor design is another general way to improve the MDA reaction, and various designs such as fluidized bed reactors [12], chemical looping processes [13], and membrane reactors [14–16] have been reported. However, a detailed discussion of MDA reactors is outside the scope of this book.

6.3 Thermodynamics of the Direct Conversion of Methane Under Non-oxidative Conditions for Hydrocarbon Production

Although the MDA reaction is typically carried out between 900 K and 1100 K, which are relatively high reaction temperatures compared to those used in the oxidative conversion of methane, non-oxidative conditions result in higher benzene selectivity. Thus, benzene and toluene are the main hydrocarbon products along with traces of ethylene and ethane [17, 18]. The MDA reaction is more suitable for the production of aromatic hydrocarbons than oxidative conversion when only methane is used as a reactant.

In this section, the characteristics of the MDA reaction are explained from a thermodynamic viewpoint and compared with those of the oxidative coupling of methane. The equilibrium conversions of methane to benzene, C_2H_6, C_2H_4, and carbon (C) together with hydrogen achieved under non-oxidative conditions at 773 K are shown below [17, 18].

$$6\,CH_4 \rightleftharpoons C_6H_6 + 9\,H_2 \quad \text{Equilibrium } CH_4 \text{ conversion: 3.1\% at 773 K} \tag{6.2}$$

$$2\,CH_4\,(g) \rightleftharpoons C_2H_6\,(g) + H_2\,(g) \quad \text{Equilibrium } CH_4 \text{ conversion: 0.45\% at 773 K} \tag{6.3}$$

$$2\,CH_4\,(g) \rightleftharpoons C_2H_4\,(g) + 2\,H_2\,(g) \quad \text{Equilibrium } CH_4 \text{ conversion: 0.62\% at 773 K} \tag{6.4}$$

$$CH_4\,(g) \rightleftharpoons C\,(s) + 2\,H_2\,(g) \quad \text{Equilibrium } CH_4 \text{ conversion: 52\% at 773 K} \tag{6.5}$$

As shown in reactions (6.2)–(6.5), the formation of carbon via methane conversion is the most favorable reaction under non-oxidative conditions. On the other hand, the transformation of methane to benzene (aromatics) is more thermodynamically favorable than its transformation to ethylene or ethane, although hydrogen formation is more favorable at higher reaction temperatures because of the endothermic nature of the dehydrogenation reaction.

On the other hand, in the oxidative conversion of methane to benzene shown in reaction (6.6), the equilibrium methane conversion is 100% at 773 K.

$$6\,CH_4 + 9/2\,O_2 \longrightarrow C_6H_6 + 9\,H_2O \qquad (6.6)$$

Even at 623 K, the conversion of methane to benzene is thermodynamically complete with 100% conversion, although the selectivity towards hydrocarbons is low.

6.4 Zeolite-Based Catalysts for the Conversion of Methane to Aromatic Hydrocarbons

Since the H-ZSM-5-supported MoO_3 zeolite catalyst for the MDA reaction was reported in 1993, many catalysts based on the dispersion of metal ions on various supports, mainly zeolites, have been investigated. In the following sections, the catalytic activities of various zeolite-based catalysts are surveyed.

6.4.1 Catalytic Activities of Mo Species Loaded on H^+-Exchanged Zeolites

The performances of a variety of catalysts in which Mo species are supported on various H^+-exchanged zeolites have been examined at ~1000 K in a fixed-bed continuous-flow reactor [5, 19–25]; data for these systems are shown in Tables 6.1 and 6.2. In almost all cases, a methane pressure of 101 kPa was used. In these tables, the single-pass conversion of methane and the selectivities towards hydrocarbons (benzene, toluene, naphthalene, C_2 hydrocarbons ($C_2H_4 + C_2H_6$), and coke) are summarized under all reaction conditions, although in some cases, the amount of coke formed on the surface of the catalyst was not provided. The selectivities were calculated on a carbon basis.

The Mo/H-MCM-22 listed in Table 6.1 was prepared by impregnation with ammonium heptamolybdate and then dried and calcined at 873 K. When the reaction was carried out at 973 K, the Mo (6 wt%)/H-MCM-22 produced benzene in 80.0% selectivity at 10.0% methane conversion. On the other hand, Mo (6 wt%)/H-ZSM-5 gave benzene with 57.8% selectivity at 10.6% methane conversion.

ZRP-1, which has a structure similar to pentasil-type zeolites, is a phosphorous- and rare earth oxide-containing material [22]. This catalyst was prepared by modifying a silica-rich H-ZSM-5 with phosphorous and rare earth oxides. The Mo/H-ZRP-1 catalyst also selectively produced aromatic hydrocarbons, with a (benzene + naphthalene) selectivity of 75.6% at 9.7% methane conversion.

Among the Mo-modified H^+-exchanged zeolites shown in Table 6.1, H-MCM-36, H-IM-5, and H-ITQ-2 also gave aromatic hydrocarbons with selectivities of 40–50% at methane conversions of 6–11%, while USY, MFS-16, and mordenite showed lower catalytic activities for the formation of aromatic hydrocarbons and produced

Table 6.1 Catalytic performance of Mo species supported on various H^+-exchanged zeolites in a fixed-bed continuous-flow reactor

H^+-exchanged zeolite (Pore diameter/Å) Ring size [Dimensionality]	Si/Al Ratio	Mo/wt%	CH_4 con./%	Selectivity/C-atom%					Reaction temperature/K	Time on stream/h	Other reaction conditions	Ref.
				Benzene	Toluene	Naphthalene	C_2 hydrocarbons	Coke				
H-MC-22 (5.5 × 4.0, 5.1 × 4.1) 10-membered ring [2]	15	6	10.0	80.0	No data	4.4	3.4	12.0	973	180	CH_4/N_2 = 90.5/9.5 Total flow rate 1500 ml g^{-1} h^{-1}	[19]
	15	2	5.7	67.8	No data	7.9	4.9	18.5	973	180		
H-ZSM-5 (5.5 × 4.0, 5.1 × 4.1) 10-membered ring [3]	25	6	10.6	57.8	No data	19.8	3.3	18.1	973	90	CH_4/N_2 = 90.5/9.5 Total flow rate 1500 ml g^{-1} h^{-1}	
	25	6	17.9	55.3	3.4	11.8	2.3	27.1	1003	60		[20]
Na-ZSM-5	12.5	2	0	0	–	–	–	No data	973			[5]
H-ZRP-1 (5.6 × 5.6, 5.1 × 5.5)	25	6	9.7	48.6	No data	27.0	No data	13.9	973	60	CH_4/N_2 = 92.5/7.5 Total flow rate 1500 ml g^{-1} h^{-1}	[22]
H-MC-36 [2]	38.2	6	11.5	47.8	No data	0.3	No data	28.5	973	60	CH_4/N_2 = 92.5/7.5	[22]

(continued)

Table 6.1 (continued)

H^+-exchanged zeolite (Pore diameter/Å) Ring size [Dimensionality]	Si/Al Ratio	Mo/wt%	CH_4 con./%	Selectivity/C-atom%					Reaction temperature/K	Time on stream/h	Other reaction conditions	Ref.
				Benzene	Toluene	Naphthalene	C_2 hydrocarbons	Coke				
H-IM-5	25	6	11.5	41.7	1.9	14.1	No data	42.3	973	60	Total flow rate 1500 ml g^{-1} h^{-1}	[23]
H-ITQ-2 [2]	25	3	6.8	31	1	9	1	44	973	110	Total flow rate 1500 ml g^{-1} h^{-1}	[24]
	25	6	9.7	48.6	No data	27.0	No data	13.9	973	60	CH_4/N_2 = 92.5/7.5 Total flow rate 1500 ml g^{-1} h^{-1}	
US-Y [3]	–	3	6.4	10.5	0.4	0	2.2	84.3	973	40	CH_4/N_2 = 98/2 Total flow rate 7.6 ml min^{-1}	[21]
H-FSM-16 16-membered ring [1]	20	3	6.6	6.5	0.2	0	2.2	87.7	973	40		[21]
	15	3	5.9	5.9	0.2	0	1.7	875	973	40		
H-Mordenite 12-membered ring [2]	22	3	7.3	4.1	0.2	0	3.5	83.1	97.3	40		[21]

coke in greater than 80% selectivity. Furthermore, as shown in Table 6.2, Mo species supported on Al_2O_3 and SiO_2 catalysts also showed high selectivities towards coke, even at methane conversions of 7.4 and 5.3%, respectively. Al_2O_3 and SiO_2 have no ordered pore structures, whereas zeolites do. Therefore, the catalytic performances of Mo-based zeolites seem to depend on their pore structure, which plays an important role in the selective production of aromatic hydrocarbons, as discussed below.

Neither H-ZSM-5 without Mo nor Mo/Na^+-exchanged ZSM-5 (Na-ZSM-5) produced aromatic hydrocarbons such as benzene. Furthermore, MoO_3 showed almost no catalytic activity in the MDA reaction, while Mo/Al_2O_3 and Mo/SiO_2 showed very low selectivities towards aromatic hydrocarbons (Table 6.2). Therefore, both the Mo species and the H^+-exchanged zeolite are essential for the conversion of methane to aromatic hydrocarbons with high selectivity, as discussed in the following sections.

6.4.2 *Effect of the Zeolite Pore Structure on the Catalytic Production of Aromatic Hydrocarbons*

As discussed in Sect. 6.4, the structure of zeolites used as Mo-functionalized catalysts seems to influence their catalytic properties, especially their selectivity towards benzene. Mo/H-MCM-22 exhibited higher selectivity towards benzene than the other Mo-modified zeolites in Table 6.1, such as Mo/H-ZSM-5. The catalytic performances of various Mo-modified H^+-exchanged zeolites, including their benzene formation rates and methane conversions, are summarized in Table 6.3 [26].

ZSM-5 has a three-dimensional pore structure that consists of networks of straight and zigzagging pores and their intersections. Both pore structures are formed by 10-membered oxygen rings with entrance diameters of ~5.5 Å. One is straight with an elliptical cross-section, while the second type has a circular cross-section. The straight pores are connected by the zigzag channels. The effective pore diameter is 5.9 Å, which is roughly equal to the kinetic diameter of benzene (5.8 Å).

As shown in Table 6.3, Mo/H-ZRP-1 showed the highest rate of benzene formation among the Mo-modified H^+-exchanged zeolite catalysts; Mo/H-ZSM-5 also showed performance comparable to that of Mo/H-ZRP-1. ZRP-1 zeolites are prepared by modification of high Si/Al ratio H-ZSM-5 with phosphorous and rare earth (RE) oxides. The chemical composition of ZRP-1 is represented by the formula $x\ RE_2O_3$: $y\ Na_2O \cdot Al_2O_3$: $z\ SiO_2$, where $x = 0.01$–0.03, $y = 0.4$–1.0, and $z = 20$–60. ZRP-1 was prepared based on the concept of blocking the acid sites on the exterior surface of ZSM-5 using phosphorous and rare earth oxides. The high catalytic performance may have resulted from the resulting dealumination (loss of aluminum from the ZSM-5 zeolite) and the formation of various P- and Al-containing compounds inside the pores, which decreased the channel size [27].

The zeolite MCM-22 has larger pores than ZSM-5, and its pore structure consists of two independent pore systems. One of these systems is composed of 12-membered

Table 6.2 Catalytic performance of Mo species supported on various supports in a fixed-bed continuous-flow reactor

Catalyst	Mo/wt%	CH_4 conversion/%	Selectivity/C-atom%					Reaction temperature/K	Time on stream/h	Ref.
			Benzene	Toluene	Naphthalene	C_2 hydrocarbons	Coke			
Mo/Al_2O_3	3	7.4	3.9	0.1	0	2.1	88.9	973	40	[21]
Mo/SiO_2	3	5.3	8.7	0.4	0.3	2.1	86.6	973	40	[21]
MoO_3	–	0.01	0	0	No data	100	44	973	110	[25]

Table 6.3 Catalytic performances of Mo and Re catalysts supported on various porous materials

H^+-exchanged zeolite or support	Si/Al ratio	Pore diameter/Å	Dimensionality of pores	Number of members in rings	3 wt% Mo		5 wt% Re	
					CH_4 conversion/%	Benzene formation rate/μmol g^{-1} s^{-1}	CH_4 conversion/%	Benzene formation rate/μmol g^{-1} s^{-1}
H-ZSM-5	19.5	5.4 × 5.6, 5.1 × 5.3	3	10	8.1	0.73	10.1	1.63
H-ZRP-1	25	5.6 × 5.6, 5.1 × 5.5	3	10	8.9	0.95	7.3	0.13
H-MCM-22	18	5.5 × 4.0, 5.1 × 4.1	2	10	8.0	0.55	5.3	0.61
H-ZSM-11	19	5.3 × 5.4	3	10	6.3	0.57	5.1	0.35
H-ß	18.5	5.6 × 5.6, 6.4 × 7.6	3	10	4.1	0.14	4.6	0.11
H-ZSM-12	20	6.0 × 5.6	1	12	4.3	0.11	2.9	0.10
H-Ferrierite	32	5.4 × 4.2, 4.8 × 3.5	2	10, 8	5.3	0.12	3.9	0.10
H-LTL	3	7.1	3	12	0.4	0.05	4.3	0.03
H-FSM-16	150	27	1		16.9[a]	0.03[a]	Not available	Not available
Al_2O_3	–	–	–	–	8.3	0.05	3.1	0.06
SAPO-5	–	7.3	1	12	4.3	0.05	3.2	0.01

Reaction conditions: 973 K; flow rate = 3000 mL g^{-1} h^{-1}, 0.3 MPa, 2% CO_2, rates of benzene formation at 2 h time on stream

[a]Without CO_2 addition

oxygen ring cages with dimensions of 18 × 7.1 × 7.1 Å. The cages are connected through 10-membered ring windows. The other system is a two-dimensional 10-membered ring pore system that does not contain any cages. There is no direct connection between the two pore systems. Thus, MCM-22 consists of an interconnected building unit forming two independent pore systems: Two-dimensional, sinusoidal 10-membered interlayer channels of 4.0 × 5.9 Å and 12-membered interlayer supercages of 7.1 × 18 Å with 4.0 × 5.9 Å entrance apertures.

ZSM-11 has a tetragonal crystal lattice and a two-dimensional pore structure consisting of straight, perpendicularly intersecting channels; the pores consist of 10-membered oxygen rings and have dimensions of 5.1 × 5.5 Å. This pore structure is almost the same as that of the ZSM-5 zeolite. ZSM-8, which also consists of 10-membered oxygen rings, has an orthorhombic crystal lattice. It has a two-dimensional system of channels with an average entrance diameter of ~5 Å. SAPO-34 has 8-membered oxygen rings and narrow pores (pore entrance diameter of 4.3 Å), which impedes the diffusion of aromatic compounds. In β-zeolite, three crystal systems (tetragonal, anorthic, and monoclinic) exist; they are highly disordered and have many defects. The pore structure of β-zeolite is two-dimensional, with pore entrance dimensions of 6.6 × 8.1 Å and 5.5 × 6.5 Å. Empirically curved channels with dimensions of 5.5 × 6.5 Å are formed by 12-membered oxygen rings.

X, Y, SAPO-5, and SAPO-11 are large-pore zeolite materials, whose pore systems are three-dimensional straight channels with supercages and large pore entrance diameters of ~7.5 Å, which is larger than the kinetic diameter of benzene molecule.

Additionally, although the selectivity towards coke and amount of coke formed during the MDA reaction were not measured in some cases, the catalytic performances of many Mo-modified H^+-exchanged zeolites have been examined [28, 29]. For example, Zhang et al. reported that the catalytic activities for the MDA reaction followed the order Mo/H-ZSM-11 > Mo/H-ZSM-5 > Mo/H-ZSM-8 > Mo/H-β > Mo/MCM-41 > Mo/H-SAPO-34 > Mo/H-mordenite > Mo/H-X > Mo/H-Y > Mo/H-SAPO-5 > Mo/H-SAPO-11 [28].

Thus, based on the results summarized in Tables 6.1 and 6.3 and the discussion above, the zeolite pore structure can be seen to play an important role in achieving high catalytic performance in the MDA reaction. Mo-modified porous materials containing pores with diameters of 5.3–5.6 Å (ZSM-5, MCM-22, ZSM-11, and ZRP-1), which is close to the kinetic diameter of a benzene molecule, and two- or three-dimensional pore systems are superior as catalysts for the MDA reaction. Mo-modified zeolites with pore diameters greater than ~6 Å exhibited low catalytic activity, with coke formation predominating.

6.4.3 *Effect of the Si/Al Ratios of H^+-Exchanged Zeolites on Their Catalytic Performances*

As mentioned above, both an H^+-exchanged zeolite structure and Mo species are crucial for achieving catalytic activity in the MDA reaction; for example, Mo/Na-ZSM-5 does not show catalytic activity in this reaction. The acidic properties of H^+-exchanged zeolite are due to their Brønsted acid sites, which originate from the acidic OH groups on the surface of the zeolite. These acidic OH groups are generated to preserve charge balance in the zeolite by the introduction of Al^{3+} cations into the zeolite lattice in the place of Si^{4+} cations, as shown in (6.7).

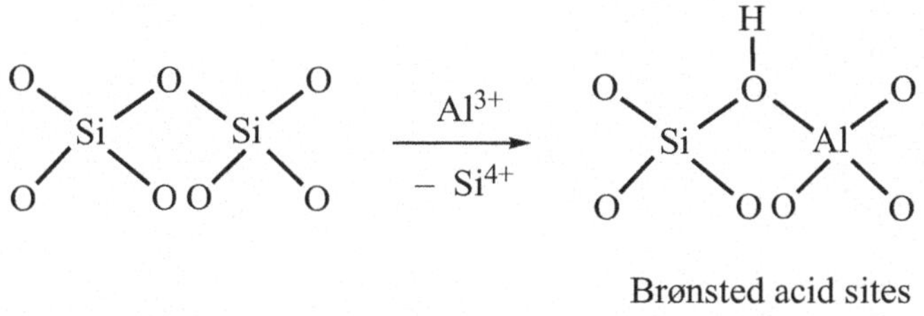

(6.7)

As shown below, H^+-exchanged zeolites are usually prepared by the exchange of Na^+ in Na^+-exchanged zeolites using acidic solutions, as shown in (6.8) or using NH_4Cl solutions (6.9) followed by heating to afford converted H^+-type zeolites. Notably, the number of Brønsted acid sites is less than the number of Al^{3+} cations introduced into the lattice. When the Na^+ cations in the Na^+-exchanged zeolite are completely replaced by protons, the number of Brønsted acid sites becomes equal to the number of Al^{3+} cations in the lattice.

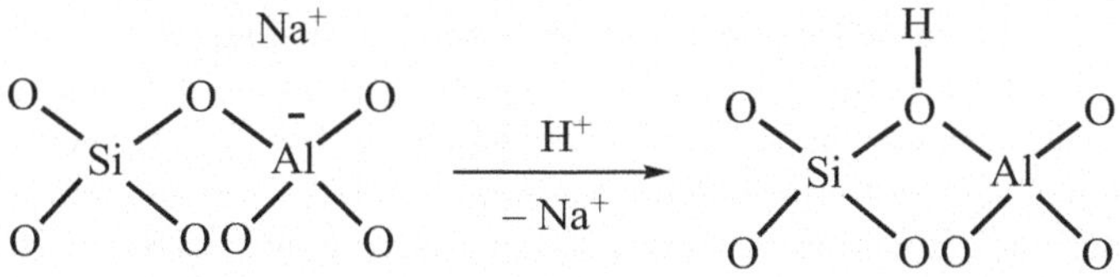

(6.8)

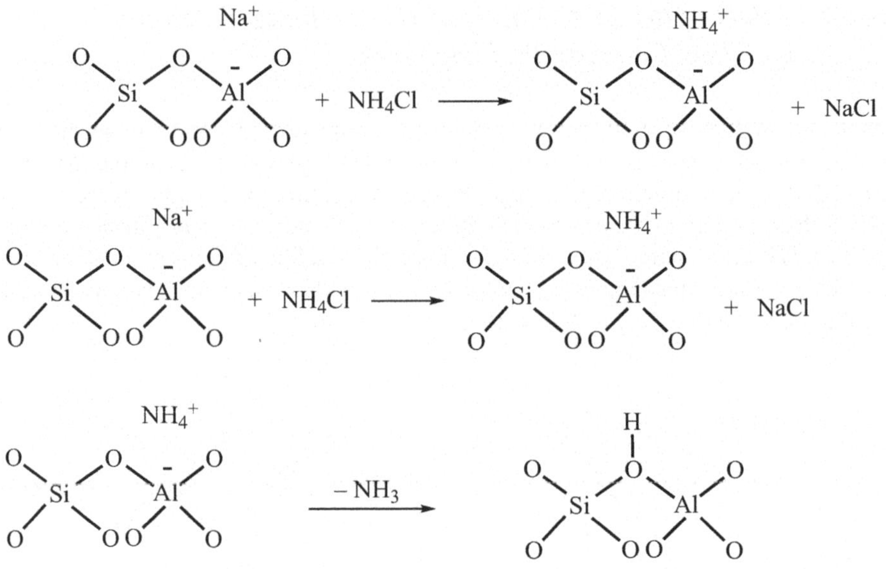

(6.9)

Since the amount of Brønsted acid sites (acidic protons) in the zeolite depends on the Si/Al ratio, the catalytic activities of H^+-exchanged zeolites modified with Mo species may be influenced by the Si/Al ratio. Liu et al. investigated the influence of the Si/Al ratio of H-ZSM-5 modified with Mo (3 wt%, 0.32 mmol/g) on the catalytic activity of the resulting zeolites in the MDA reaction [21, 30]; the results are summarized in Table 6.4. For instance, the high catalytic performance was observed for the 3 wt% Mo catalysts with a Si/Al ratio of approximately 20. The concentration of acidic protons (Brønsted acid sites) in H-ZSM-5 can also be estimated from the Si/Al ratio. As shown in Table 6.4, the conversion of methane remained relatively constant at Si/Al ratios of up to 36.7 (0.43 mmol/g protons), i.e., when the acidic proton concentration was greater than that of Mo (0.32 mmol/g). However, when the concentration of acid protons was decreased below that of Mo via increasing the Si/Al ratio, the catalytic activity for the formation of aromatic hydrocarbons decreased, as shown in Table 6.4.

Sriglan et al. reported that the formation rate of aromatic hydrocarbons over Mo (3.9 wt%)/H-ZSM-5 catalysts depended on their Si/Al ratio [31], with the aromatic hydrocarbon formation rate increasing as the Si/Al ratio was increased from 14 to 28. However, hydrocarbon formation decreased when the Si/Al ratio was further increased to 54. The results discussed above demonstrate that the Si/Al ratio of Mo/H^+-exchanged zeolite catalysts influences their catalytic activities in the MDA reaction.

Similarly, Mátus et al. found that the catalytic activity of Mo-modified H-ZSM-5 depended on the concentration of Mo when the Si/Al ratio of the zeolite was held

Table 6.4 Catalytic performance of 3 wt% (0.32 mmol g^{-1}) Mo H-ZSM-5 catalysts with different Si/Al ratios for the methane aromatization reaction at 973 K and 101 kPa

Si/Al ratio	Al atoms per unit cell	Al^{3+} concentration mmol g^{-1}	CH_4 conversion/%	Selectivity/C-atom %		Hydrocarbon product distribution/%			
				Hydrocarbon	Coke	Benzene	Toluene	Naphthalene	C_2
11.9	7.4	1.3	8.8	55.2	41.6	65.4	3.4	26.5	4.7
13.0	6.9	1.2	8.0	54.1	43.6	67.0	3.7	23.4	5.6
19.8	4.6	0.80	10.2	62.8	35.3	59.4	3.1	33.2	3.9
26.5	3.5	0.61	7.2	48.1	47.3	62.9	3.3	26.3	7.3
36.7	2.5	0.43	8.3	35.6	60.9	66.5	3.5	22.8	7.1
108	0.88	0.15	7.5	16.7	79.3	59.0	4.0	14.5	21.5
400	0.24	0.04	5.1	30.8	61.4	64.0	3.6	13.5	18.3
950	0.10	0.02	6.3	19.7	76.9	57.1	4.2	10.6	26.9

Reprinted from ref. [21], Copyright 2020, with permission from Elsevier

constant at 17 [32]. This result shows that the ratio of the Mo/Brønsted acid sites affects the production of aromatic hydrocarbons.

The results discussed above suggest that the Mo species interact with the acidic protons (Brønsted acid sites) in the zeolites to produce active Mo species that act as the catalytically active sites. This point will be discussed further in Sect. 6.5.

6.4.4 Catalytic Activities of H^+-Exchanged Zeolites Modified with Various Metal Species

In addition to Mo/H^+-exchanged zeolites, H^+-exchanged zeolites modified with other metal species (Fe [33–36], V [33], Cr [33], W [33, 37, 38], Re [33, 39–41], Ru [42], Zn [43, 44], Pt [45], and Mn [46]) have also been reported to catalyze the MDA reaction to produce aromatic hydrocarbons. Many metal species (Fe [47–49], Ni [50], Co [51], Cu [52, 53], Zn [54], Ga [55, 56], Pt [57, 58], Rh [58], W [59], Cr [60], Ru [61], Zr [62], V [62], Ag [63], In [64], Li [65], and P [65]) have also been added as promoters to improve the catalytic performances of Mo/H^+-exchanged zeolites (mainly H-ZSM-5).

The catalytic activities of these materials were strongly dependent on the identity of the metal ion and pretreatment, which allowed catalyst performance to be adjusted by control of these parameters. For example, Weckhuysen et al. prepared H-ZSM-5 zeolite catalysts containing 2.0 wt% of different transition metal ions via the impregnation method and then pretreated the catalysts with CO at 773 K for 6 h. The activities of the resulting catalysts were found to decrease in the order (Table 6.5) [33]:

Mo (18.3) > W (10.8) > Fe (5.7) > V (3.9) > Cr (1.5)

Table 6.5 Catalytic performances of 2.0 wt% transition metal ion-loaded H-ZSM-5 zeolites at 1023 K

Transition metal ion	CH_4 conversion/%[a]	Methane consumption rate[b]	Maximum selectivity/%		
			Benzene	Naphthalene	Aliphatic hydrocarbons
Mo	7.9	18.3	72.2	13.6	12.8
W	2.4	10.8	50.8	0	20.1
Fe	4.1	5.7	73.4	16.1	22.1
V	3.2	3.9	31.6	6.3	20.4
Cr	1.1	1.5	72.0	3.7	26.7

[a] CH_4 conversion was measured after 3 h of reaction. Catalysts were prepared using the impregnation method and then pre-reduced with CO at 773 K for 6 h

[b] CH_4 consumption rate given in (Molecules CH_4 reacted)/((Metal atom) h)

The values in parentheses are the rates of methane consumption in units of methane molecules reacted per metal atom per hour. As shown in Table 6.5, the Mo-loaded H-ZSM-5 showed the highest overall catalytic performance for the production of aromatic hydrocarbons among the different metals tested.

The catalytic performances Re (5 wt%)/H^+-exchanged zeolites are also shown in Table 6.3. The trend in the catalytic performances of the Re-impregnated H^+-exchanged zeolites was the same as those for their Mo-impregnated counterparts in all respects, such as the methane conversion and the benzene formation rate.

6.5 Interaction of Mo Species with Brønsted Acid Sites on H^+-Exchanged Zeolites

As discussed in Sect. 6.4, Mo species are believed to interact with H^+ (acidic protons) because of the presence of Brønsted acid sites on and/or in the zeolite. The catalytic activity of Mo/H^+-exchanged zeolite catalysts in the formation of aromatic hydrocarbons has been demonstrated to depend on the ratio of Mo to Brønsted acid sites. To understand the role of the Mo species generated on H^+-exchanged zeolite, first, the Mo species present during the preparation and pretreatment of the Mo-modified zeolite catalyst, as well as during the MDA reaction, should be investigated. In the following sections, the formation and generation of Mo species on and/or in H^+-exchanged zeolites during these three steps is discussed, mainly in terms of the zeolite H-ZSM-5.

6.5.1 Methods for the Preparation of Mo-Modified H-ZSM-5 Catalysts

Mo-modified H-ZSM-5 (Mo/H-ZSM-5) catalysts for the MDA reaction are usually prepared by one of the following two methods [21, 48, 65–67].

Method 1: A zeolite is impregnated using a slurry or the incipient wetness method using a basic solution of a molybdenum compound (ammonium heptamolybdate ($(NH_4)_6Mo_7O_{24}$)) or ammonium paramolybdate), followed by treatment in air at a temperature of 700–1000 K.

Method 2: A molybdenum compound such as MoO_3 is mechanically mixed with an H^+-exchanged zeolite (H-ZSM-5) and then heated at the reaction temperature. The molybdenum species are thus anchored on the zeolite as the result of a solid-phase reaction.

In both **Method 1** and **Method 2**, an H^+-exchanged zeolite or $NH_4{}^+$-exchanged zeolite must be used. Furthermore, in **Method 1**, changing the pH of the molybdenum solution affects the properties of the resulting catalysts [68].

6.5.2 Formation of Various Mo Species in H^+-Exchanged Zeolites

The catalytic performances of Mo-modified H-ZSM-5 (Mo/H-ZSM-5) for the MDA reaction strongly depended on catalyst preparation conditions and/or preparation methods. Furthermore, the formation of various Mo species could be observed by analytical methods such as NMR, TPD (temperature-programed desorption) TG (thermogravimetry), and UV-Raman spectroscopy [67, 69–74]. These results show that Mo species probably interact with Brønsted acid sites on ZSM-5 and vary during catalyst perpetration.

Xu et al. examined the Mo species on H-ZSM-5 during calcination treatment in the air using infrared (IR) spectroscopy and differential thermal analysis (DTA). They detected small MoO_3 crystallites on the exterior surface of ZSM-5 zeolites during calcination at around 500–650 K [67, 70]. They further showed that the IR bands associated with the MoO_3 crystallites on the external surface of the ZSM-5 crystals disappeared when the sample was further calcined at 773 K, followed by the migration of $(MoO_3)_n$ oligomers into the zeolite channels (pores).

The Tamman temperature of MoO_3 is 534 K, and the sublimation of MoO_3 becomes detectable above 623–673 K. Thus, the surface migration of MoO_3 onto H-ZSM-5 occurs at high temperature, i.e., catalyst preparation temperatures of around 700–1000 K. Furthermore, the vapor pressure of MoO_3 as $(MoO_3)_n$ oligomers is 56 Pa at 973 K. Therefore, it is possible that MoO_3 molecules diffuse into the cavities (pores) in zeolite and can interact with the protons at the Brønsted acid sites in the cavities.

Borry et al. proposed that MoO_3 species that migrate into zeolite cavities react with the protons at the Brønsted acid sites to form $MoO_2(OH)^+$ and Species (**I**) as follows [71]:

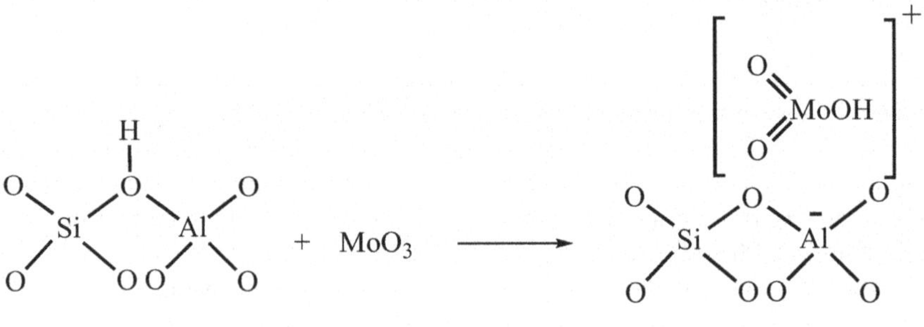

Species (**I**)

(6.10)

X-ray diffraction, ^{27}Al magic angle spinning nuclear magnetic spectroscopy (^{27}Al MAS-NMR), and mass spectral measurements showed that $(Mo_2O_5)^{2+}$ and Species (**II**) were generated in H-ZSM-5 by the condensation of Species (**I**).

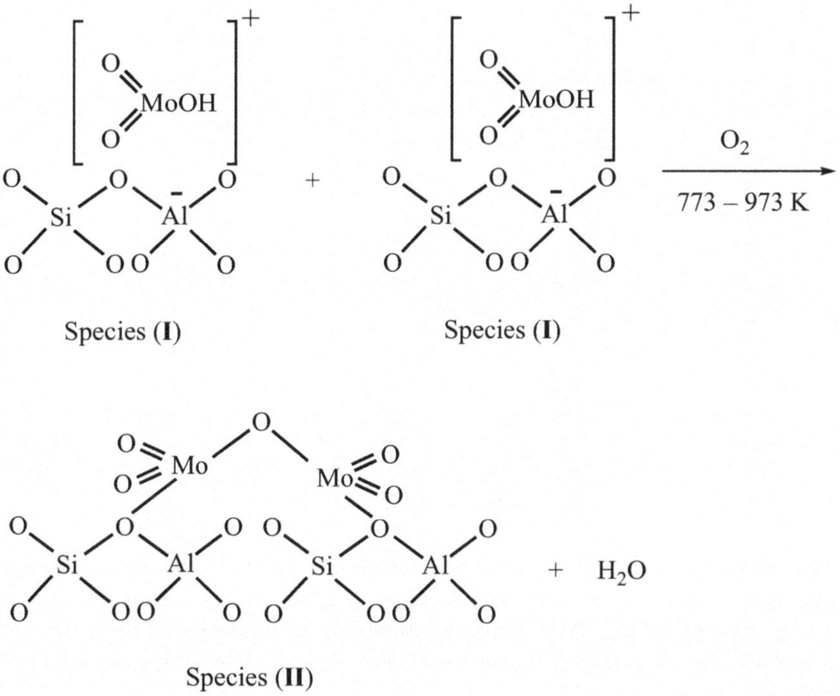

(6.11)

In reference to $Mo_2O_5^{2+}$, the authors mentioned that despite the fact that $Mo_2O_5^{2+}$ species are not known in solution, the exchanged species in a zeolite should be cationic. In 2000, Li et al. used Raman and X-ray near-edge spectroscopy to confirm the formation of Species (**II**) on H-ZSM-5 when ammonium hexamolybdate–impregnated H-ZSM-5 with a Si/Al ratio of 14.3 was calcined at 973 K [72].

Species (**I**), $Mo_2(OH)^+$, can also react with the protons at the Brønsted acid sites to form MoO_2^{2+} cations that bridge two acid sites and water in zeolites with Si/Al ratios much lower than ZSM-5 zeolites. For example, in the zeolite H-Y (Si/Al ratio = 2.6), Species (**III**) was produced by the reaction of acid protons with MoO_2Cl_2 as follows [74].

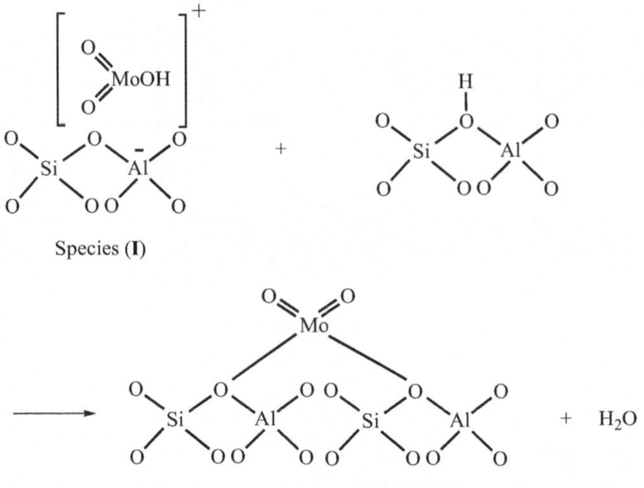

Species (**I**)

Species (**III**)

(6.12)

The formation of Species (**II**) and Species (**III**) depends on the Si/Al ratio of the zeolite; that is, they are strongly influenced by the distance between neighboring Al sites in the zeolite lattice. The relationship between the formation of these molybdenum species and the distance between neighboring Al sites will be discussed in the following section.

Vahel et al. reported the chemical properties of MoO_3, including its reversible reaction with H_2O to form $MoO_2(OH)_2$ [75].

$$MoO_3 + H_2O \rightleftharpoons MoO_2(OH)_2 \quad (6.13)$$

In the preparation of Mo/H-ZSM-5, H_2O is formed during the calcination of the Mo-modified H-ZSM-5, and $MoO_2(OH)_2$ has a vapor pressure of 4.9 Pa at 973 K. Therefore, $MoO_2(OH)_2$ can react with the acidic protons in H^+-exchanged zeolites as follows:

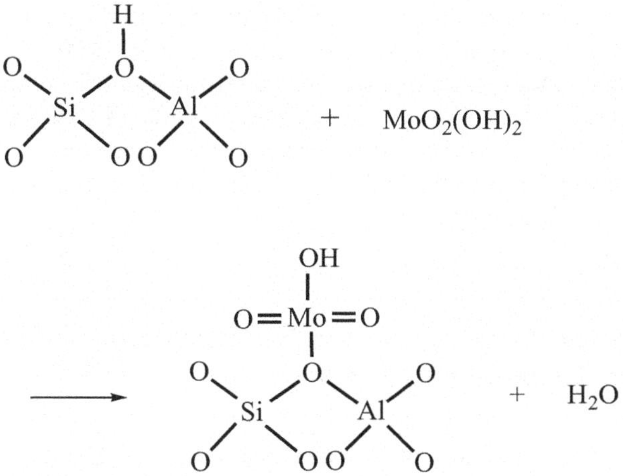

Species (**IV**)

(6.14)

Species (**IV**) further reacts with acidic protons to produce Species (**II**) and Species (**III**).

A similar condensation reaction of the surface OH groups in zeolites leads to the extraction of Al^{3+} ions from the framework, accompanied by the formation of H_2O and the disappearance of two Brønsted acid sites; this process is typically referred to as dealumination. The tendency for Al atoms to be extracted from the framework became more significant with increasing Mo loading [67]. The extra framework Al atoms form small domains of not only Al_2O_3, but also $Al_2(MO_4)_3$ via the reaction of MoO_3 with Al_2O_3.

Although Mo supported on Al_2O_3 showed low catalytic activity in the MDA reaction (Table 6.2), the structure of Mo species on various Al_2O_3 structures, such as γ-Al_2O_3, has been studied. For example, Giordano et al. proposed the formation of aluminum-molybdic complexes based on IR spectra and ultraviolet reflectance spectra [76]. Thus, the condensation of molybdate ions with the surface OH groups of Al_2O_3 proceeded to produce Mo species, namely, the monomeric and dimeric surface structures of Species (**V**) and Species (**VI**).

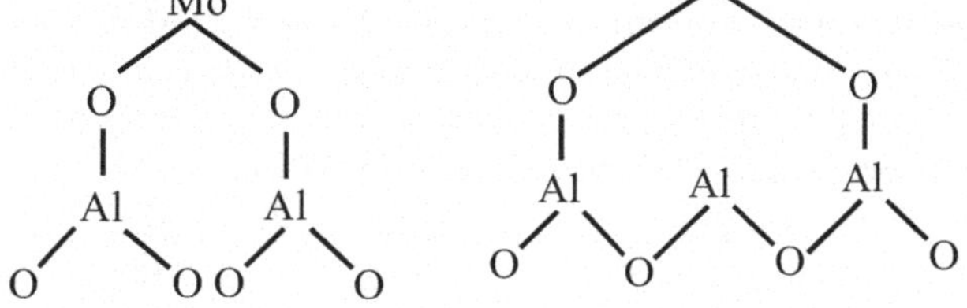

Species (**V**) Species (**VI**)

(6.15)

Therefore, the impregnation of Mo compounds, such as ammonium heptamolybdate ($(NH_4)_6Mo_7O_{24}$) initially leads to the formation of extra MoO_3 crystals during calcination. These MoO_3 crystals may react with surface OH groups, such as the acidic protons of Brønsted acid sites, to form Species (**II–VI**). The structural details of Species (**II**) and Species (**III**) will be further discussed in Sect. 6.5.4, which deals with the relationship between the Si/Al ratios in H-ZSM-5 zeolites and the formation of Species (**II**) and Species (**III**).

6.5.3 Chemical Properties of the Mo Species on H-ZSM-5 After Calcination at 773 K

As mentioned above, a variety of Mo species (Species **II–VI**) may be formed on H-ZSM-5 zeolites after calcination at around 700–1000 K. Xu et al. attempted to use an ammoniacal solution to extract the Mo species on Mo-impregnated H-ZSM-5 with a Si/Al ratio of 25 that had been calcined for 773 K for 6 h and then dried at 383 K overnight [77]. They reported that two kinds of Mo species were present on the Mo/H-ZSM-5 catalysts: a soluble species (Mo extracted in NH_3 solution) and species that were not dissolved into the NH_3 solution (residual Mo). They also showed that the ratio of soluble Mo species to residual Mo species depended on the original Mo loading. For example, for Mo loadings of less than 3.0 wt%, the amount of residual Mo was almost constant. However, at Mo loadings greater than 3.0 wt%, the amount of residual Mo increased with the Mo loading. The non-linear dependence of the amount of residual Mo on the original Mo loading implied that some Mo species were anchored on the surface and/or inside the micropores of the H-ZSM-5.

6.5.4 Influence of the Si/Al Ratio on the Structure of Mo Species on H^+-Exchanged Zeolites After Calcination

As described in Sect. 6.3, after the calcination of Mo-modified H^+-exchanged zeolites, Mo may possibly exist as Species (**II**), $Mo_2O_5^{2+}$, and Species (**III**), MoO_2^{2+}, while dealuminated Al species may react with MoO_3 to produce the aluminum-molybdic complexes Species (**IV**) and Species (**V**). Thus, a variety of Mo oxide species are formed on H^+-exchanged zeolites such as H-ZSM-5. For example, Gao et al. reported the formation of nanostructured Mo oxide on and/or in H-ZSM-5 [78].

According to ref. [79], the formation of the anchored structures of Species (**II**) and Species (**III**) depends on the distance between an Al atom and the first neighboring Al atom (next-nearest neighbor Al atom) in the zeolite lattice, and the distances between two oxygen atoms anchored to Al atoms on the zeolite in Species (**II**) and Species (**III**) were estimated.

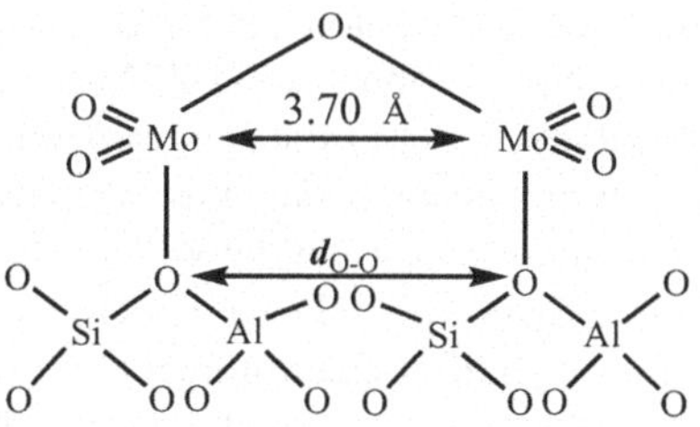

Fig. 6.1 Structure of Mo species (**II**) (Reprinted from ref. [72], Copyright 2020, with permission from Elsevier)

Li et al. estimated the distance between two oxygen atoms in Species (**II**) (Fig. 6.1, $d_{O\text{-}O}$) as 4.92 Å from multiple scattering simulations of $Mo_2O_7^{2-}$; the structure of $Mo_2O_7^{2-}$ resembled that of the previously reported $MgMo_2O_7$ dimer, which can be considered as a model of the zeolite framework [72]. They also estimated the distance between two Mo atoms to be 3.7 Å, which was similar to the Mo–Mo distances in $MgMo_2O_7$, which were determined as 3.687 Å and 3.685 Å in the crystal structures reported in ref. [80] and [81], respectively.

The crystal structure of $MgMoO_4$, which was considered as a model structure for Species (**III**), was also determined by X-ray diffraction spectroscopy and extended X-ray fine structure spectroscopy (EXAFS) [82, 83]. In ref. [83], Amberg et al. reported the crystal structure of $MgMoO_4 \cdot H_2O$, in which the Mg–O distances ranged from 2.034 to 2.130 Å, the Mg–Mg distance was 3.191 Å, and the Mg–Mo and Mo–Mo distances were 3.441 Å and 4.061 Å, respectively. Therefore, if Species (**III**) were bonded to acid sites at next-nearest neighbor Al atoms, the distance between two oxygen atoms in Fig. 6.1 should be about 3.2 Å to anchor the Mo species to two Brønsted acid sites. Since the Mg–O distance was about 2.1 Å in $MgMoO_4 \cdot H_2O$, Species (**III**) could be formed at next-neighbor Al distances of as great as 5.3 Å (3.2 + 2.1 = 5.3 Å) in the zeolite.

The structure of the Mo species on H-ZSM-5 was investigated using density functional theory (DFT), and MoO_2^{2+} (Species (**III**)) was found to be the most likely structure [84]. Furthermore, the theoretical predictions of the model of Species (**III**) agreed with the experimental results from MAS-NMR, ESR, and Fourier transform infrared (FTIR) measurements.

ZSM-5 zeolites are usually synthesized with Si/Al ratios in the range 12–1000. H-ZSM-5 zeolites with Si/Al ratios in this range have often been prepared and used as Mo/H-ZSAM-5 catalysts for the MDA reaction. The unit cell composition of Na-ZSM-5 is $Na_n(Al_nSi_{(96-n)}O_{192}) \cdot 6H_2O$. Thus, the number of T (Al or Si) atoms in the unit cell is 96. Therefore, the number of Al atoms per unit cell can only be larger than 1 when the Si/Al ratio is smaller than 95. The sites occupied with Al or Si atoms are denoted as T sites.

Rice et al. calculated the distance between two Al atoms at next-nearest neighbor T sites occupied with Al atoms or next-next-nearest neighbor T sites to be 4.2–6.5 Å [79]; these distances correspond to two or more Al atoms within the sphere. They also estimated the occupancy of T sites by Al atoms as a function of the Si/Al ratio, and showed the necessity of the formation of Species (**II**) to overcome the long distances between two Al atoms at Si/Al ratios lower than 25. Goodman et al. reported that in the straight channels of H-ZSM-5, all T sites were within 4.75 Å of the axis running down the center of the zeolite channel (Si/Al ratio < 25) [85].

On the basis of the above estimations of the distance between two Al atoms in next-next-nearest neighbor T sites in the zeolite ZSM-5, the formation of Species (**II**) and Species (**III**) was concluded to occur in H-ZSM-5 zeolites with a Si/Al ratio lower than 95 if dealumination did not occur.

Borry et al. examined the ratio of the number of acidic protons to the number of Mo atoms introduced into H-ZSM-5 (H^+/Mo) at a Si/Al ratio of 14.3. They experimentally demonstrated that the H^+/Mo ratio was constant at approximately 1.24 when the Mo content was lower than 3.6 wt% [71]. This indicated that Species (**II**) was mainly produced as the Mo species on and/or in the H-ZSM-5. However, when the Mo content was increased to 6.3 wt%, the H^+/Mo ratio decreased to 0.64. This means that the formation of Species (**III**) predominates over that of Species (**II**).

Tessonnier et al. also examined the H^+/Mo ratios in Mo-modified H-ZSM-5 zeolites [86]. When 2 wt% Mo was supported on H-ZSM-5 with a Si/Al ratio of 40, the H^+/Mo ratio was almost unity, indicating the formation of Species (**II**). On the other hand, when Mo (2 wt% or 4 wt%) was supported on H-ZSM-5 with a Si/Al ratio of 15, the H^+/Mo ratio was almost two in both cases. This result indicated that the monomeric bidentate Mo species (Species (**III**)) was formed. As mentioned previously, a species similar to Species (**III**) was formed in the H-Y zeolite, which has a low Si/Al ratio of 2.65 [74]. At higher Si/Al ratios (low Al contents), the distance between two Brønsted acid sites becomes too long for this bidentate Mo complex to be formed. Therefore, the dimeric Species (**II**) is formed via the condensation of two monomeric Mo species to overcome the distance between two Brønsted acid sites.

6.6 Mo Species of Mo/H-ZSM-5 Zeolite Catalysts During the Working Stage

Section 6.5 discussed the anchored Mo species (**II–VI**) that can potentially form via the interaction of MoO_3 with the surface of the H^+-exchanged zeolites. However, these Mo species are not necessarily the catalytically active species in the MDA reaction. When the Mo-modified H^+-exchanged zeolites such as Mo/H-ZSM-5 are used as catalysts in the MDA reaction, the production of hydrocarbons, such as ethylene and benzene, is usually observed in the induction period and does not depend on the type of Mo-modified catalyst [66, 87–93]. This induction period has also been observed when unsupported Mo_2O_3 is used as a catalyst.

In both cases, the formation of CO_2 has been observed upon contact of the catalyst with methane, followed by the production of CO and H_2O. After the formation of CO_2, CO, and H_2O, the hydrocarbons C_2H_6, C_2H_4, and benzene were produced. These phenomena indicate that the catalytic active species (active sites) were generated during the induction period. The details of the generation of the catalytically active species, the mechanisms of the C–C bond formation reaction, i.e., the formation of hydrocarbons such as C_2H_4 and benzene, and the roles of the catalysts will be discussed in the following sections.

6.6.1 Changes in the Reaction Products During the Induction Period

As discussed above, the types of Mo species on H^+-exchanged zeolites such as H-ZSM-5 may change during the induction period. The change in the Mo species would be expected to affect the types or amounts of reaction products produced during this induction period. Ding et al. examined the changes in the products during the initial reaction stage [93]. They observed changes in the concentrations of unreacted CH_4 and the reaction products (CO, CO_2, H_2O, and hydrocarbons) as the reaction progressed at 950 K using a Mo (4 wt%)/H-ZSM-5 (Si/Al = 25) catalyst and a 1:1:1.3 CH_4/Ar/He mixtures at a flow rate of 100 ml min^{-1}. The results are shown in Fig. 6.2 as plots of the conversion of CH_4 and the formation rates of CO, CO_2, H_2O, and hydrocarbons (C_2H_6, C_2H_4, benzene, and naphthalene) against the reaction time. After an induction period of about 200 s, the consumption rate of CH_4 increased sharply and reached its maximum at 310 s. During this ~200 s induction period, the conversion of CH_4 led predominantly to CO_2, H_2O, and CO, along with H_2, without the concurrent formation of hydrocarbons. After ~200 s, the formation of C_2H_4 and benzene began; the rate of C_2H_6 formation increased gradually and reached maximum at 310 s. The formation of naphthalene began later than that of benzene at about 300 s.

6.6.2 Changes in the Mo Species During Induction Period

As discussed in Sect. 6.6.1, during the initial stage of the reaction, CO_2, H_2O, and CO were formed when Mo/H-ZSM-5 was brought into contact with CH_4. The formation rates of these compounds varied strongly with the reaction time until the formation of C_2H_6, C_2H_4, and benzene began [93]. The formation of these initial products was attributed to the partial removal of oxygen from the Mo species on H-ZSM-5. Thus, the reduction of Mo species such as Species (**II**) occurred, with CO and H_2 as the predominant products. The amount of H_2 produced was approximately twice that of CO. In ref. [93], the authors showed that carbon species were retained on the catalyst surface as Mo carbidic carbon, MoC_x. Other groups have also reported the reduction of Mo species to lower-valent molybdenum species including Mo^0 metal and the formation of Mo carbidic carbon (MoC_x and Mo_2C) and oxycarbidic carbon (MoC_xO_y) [21, 88–90, 94–99].

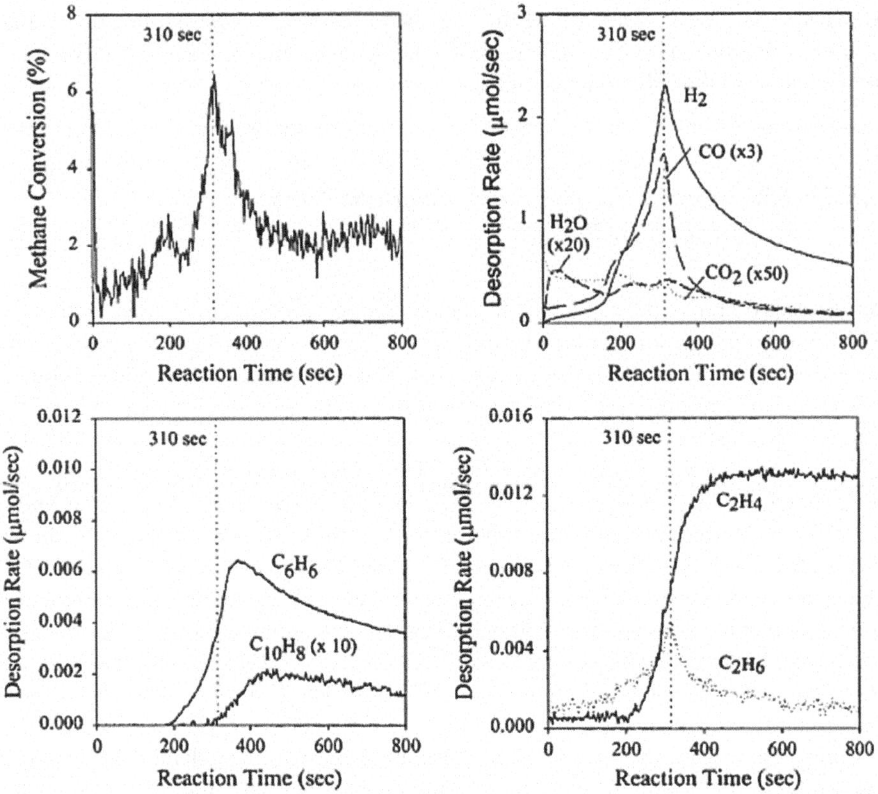

Fig. 6.2 Transient reaction of 4 wt% Mo/H-ZSM-5 with methane at 950 K (Reprinted with permission from ref [93]. Copyright 2020 American Chemical Society)

However, unsupported molybdenum compounds (Mo, MoO_2, MoO_3, Mo_2C, and $MoC_{(1-x)}$) gave H_2, H_2O, and CO_2 as products were detected, Only a trace amount of ethane was formed [100]. Therefore, Mo species such as Species (**II–VI**) are reduced to lower-valent Mo species during the induction period and then further reduced to Mo carbidic carbon, which must be supported on an H^+-exchanged zeolite such as H-ZSM-5 to obtain catalytic activity for the formation of aromatic hydrocarbons, as will be discussed in the next section.

6.7 Active Species and Reaction Mechanism for the Formation of Aromatic Hydrocarbons

The most investigated metal for the MDA reaction is Mo, because of its superior activity. However, as mentioned in Sect. 6.4.4, in addition to Mo, several other metals can be used to modify H^+-exchanged zeolites for the catalysis of the MDA reaction.

These metal species can be separated into two categories [11]. The first category comprises Mo, W, Fe, V, Cr, Re, and Mn, which undergo an activation (induction) period before the production of hydrocarbon products. Thus, during this period, the metal ion is reduced to its active phase, most likely via carburization. The second category is made up of Ag, In, and Zn, in which the active site has been reported to be a cation that acts as a proton abstraction site for CH_4. Details about this type of catalyst are given in Chap. 5.

In the MDA reaction, aromatic hydrocarbons are not the initial products. Thus, the first C–C bond formation to produce C2 hydrocarbons (i.e., ethylene and/or ethane) should be distinguished from subsequent C–C bond formation to produce aromatic hydrocarbons.

6.7.1 Active Mo Species and C–C Bond Formation to Give C_2H_6 and C_2H_4 During the Initial Stage

Since CO, CO_2, H_2, and H_2O are produced during the induction period when a Mo/H^+-exchanged zeolite is brought into contact with methane at approximately 900 K, various molybdenum species can potentially be formed, including the molybdenum carbides MoC_x [21, 93, 101], $MoC_{(1-x)}$ [97], Mo_2C [88, 93, 96, 100, 102, 103], and/or MoC_xO_y [98], as well as Mo [90]. When Mo, MoO_2, MoO_3, Mo_2C, and $MoC_{(1-x)}$ were used as catalysts in the conversion of methane at 973 K, only trace C_2H_6 was observed after the induction period [90]. Furthermore, these compounds interacted with CH_4 to produce H_2 over Mo, while H_2O and CO_2 were formed over MoO_2 and MoO_3. When Mo_2C and $MoC_{(1-x)}$ were used as catalysts at the same reaction temperature, small amounts of C_2H_4 and C_2H_6 were observed together with CO and hydrogen (H_2). These results strongly suggest that carbonized Mo species such as MoC_x take part in the first C–C bond formation to produce C_2H_6 and C_2H_4 from CH_4.

Several reaction mechanisms for the formation of these C_2 hydrocarbons have been proposed. However, the experimental evidence up to the present has not always been clear. The reaction mechanisms considered to be most likely are listed below.

(1) Catalysis of the formation of C_2H_6 by MoC_x

According to ref. [6], during the initial stage, MoC_x species react with methane to produce a specific carbonaceous intermediate (Intermediate (**I**)) on the catalyst surface, which is possibly converted to C_2H_6 by the further reaction of methane with the CH_y and/or CH_z species as follows:

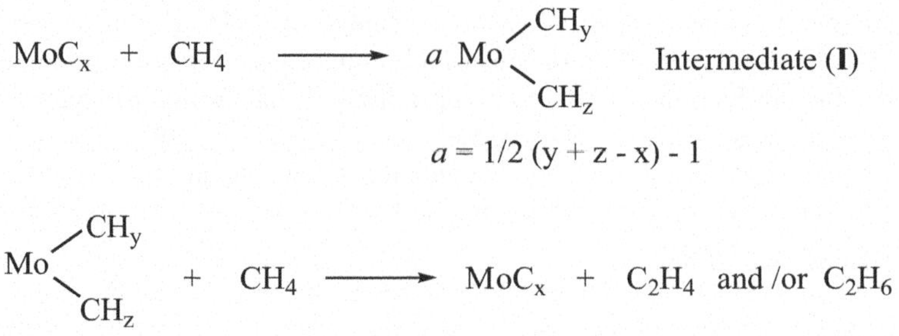

(6.16)

(2) C_2H_6 formation via molybdenum–carbene intermediates

Xu et al. [67] and Pierella et al. [104] proposed the molybdenum–carbene mechanism. In this mechanism, CH_4 reacts with Mo^{6+} on H-ZSM-5 to produce CH_3^+ (a methoxy species on the Brønsted acid sites of the zeolite) and $[Mo–H]^{5+}$. These are subsequently converted into a molybdenum–carbene species, (Mo = CH_2), which further reacts with CH_4 to produce C_2H_6 as shown below.

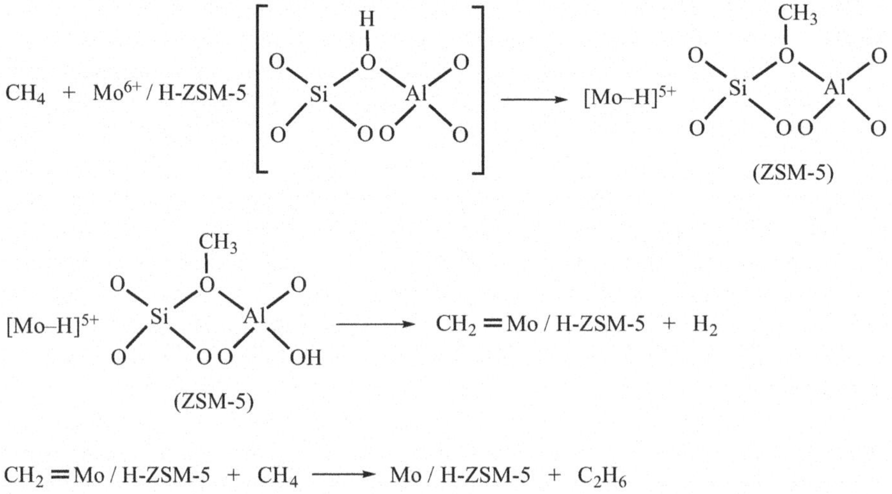

(6.17)

When a Fe/H^+-exchanged zeolite was used as a catalyst for the MDA reaction, carburized Fe species were produced and gave C2 hydrocarbons (C_2H_4 and C_2H_6) together with aromatic hydrocarbons [34]. Since Fe ions usually form Fisher-type metal–carbene complexes, Tan proposed that a Fe = CH_2 complex was formed on the iron carbide formed during the induction period and then catalyzed the production of C_2H_6 from methane.

(3) Coupling of methyl radicals (•CH_3) to produce C_2H_6

The methyl radical (•CH_3) is another reaction intermediate that can form on the surface of Mo species such as MoO_3. This species can then undergo oxidative coupling to form C_2H_6, as discussed in detail in Chap. 5.

$$\begin{aligned} CH_4 + MoO_x &\longrightarrow \bullet CH_3 + MoO_{(x-1)}OH \\ \bullet CH_3 + \bullet CH_3 &\longrightarrow C_2H_6 \end{aligned} \tag{6.18}$$

In cases (1)–(3), at high reaction temperatures of approximately 900 K, C_2H_6 is easily dehydrogenated to C_2H_4 and H_2 in the absence of a catalyst.

6.7.2 Reaction Mechanism and Role of the Catalyst in the Formation of Benzene from Ethylene

As discussed above, free Mo compounds such as Mo_2C, which were not supported (and/or modified) on zeolites, give C_2 hydrocarbons but do not produce aromatic hydrocarbons, and their catalytic activities are very low compared to those of Mo/H$^+$-exchanged zeolite catalysts. On the other hand, Mo/H$^+$-exchanged zeolite catalysts often yield not only C2 hydrocarbons, but also aromatic hydrocarbons such as benzene. Furthermore, their catalytic activities are greatly enhanced. Thus, to produce aromatic hydrocarbons catalytically, it is crucial to modify H$^+$-exchanged zeolites with Mo species. These results suggest that the Brønsted acid sites of the zeolite may play a role not only in the generation of the catalytically active sites of the Mo species, but also in the catalysis of C–C bond formation to produce benzene.

To determine the role of the Brønsted acid sites, Sheng et al. investigated the chemical properties of Mo_2C/H-[Al]-ZSM-5 and Mo_2C/H-[B]-ZSM-5 catalysts [105]. Here H-[Al]-ZSM-5 represents typical aluminosilicate H-ZSM-5, while H-[B]-ZSM-5 is borosilicate zeolite, in which the zeolite lattice contains B atoms rather than Al atoms. The acid strength of H-[B]-ZSM-5 is much weaker than that of H-[Al]-ZSM-5. For example, H-[Al]-ZSM-5 can convert ethane to higher hydrocarbons such as benzene around 800 K, however, H-[B]-ZSM-5 cannot achieve this conversion under any conditions. The different acid strength of the acidic O–H groups (Brønsted acid sites) is reflected in their different IR stretching bands, which were observed at 3610 cm^{-1} and 3709 cm^{-1} for H-[Al]-ZSM-5 and H-[B]-ZSM-5, respectively [105].

As discussed previously, when Mo/H-[Al]-ZSM-5 was used as the catalyst, C_2H_4 and benzene were produced simultaneously after an induction period. This result indicates that the initial reaction product C_2H_4 is quickly converted to benzene over Mo_2C/H-[Al]-ZSM-5. C_2 hydrocarbons such as C_2H_4 were also demonstrated to be produced on the Mo species rather than the Brønsted acid sites. For example, free MoO_3 was able to produce C_2H_6, but not aromatic hydrocarbons. To further examine the catalytic performance of Mo_2C/H-[Al]-ZSM-5 and Mo_2C/H-[B]-ZSM-5, Sheng et al. also carried out the conversion of methane at 773 K [105]. At this reaction temperature, the equilibrium methane conversion was 2.4%. In both catalysts, the conversion of CH_4 was the same, 0.8%. As shown in Table 6.6, however,

Table 6.6 Methane activation selectivity using Mo_2C/H-[Al]-ZSM-5 and Mo_2C/H-[B]-ZSM-5 at approximately 1% methane conversion at 923 K in a flow reactor

Catalyst	Product Selectivity/%	
	C_2H_4	Benzene
Mo_2C/H-[Al]-ZSM-5	5	91
Mo_2C/H-[B]-ZSM-5	90	7

Reported in ref. [105]
Catalyst: 50 mg, CH_4/He: 5/95

Mo_2C/H-[Al]-ZSM-5 selectively produced benzene, while Mo_2C/H-[B]-ZSM-5 produced C_2H_4. These results indicate that the strongly acidic Brønsted acid sites (acidic protons) catalyze the conversion of C_2H_4 to benzene (the oligomerization of C_2H_4), while the Mo species catalyze the formation of C_2 hydrocarbons from methane. Thus, the MDA reaction proceeds via bifunctional catalysis.

Liu et al. also reported the bifunctional catalytic activity of Mo/H-ZSM-5 in the MDA reaction [21]. They demonstrated a relationship between the rate of benzene formation and the amount of Brønsted acid sites. A close correlation between the amount of Brønsted acid sites and the benzene formation rate was observed. However, the amount of Lewis acid sites was not correlated to the rate of benzene formation.

6.8 Various Attempts to Improve the Production of Aromatic Hydrocarbons by Mo/H-ZSM-5

In the MDA reaction, which is usually carried out using a pure methane feed and a Mo/H-ZSM-5 zeolite catalyst, the conversion of methane and rate of benzene formation often decreases after a few hours because of significant coke formation. Various strategies have been attempted to suppress the deactivation of the catalyst and improve its catalytic performance, as described in the sections below.

6.8.1 Promotion of the Catalytic Performance of Mo/H⁺-Exchanged Zeolite Catalysts Using Various Additives

Various metals have been added to Mo/H^+-exchanged zeolites to improve their catalytic activity, benzene selectivity, and catalyst stability. For example, a Cu(II)-ion exchanged H-ZSM-5 (Cu/H-ZSM-5) was prepared by exchanging 37.8% of the H^+ in H-ZSM-5 with Cu^{2+} and then mechanically mixing the Cu^{2+}-exchanged zeolite with MoO_3 (3 wt%). This mixture was then calcined at 773 K for 4 h to produce (Mo (3 wt%)/CuH-ZSM-5). The Mo (3 wt%)/CuH-ZSM-5 exhibited higher catalytic activity than Mo(3 wt%)/H-ZSM-5, with the maximum methane conversion increasing from 7.4 to 10.1% [52]. The selectivity towards benzene also increased from 92.7 to 94.8%. Furthermore, the deactivation of the catalyst was suppressed because of the resulting change in the character of the carbonaceous deposits. Zeolite dealumination was also suppressed in the presence of the Cu^{2+} ions, which increased the concentration of Mo^{5+} in Mo/CuH-ZSM-5 compared to that in Mo/H-ZSM-5.

The addition of Ru [61], Co [47], W [62], and Zr [62] to H-ZSM-5 has also been found to result in improved catalytic performance. For example, the improved catalytic properties of Ru-modified Mo/H-ZSM-5 were attributed to its decreased concentration of Brønsted acid sites with strong acid strength and increased concentration of sites with weak or medium acid strength, as well as the easier reduction of the initially formed molybdenum oxides. Pt-modified Mo/H-ZSM-5 showed increased catalyst stability due to the suppression of carbonaceous deposits during the MDA reaction [106]. Furthermore, Zr and La modification also decreased the carbonization rate of W/H-ZSM-5 catalysts [38]. W/H-ZSM-5 was prepared by dissolving $(NH_4)_2WO_4$ in water and adding a small amount of H_2SO_4 to adjust the pH value of the solution to 2–3, followed by drying at 383 K for 2 h and calcination at 773 K for 4 h. Subsequently, Zn ($ZnSO_4$) or La ($La(NO_3)_3$) was impregnated into the W-H_2SO_4/H-ZSM-5 in NH_3 solution, and the samples were dried at 383 K for 2 h and finally calcined at 673 K for 4 h. When W (2.5 wt%)–Zn (1.5 wt%)–H_2SO_4/ZSM-5 was used as a catalyst at 1123 K, 0.1 MPa, and a gas hourly space velocity (GHSV) of 1500 mL g^{-1} h^{-1}, the methane conversion was 23% with ~96% selectivity towards benzene, while only 0.02% coke relative to the catalyst weight was formed after 3 h.

To reduce the formation of carbonaceous residues, dealumination of the parent zeolite has been examined [107–109]. For example, the selectivity of coke formation decreased from 37.9% over Mo/H-ZSM-5 to 18.8% over dealuminated Mo/H-ZSM-5; the methane conversion was maintained at ~11% while the selectivity towards benzene increased from 46.1 to 63.5% [107].

6.8.2 Suppression of Brønsted Acid Sites on the External Surface of H^+-Exchanged Zeolites

To suppress coke formation, silanation treatment was used to decrease the amount of Brønsted acid sites on the outer (external) surface of the parent zeolite H-ZSM-5; this method had almost no effect on the acidic O–H groups (Brønsted acid sites) located in the zeolite channels [110]. For example, 3-aminopropyl triethoxysilane was used to silanate the parent H-ZSM-5 zeolite. The addition of the optimum amount (0.5 wt% as SiO_2) of the silanizing agent to H-ZSM-5 relative to the final weight of Mo/H-ZSM-5 (0.5 wt% SiO_2/6 wt% Mo/H-ZSM-5) tuned the apertures of the external micropores of H-ZSM-5 to a diameter of ~5 Å, which was similar to that of MCM-22. The SiO_2(0.5 wt%)/Mo(6 wt%)/H-ZSM-5 exhibited an improved benzene selectivity of ~90% in the MDA reaction because of the suppression of both naphthalene and coke formation.

6.8.3 Effect of the Addition of Small Amounts of CO and CO_2 to the Methane Feed

Adding small amounts of CO and CO_2 to the methane reactant feed is another approach that can be used to improve the catalytic performance of modified H-ZSM-5. These compounds have been found to promote the formation of benzene and improve the stability of Mo/H-ZSM-5 catalysts [111–114]. For example, Ohnishi et al. examined the effect of the addition of CO and CO_2 on the catalytic performances of H-ZSM-5 zeolite catalysts modified with various metal species (Mo, Re, and Fe) [30, 111, 112]. When Fe-modified Mo (3 wt%)/H-ZSM-5 (Fe/Mo (3 wt%)/H-ZSM-5) and Co-modified Mo (3 wt%)/H-ZSM-5 (Co/Mo (3 wt%)/H-ZSM-5) were used as catalysts for the conversion of methane containing 1.8% (v/v) CO at 973 K, benzene was produced at a rate of ~1000 mmol g^{-1} s^{-1} on a carbon basis. Catalytic stability was observed for >100 h over both catalysts because of the minimization of the coke formation (<20%). Similarly, the catalytic stability of 3 wt% Mo/H-ZSM-5 was also improved by adding CO_2 to the methane feed gas instead of CO at the same reaction temperature [30].

The CO_2 reforming reaction and reverse Boudouard reaction are shown below as reaction (6.19) and (6.20).

$$CO_2 + CH_4 \rightleftharpoons 2\,CO + 2\,H_2 \tag{6.19}$$

$$CO_2 + C \rightleftharpoons 2\,CO \tag{6.20}$$

As shown in reaction (6.19), when CO_2 is added to methane, it is converted to CO, therefore, the effect of CO_2 is analogous to that of CO. Furthermore, Ohnishi et al. carried out a temperature-programed oxidation (TPO) reaction to measure the amount of coke deposited on Mo/H-ZSM-5 and Fe/Co-modified Mo/H-ZSM-5 catalysts [111]. This TPO experiment showed that in contrast to CO, the addition of CO_2 to the methane feed reduced not only the inert coke, but also the reactive coke, which was converted to aromatics such as benzene and naphthalene. ^{13}CO isotopic labeling tracer studies also showed that the ^{13}C from ^{13}CO was easily incorporated into the methane and products such as benzene and ethylene. On the basis of these results, the authors speculated that the role of CO addition to methane feed was based on the formation of minute amounts of CO_2 and C by the Boudouard reaction, in which C represents hydrogenated carbon species (CH_x), resulting in improved catalyst stability due to the suppression of coke formation on the catalyst surface.

However, Liu et al. reported that the CH_4 conversion rates were not influenced by the addition of CO and CO_2 as co-reactants [115]. They explained that CO_2 led to the oxidation of the MoC_x on H-ZSM-5, but the CH_x reaction intermediates formed in the rate-determining C–H bond activation step increasingly formed CO rather than hydrocarbons.

6.8.4 *Effect of the Addition of Small Amounts of O_2, H_2, and H_2O*

The presence of very small amounts of O_2 [116], H_2 [117], and H_2O [118] have been found to have a positive effect on the MDA reaction. The critical oxygen concentration in the feed, which should not be exceeded to avoid the deep oxidation of methane, has been found to increase with reaction temperature. At 973 K and 1073 K, the O_2/CH_4 molar ratio should not exceed 6.5×10^{-3} and 2.4×10^{-2}, respectively. The addition of small amounts of H_2O (1.7–2.2%) to the reactant improved the catalytic activity and stability of Mo (6 wt%)/H-ZSM-5 catalysts.

6.9 Summary of the Catalytic Features of Mo/H^+-Exchanged Zeolite Catalysts for MDA Reaction

The MDA reaction can potentially afford both high-value aromatic compounds, as well as pure hydrogen simultaneously. The most important features of Mo/H^+-exchanged zeolite catalysts for the MDA reaction can be summarized as follows:

(1) Mo/H^+-exchanged zeolites are bifunctional catalysts. Thus, the Mo species should be modified onto an H^+-exchanged zeolite to achieve good catalytic activity.
(2) There is an induction period in the early stage of the reaction, during which the active Mo species, such as MoO_x, are reduced into molybdenum carbonaceous species with CH_4.
(3) Various interactions, such as the interaction between the Mo species and the Brønsted acid sites on H^+-exchanged zeolite, may be involved in the catalysis and have a crucial influence on the catalytic performance.
(4) Carbonaceous compounds are formed during the MDA reaction and lead to severe deactivation of the Mo/H^+-exchanged zeolite catalysts.

However, catalyst deactivation remains a serious problem, and the reaction inevitably produces carbon deposits on the catalyst, which precludes industrial applications. To increase catalyst lifetime and optimize the condition of continuous catalyst regeneration, one can resort to computational studies [84] and perform further catalyst characterization [119]. A few studies also employ DFT methods to follow the reaction mechanism and carbon deposition. Developments of these studies possibly bring proper technological implementation that will make the synthesis of benzene and hydrogen by MDA economically profitable.

References

1. Guo X, Fang G, Li G, Ma H, Fan H, Yu L, Ma C, Wu X, Deng D, Wei M, Tan D, Si R, Zhang S, Li J, Sun L, Tang Z, Pan X, Bao X (2014) Direct non oxidative conversion of methane to ethylene, aromatics and hydrogen. Science 344:616–619
2. Bragin O, Vasina T, Isakov YI, Prebrazhenskii A, Prebrazhenskii A, Palishkina N, Minachev KM (1982) Catalytic aromatization of methane and ethane. Russ Chem Bull 31:847
3. Shepelev SS, Inoe KG (1983) Catalytic properties of zeolite with structure and chemical compositions in preparation of hydrocarbons from methane. React Kinet Catal Lett 23:319–322
4. Anderson JR, Tsai P (1985) Oxidation of methane over H-ZSM-5 and other catalysts. Appl Catal 191:141–152
5. Wang L, Tao L, Xie M, Xu G (1993) Dehydrogenation of aromatization of methane under non-oxidative conditions. Catal Lett 21:35–41
6. Xu Y, Lin L (1999) Recent advances in methane dehydroaromatization over transition metal ion-modified zeolite catalysts under non-oxidative conditions. Appl Catal A: General 188:53–67
7. Ismagilov ZR, Matus EV, Tsikoza LT (2008) Direct conversion of methane on Mo/ZSM-5 catalysts to produce benzene and hydrogen: achievements and perspectives. Energy Environ Sci 1:526–541
8. Ma S, Guo X, Zhao L, Scott S, Bao X (2013) Recent progress in methane dehydroaromatization: from laboratory curiosities to promising technology. J Energy Chem 22:1–20
9. Mamonov NA, Fadeeve EV, Grigoriev DA, Mikhailov MN, Kustov LM, Alkhimov SA (2013) Metal/Zeolite catalysts of methane dehydroaromatization. Russ Chem Rev 82:567–585
10. Spivey JJ, Hutchings G (2014) Catalytic aromatization of methane. Chem Soc Rev 43:792–803
11. Vollmer I, Yurulina I, Kapteijn F, Gascon J (2018) Progress in developing a structure-activity relationship for the direct aromatization of methane. ChemCatChem 10:1–15
12. Xu Y, Lu J, Wang J, Suzuki Y, Zhang ZG (2011) The catalytic stability of Mo/HZSM-5 in methane dehydroaromatization at sever and periodic CH_4-H_2 switch operation conditions. Chem Eng J 168:390–402
13. Brady C, Murphy B, Xu B (2017) Enhanced methane dehydroaromatization *via* coupling with chemical looping. ACS Catal 7:3924–3928
14. Cao Z, Jiang H, Luo H, Baumann S, Meulenberg WA, Assmann J, Mleczko L, Liu Y, Caro J (2013) Natural gas to fuels and chemicals: improved methane aromatization in as oxygen-permeable membrane reactor. Angew Chem Int Ed 52:13794–13797
15. Yuste-Tirados I, Maleró-Fjeld H, Vestre PK, Coors WG, Martinez A, Norby T, Serra JM, Kjólseth C (2016) Direct conversion of methane to aromatics in catalytic co-ionic membrane reactor. Science 353:563–566
16. Xue J, Chen Y, Wei Y, Feldhoft A, Wang H, Caro J (2016) Gas to liquids: natural gas conversion to aromatic fuels and chemicals in hydrogen permeable ceramic hollow fiber membrane reactor. ASC Catal 6:2448–2451
17. Xu Y, Bao X, Lin L (2003) Direct conversion of methane under nonoxidative conditions. J Catal 216:386–395
18. Okolie C, Lyu Y, Kovarik LV, Stavitski E, Sievers C (2018) Coupling of methane to ethane, ethylene and aromatics over nickel on ceria-zirconia at low temperatures. ChemCatChem 10:2700–2708
19. Shu Y, Ma D, Xu L, Xu Y, Bao X (2000) Methane dehydro-aromatization over Mo/MCM-22 catalysts: a highly selective catalysts for the formation of benzene. Catal Lett 70:67–73
20. Li Y, Su L, Wang H, Li H, Shen W, Bao X, Xu Y (2003) Combined single-pass conversion of methane *via* oxidative coupling and dehydroaromatization. Catal Lett 89:275–279
21. Liu S, Wang L, Ohnishi R, Ichikawa M (1999) Bifunctional catalysis of Mo/HZSM-5 in the dehydroaromatization of methane to benzene and naphthalene XAFS/TG/DTA/MAS/FTIR characterization and supporting effects. J Catal 181:175–188

22. Shu Y, Ma D, Tian B, Xu Y, Bao X (1999) Methane dehydrogenation and aromatization over Mo/HZRP-1 catalyst in the absence of oxygen, I. Catalytic evaluation including formation of coke over Mo/HZRP-1 catalyst. Chin J Catal 20:279–284
23. Liu H, Kan Q (2017) Improved performance of hierarchical porous Mo/H-IM-5. Appl Petrochem Res 7:97–105
24. Martìnez A, Peris E, Sastre G (2005) Dehydroaromatization of methane under non-oxidative conditions over bifunctional Mo/ITQ-2 catalysts. Catal Today 107–108:676–684
25. Wang ST, Xu Y, Liu W, Wang L, Guo X (1996) Methane activation without using oxidants over supported Mo catalysts. Appl Catal A: General 136:7–17
26. Ohnishi R, Xu X, Issoh K, Ichikawa M (2001) Catalytic dehydroaromatization of methane towards benzene and naphthalene on zeolite-supported Re and Mo-Templating roles of micropore and novel mechanism. Stud Surf Sci Catal 136:393–398
27. Shu Y, Ma D, Bao X, Xu Y (2000) Methane dehydro-aromatization over a Mo/Phosphoric rare earth oxide-containing penta-sil type zeolite in the absence of oxygen. Catal Lett 66:161–167
28. Zhang CL, Li S, Yuan Y, Zhang WX, Wu TH, Lin LW (1998) Aromatization of methane in the absence of oxygen over Mo-based catalysts supported on different types of zeolites. Catal Lett 56:207–213
29. Wu Y, Emdadi L, Schulman E, Shu T, Tran DT, Wang X, Liu D (2018) Overgrowth of lamellar silicate-1 on MFI and BEA zeolites and its consequence on non-oxidative methane aromatization reaction. Microporous Mesoporous Mater 263:1–10
30. Liu S, Wang L, Ohnishi R, Ichikawa M (2000) Bifunctional catalysis of Mo/HZSM-5 in the dehydroaromatization of methane with CO/CO_2 to benzene and naphthalene. Kinet Catal 41:132–144
31. Sriglan A, Savasci ÖT, Erdem-Senatalar A, Tuel A, Sapaly G, Taârit YB (2007) The effect of support morphology on the activity of HZSM-5 supported molybdenum catalysts fro aromatization of methane. J Catal 246:35–39
32. Mátus EV, Ismagilov IZ, Sukhova OB, Zaukouskii IV, Tsikoza LT, Ismagilov ZR (2007) Study of methane dehydroaromatization on impregnated Mo/ZSM-5 catalysts and characterization of nanostructured molybdenum phase and carbonaceous deposit. Ind Eng Chem Res 46:4063–4074
33. Weckhuysen BM, Wang D, Rosynek MP, Lunsford JH (1998) Conversion of methane to benzene over transition metal ion ZSM-5 zeolites. I. Catalytic characterization. J Catal 175:338–346
34. Tan P (2016) Active phase, catalytic activity, and induction period of Fe/Zeolite material in nonoxidative aromatization of methane. J Catal 338:21–29
35. Li Y, Veser G (2016) The nature of the selective species in Fe-HZSM-5 for non-oxidative methane dehydroaromatization. Catal Sci Technol 6:5440–5452
36. Weckhuysen BM, Wang D, Rosynek MP, Lunsford JH (1997) Catalytic conversion of methane into aromatic hydrocarbons over iron oxide loaded ZSM-5 zeolite. Angew Chem Int Ed Engl 36:2374–2376
37. Ding W, Meitner GD, Marler DO, Igleshia E (2001) Synthesis, structural characterization, and catalytic properties of tungsten-exchanged H-ZSM-5. J Phys Chem B 105:3928–3936
38. Zeng JL, Xiong ZT, Zhang HB, Lin GD, Tsai KR (1998) Nonoxidative dehydrogenation and aromatization of methane over W/HZSM-5 based catalyst. Catal Lett 53:119–124
39. Ohnishi R, Ichikawa M (2002) Effect of CO_2/CO addition in dehydroaromatization of methane on zeolite-supported Mo and Re catalysts towards hydrogen, benzene, and naphthalene. Catal Surv Jpn 7:103–110
40. Wang L, Ohnishi R, Ichikawa M (2000) Selective dehydroaromatization of methane towards benzene on Re/HZSM-5 catalysts and effects of CO/CO_2 addition. J Catal 190:276–283
41. Su L, Ma D, Kiu X, Xu Y, Bao X (2002) Methane aromatization on Re/HZSM-5 system under non-oxygen condition. Chin J Catal 23:41–45
42. Nahreen S, Praserthdam S, Beltran SP, Ballbuena PB, Adhikari S, Gupta RB (2016) Catalytic upgrading of methane to higher hydrocarbons in a nonoxidative chemical conversion. Energy Fuels 30:2584–2593

43. Abdelsayed V, Smoith MW, Shekhuwat D (2015) Investigation of the stability of Zn-based HZSM-5 catalysts for methane dehydroaromatization. Appl Catal A: General 505:365–374
44. Zhang H, Lou H, Zheng X (2004) Catalytic conversion of methane at low temperature over 6% Zn/HZSM-5 zeolite. Chin J Catal 25:255–256
45. Gerceker D, Motagamwala AH, Rivera-Dones KR, Miller JB, Huber GW, Mavrikakis M, Dumesic JA (2017) Methane conversion to ethylene and aromatics on PtSn catalysts. ASC Catal 7:2088–2100
46. Tan PL, Au CT, Li SY (2006) Methane dehydrogenation and aromatization over 4wt% Mn/HZSM-5 in the absence of an oxidant. Catal Lett 112:239–245
47. Burna SJ, Hargreaves JS, Pal P, Parida KM, Parija S (2006) The effect of dopants on the activity of MoO_3/ZSM-5 catalysts for the dehydroaromatization of methane. Catal Today 114:383–387
48. Liu S, Dong Q, Ohnishi R, Ichikawa M (1997) Remarkable non-oxidative conversion of methane to naphthalene and benzene on Co and Fe modified Mo/HZSM-5 catalysts. Chem Commun 1455–1456
49. Cheng X, Yan P, Zhang X, Yang F, Pai C, Li D, Ma XX (2017) Enhanced methane dehydroaromatization in the presence of CO_2 over Fe- and Mg-modified Mo/ZSM-5. J Mol Catal 437:114–120
50. Vosmerikov AV, Zaikovskii VI, Korobitsyna LL, Echevskii GV, Barbashin YE, Zhuravkov ZP (2009) Nonoxidative conversion methane into aromatic hydrocarbons on Ni-Mo/ZSM-5 catalysts. Kinet Catal 50:725–733
51. Liu HM, Tian BL, Li YG, Bao XH, Xu YD (2003) Effect of Co additive on the formation of coke deposits in methane dehydro-aromatization over Mo/HZSM-5 catalyst. Chin J Catal 24:123–128
52. Li S, Zhang C, Kan Q, Wang D, Wu T, Lin L (1999) The function of Cu(II) ions in the Mo/Cu/H-ZSM-5 catalyst for methane conversion under non-oxidative condition. Appl Catal A: General 187:199–206
53. Qi S, Yang B (2004) Methane aromatization using Mo-based catalysts prepared by microwave heating. Catal Today 98:639–645
54. Zhang YP, Wang DJ, Fei JH, Zheng XM (2001) Methane aromatization under O_2-free conditions on zinc modified Mo/HZSM-5 catalyst. React Kinet Catal Lett 74:151–161
55. Liu BS, Yang Y, Sayari A (2001) Non-oxidative dehydroaromatization of methane over Ga-promoted Mo/HZSM-5 based catalysts. Appl Catal A: General 214:95–102
56. Liu BS, Jiang L, Sun H, Au CT (2007) XPS, XAFS and TG/DTA characterization of deposited carbon in methane dehydroaromatization over Ga-Mo/ZSM-5 catalyst. Appl Surf Sci 253:5092–5100
57. Tshabalala TE, Coville NJ, Scurrell MS (2014) Dehydroaromatization of methane over doped Pt/Mo/H-ZSM-5 zeolite catalysts: the promotion effect of tin. Appl Catal A: General 485:238–244
58. Kojima R, Kikuchi S, Ma H, Bai J, Ichikawa M (2006) Promotion effects of Pt and Rh on catalytic performance of Mo/HZSM-5 and Mo/HMCM-22 in selective methane-to-benzene reaction. Catal Lett 110:15–21
59. Wang ST, Xu Y, Wang L, Liu S, Li G, Xie M, Guo X (1996) Methane and ethane activation without adding oxygen: promotional effect of W in Mo-W/H-ZSM-5. Catal Lett 38:39–43
60. Zhang Y, Wang W, Fei J, Zheng X (2002) Effect of Cr addition on the methane aromatization performance of the Mo/HZSM-5. Aust J Chem 55:531–534
61. Shu Y, Xu Y, Wang ST, Wang L, Guo X (1997) Promotion effect of Ru on the dehydrogenation and aromatization of methane in the absence of oxygen over Mo/HZSM-5 catalysts. J Catal 170:11–19
62. Wang L, Xu Y, Wong ST, Cui W, Guo X (1997) Activity and stability enhancement of Mo/HZSM-5 based catalysts for methane non-oxidative transformation to aromatics and C_2 hydrocarbons: effect of additives and pretreatment conditions. Appl Catal A: General 152:173–182

63. Ngobeni MW, Carley AF, Scurrell MS, Nicolaides CP (2009) The effect of boron and silver on the oxygen-free conversion of methane over Mo/H-ZSM-5 catalysts. J Mol Catal A: Chemical 305:40–46
64. Zhao J, Wang X, Zhang T, Tang J, Ren L, Li L, Lin L (2002) Promotion effect of Indium on dehydroaromatization of methane over Mo/HZSM-5 catalyst without adding oxygen. Chin J Catal 23:197–198
65. Chen L, Lin L, Xu Z, Li X, Zhang T (1995) Dehydro-oligomerization of methane to ethylene and aromatics over Molybdenum/HZSM-5 catalyst. J Catal 157:190–200
66. Wang D, Lunsford JH, Rosynek MP (1996) Catalytic conversion of methane to benzene over Mo/ZSM-5. Top Catal 3:289–297
67. Xu Y, Liu S, Wang L, Xie M, Guo X (1995) Methane activation without using oxidant over Mo/HZSM-5 zeolite catalysts. Catal Lett 30:135–149
68. Tan PL, Au CT, Lai SY (2007) Effect of acidify ion and basify ion of impregnating solution on the performance of Mo/HZSM-5 in methane aromaticzation. Appl Catal A: General 324:36–41
69. Ma D, Lu Y, Su L, Xu Z, Tian Z, Xu Y, Lin L, Bao X (2002) Remarkable improvement on the methane aromatization reaction: a highly selective and coking-resistant catalyst. J Phys Chem B 106:8524–8530
70. Xu Y, Shu Y, Liu S, Huang J, Guo X (1995) Interaction between ammonium heptamolybdate and NH_4ZSM-5 zeolite: the location of Mo species and acidity of Mo/HZSM-5. Catal Lett 35:233–243
71. Borry RW III, Kim YH, Huffsmith A, Reimer JA, Iglesia E (1999) Structure and density of Mo and acid sites in Mo-exchanged H-ZSM-5 catalysts for nonoxidative methane conversion. J Phys Chem B 103:5787–5796
72. Li W, Meitzner GD, Borry RW III, Iglesia E (2000) Raman and X-ray absorption studies of Mo species in Mo/H-ZSM-5 catalyst for non-oxidative CH_4 reaction. J Catal 191:373–383
73. Kim YH, Borry RW III, Iglesia E (2000) Genesis of methane activation sites in Mo-exchanged H-ZSM-5 catalysts. Microporous Mesoporous Mater 35–36:495–509
74. Minming H, Howe H (1987) Characterization of MoY zeolites prepared by aqueous ion exchange. J Catal 108:283–293
75. Vahel A, Huebener S, Eichler B (1995) Thermochromatographic studies of oxide and hydroxide species of molybdenum-model experiments with respect to the physico-chemical characterization of element 106. Radiochi Acta 69:233–239
76. Giordano N, Bart JC, Vaghi A, Castellan A, Maritiotti G (1975) Structure and catalytic activity of MoO3 Al2O3 systems I. Solid-state properties of oxidized catalysts. J Catal 36:81–92
77. Xu Y, Liu W, Wang ST, Wang L, Guo X (1996) Dehydrogenation and aromatization of methane in the absence of oxygen on Mo/HZSM-5 catalysts before and after NH_4OH extraction. Catal Lett 40:207–214
78. Gao J, Zheng Y, Jehng JM, Tang Y, Wachs IE, Podkolzin SG (2015) Identification of molybdenum oxide nanostructures on zeolite for natural gas conversion. Science 348:686–690
79. Rcie MJ, Chakraborty AK, Bell AT (1999) Al next nearest neighbor, ring occupant and proximity statistics in ZSM-5. J Catal 186:222–227
80. Standicka K, Haber J, Kozlowski R (1977) The crystal structure of magnesium dimolybdate. Acta Cryst B33:3859–3862
81. Savinelli RO, Scott SL (2010) Wavelet transform EXAFS analysis of mono- and dimolybdate model compounds and a Mo/H-ZSM-5 dehydroaromatization catalyst. Phys Chem Chem Phys 12:5660–5667
82. Bars PO, Le JY, Grandjean MD (1977) Etude de chromates, molybdates et tungstates hydrate's I. Etude structurale de MgMoO4 5H2O. Acta Crst B33:1155–1157
83. Amberg M, Günter JR, Schmalle H (1988) Preparation, crystal structure, and luminescence of magnesium molybdate and tungstate monohydrate, MgMoO4 · H2O and MgWO4 · H_2O. J Solid State Chem 77:162–169
84. Zhou D, Ma D, Liu X, Bao X (2001) Study with density functional theory method on methane dehydro-aromatization over Mo/HZSM-5 catalysts I: optimization of active Mo species bonded to ZSM-5 zeolite. J Chem Phys 114:9125–9129

85. Goodman BR, Hass KC, Schneider WF, Adams JB (2000) Statistical analysis of Al distributions and metal ion pairing probabilities in zeolites. Catal Lett 68:85–93
86. Tessonnier JP, Louis B, Walspurger S, Sommer J, Ledoux MJ, Pham-Huu C (2006) Quantitative measurement of the Brönsted acid sites in solid acids: toward a single-site design of Mo-modified ZSM-5 zeolite. J Phys Chem B 110:10390–10395
87. Solymosi F, Erdöhelyi A, Szöke A (1995) Dehydrogenation of methane on supported molybdenum oxide. Formation of benzene from methane. Catal Lett 32:43–53
88. Wang D, Lunsford JH, Rosynek MP (1997) Characterization of a Mo/ZSM-5 catalyst for the conversion of methane to benzene. J Catal 169:347–358
89. Eeckhuysen BM, Wang D, Rosynek MP, Lunsford JH (1998) Conversion of methane to benzene over transition metal ion ZSM-5 zeolites. II Catalyst characterization by X-ray photoelectron spectroscopy. J Catal 175:347–351
90. Solymois F, Cserényi J, Szöke A, Bánsági T, Oszkó A (1997) Aromatization of methane over supported and unsupported Mo-based catalysts. J Catal 165:150–161
91. Jiang H, Wang L, Cui W, Xu Y (1999) Study on the induction period of methane aromatization over Mo/HZSM-5: partial reduction of Mo species and formation of carbonaceous deposit. Catal Lett 57:95–102
92. Ma D, Shu Y, Cheng M, Xu Y, Bao X (2000) On the induction period of methane aromatization over Mo-based catalysts. J Catal 194:105–114
93. Ding W, Li S, Meitzner GD, Iglesia E (2001) Methane conversion to aromatics on Mo/H-ZSM-5: structure of molybdenum species in working catalysts. J Phys Chem B 105:506–513
94. Li B, Li S, Li N, Chen H, Zhang W, Bao X, Lin B (2006) Structure and acidity of Mo/ZSM-5 synthesized by solid state reaction for methane dehydrogenateion and aromatization. Microporous Mesoporous Mater 88:244–253
95. Liu H, Bao X, Xu Y (2006) Methane dehydroaromatization under nonoxidative conditions over Mo/HZSM-5 catalysts: identification and preparation of the Mo active species. J Catal 239:441–450
96. Vollmer I, van der Linden B, Ould-Chikh S, Aguilar-Tapia A, Yarulina I, Abou-Hamad E, Sneider YG, Suarez AIO, Hazemann JL, Kaptteijin F, Gascon G (2018) On the dynamic nature of Mo sites for methane dehydroaromatization. Chem Sci 9:4801–4807
97. Budde PK, Singh AK, Upadhyayula S (2018) Non-oxidative methane dehydroaromatization reaction over highly active α-MoC1-x ZSM-5 derived from pretreatment. J Chem Sci (Berlin Germany) 130:1–6. https://doi.org/10.1007/s12039-018-1432-5
98. Lezcano-González I, Oord R, Rovezzi M, Glatzel P, Botchway SW, Weckhuysen BM, Beale AM (2016) Molybdenum speciation and its impact on catalytic activity during methane dehydroaromatization in zeolite ZSM-5 as revealed by operand X-ray method. Angew Chem Int Ed 55:5215–5219
99. Ma D, Shu Y, Bao X, Xu Y (2000) Methane dehydro-aromatization under nonoxidative conditions over Mo/HZSM-5 catalysts: ESR study of the Mo species on/in the ZSM-5 zeolite. J Catal 189:314–325
100. Solymois F, Szöke A, Cserényi J (1996) Conversion of methane to benzene over MO_2C and Mo_2C/ZSM-5 catalysts. Catal Lett 39:157–161
101. Liu H, Shen Bao WX, Xu Y (2005) Methane dehydroaromatization over Mo/H-ZSM-5 catalysts: the reactivity of MoC_X species formed from Mo_X associated and non-associated with Brönsted acid sites. Appl Catal A: General 295:79–88
102. Zhang JZ, Log MA, Howe RF (1998) Molybdenum ZSM-5 zeolite catalysts fir the conversion of methane to benzene. Catal Today 44:293–300
103. Liu H, Shen W, Bao X, Xu Y (2006) Identification of Mo active species for methane dehydro-aromatization over Mo/HZSM-5 catalyst in the absence of oxygen: ^1H MAS NMR and EPR investigations. J Mol Catal A Chemical 244:229–236
104. Pierella LB, Wang L, Anunziata OA (1997) Methane direct conversion to aromatic hydrocarbons at low reaction temperature. React Kinet Catal Lett 60:101–106
105. Sheng H, Schreiner EP, Zheng W, Loba RF (2018) Non-oxidative coupling of methane to ethylene using Mo_2C/[B]-ZSM-5. ChemPhysChem 19:504–511

106. Chen L, Lin L, Xu Z, Zhang T, Li X (1996) Promotional effect of Pt on non-oxidative methane transformation over Mo-HZSM-5 catalyst. Catal Lett 39:169–172
107. Lu Y, Ma D, Xu Z, Tian Z, Bao X, Lin L (2001) A high coking-resistance catalyst for methane aromatization. Chem Commun 2048–2049
108. Shu Y, Ohnishi R, Ichikawa M (2003) Improved methane dehyfrocondensation reaction on HMCM-22 and HZSM-5 supported rhenium and molybdenum catalysts. Appl Catal A: General 252:315–329
109. Ding X, Song Y, Lin W (2007) A new way to enhance the coke-resistance of Mo/HZSM-5 catalyst for methane dehydroaromatization. Catal Commun 8:539–542
110. Kikuchi S, Kojima R, Ma H, Bai J, Ichikawa M (2006) Studies on Mo/HZSM-5 catalysts modified by bulky aminoalkyl-substituted silyl compounds for the selective methane to benzene (MTB) reaction. J Catal 242:349–356
111. Ohnishi R, Liu S, Ding Q, Wang L, Ichikawa M (1999) Catalytic dehydration of methane with CO and CO_2 toward benzene and naphthalene on Mo/HZSM-5 and Fe/Co-modified Mo/HZSM-5. J Catal 182:92–103
112. Shu Y, Ohnishi R, Ichikawa M (2002) Pressurized dehydrocondensation of methane toward benzene and naphthalene on Mo/HZSM-5 catalyst: optimization of reaction parameters and promotion by CO_2 addition. J Catal 206:134–142
113. Bradford MCJ, Te M, Konduru M, Fuentes DX (2004) CH_4-C_2H_6-CO_2 conversion to aromatics over Mo/SiO_2/H-ZSM-5. Appl Catal A: General 266:55–66
114. Skutil K, Taniewski M (2006) Some technological aspects of methane aromatization (direct and *via* oxidative coupling). Fuel Process Technol 87:511–521
115. Liu Z, Nutt MA, Iglesia E (2002) The effects of CO_2, CO and H_2 co-reactants on methane reactions catalyzed by Mo/H-ZSM-5. Catal Lett 81:271–279
116. Yuan S, Li J, Hao Z, Feng Z, Xin Q, Ying P, Li C (1999) The effect of oxygen on the aromatization of methane over the MO/HZSM-5 catalyst. Catal Lett 63:73–77
117. Ma H, Ohnishi R, Ichikawa M (2003) Highly stable performance of methane dehydroaromatization on Mo/HZSM-5 catalyst with a small amounts of H_2 addition into methane feed. Catal Lett 89:143–146
118. Liu S, Ohnishi R, Ichikawa M (2003) Promotional role of water added to methane feed on catalytic performance in the methane dehydroaromatization reaction on Mo/HZSM-5 catalyst. J Catal 220:57–65
119. Zhang W, Xu S, Han X, Bao X (2012) *In situ* solid-state NMR for heterogeneous catalysis: a joint experimental and theoretical approach. Chem Soc Rev 41:192–210

Chapter 7
C–C Bond Formation via Carbocations in the Methane Conversion Under Non-oxidative Conditions

7.1 Introduction

Chapter 6 discussed the conversion of methane to aromatic hydrocarbons such as benzene, which is a thermodynamically unfavorable reaction under non-oxidative reaction conditions. However, Olah et al. reported that even under such circumstances, superacids such as $FSO_3H–SbF_5$ catalyze the poly-condensation of methane to produce higher molecular weight alkanes, such as ethane (C_2H_6) and propane (C_3H_8) at approximately 400 K [1]. They also demonstrated the reaction of alkenes such as C_2H_4 (ethylene) with methane to produce alkanes such as C_3H_8 via superacid catalysis. Therefore, superacid catalysts do not produce alkenes during the conversion of alkanes such as methane, but instead are selective towards carbon–carbon (C–C) bond formation at lower reaction temperatures than heterogeneous catalysts such as zeolites.

In 1982, Sommer et al. reported the alkylation of CH_4, C_2H_6, C_3H_8, and n-C_4H_{10} (n-butane) with C_2H_4 to produce various alkanes in an HF-containing TaF_5 solution [2]. Furthermore, in 1997 Choudhary et al. reported that an H-galloaluminosilicate zeolite could heterogeneously catalyze the reaction of CH_4 with lower alkenes such as C_2H_4 and C_3H_6 to produce mainly aromatic hydrocarbons around 800 K [3]. In 2002, Baba et al. have found that Ag^+-exchanged zeolites (Ag-zeolites) produce propylene (C_3H_6) and hydrogen (H_2) via the reaction of CH_4 with C_2H_4. Thus, in this reaction, propylene and hydrogen are produced, while superacid catalysts give C_3H_8 rather than C_3H_6 [4].

$$CH_4 + CH_2{=}CH_2 \longrightarrow CH_3CH{=}CH_2 + H_2 \qquad (7.1)$$

These results suggest that the reaction of methane with alkenes can be a useful way to produce higher molecular weight hydrocarbons (higher hydrocarbons), and thus, another method of CH_4 utilization to provide useful raw materials, in addition to the polycondensation of methane via the methane dehydroaromatization (MDA) reaction mentioned in Chapter 6. Furthermore, as discussed in Chapter 1, the reactions

T. Baba and A. Miyaji, *Catalysis and the Mechanism of Methane Conversion to Chemicals*,
https://doi.org/10.1007/978-981-15-4132-2_7

of CH_4 with other hydrocarbons (alkenes, alkanes, and aromatics) are thermodynamically more favorable than polycondensations of methane, such as MDA, under non-oxidative reaction conditions [5].

This chapter focuses on the activation of methane for the formation of C–C bonds via the reaction of methane with itself or various hydrocarbons such as ethylene and benzene. The mechanisms for the activation of methane and the subsequent C–C bond formation, along with the role of the catalysts in these reactions, will be the main topics of discussion.

7.2 Superacid Catalysts for the Activation of CH_4 and Other Lower Alkanes for C–C Bond Formation

Olah et al. reported the conversion of methane to higher molecular weight hydrocarbons (higher hydrocarbons) via the treatment of CH_4 either at 423 K in an autoclave with a ten-fold excess of 1:1 FSO_3H-SbF_5 solution or at 413 K under atmospheric pressure [1]. In the following sections, the catalytic properties of superacids and their role in the catalytic activation of C–H bonds in CH_4 and other alkanes are discussed.

7.2.1 Definition of Brønsted Superacids

As mentioned above, superacids have been shown to catalyze the conversion of methane into higher hydrocarbons. In this section, the chemical properties of superacids will first be discussed. A Brønsted superacid is defined as an acid whose acid strength is at least 100 times stronger than that of 100% sulfuric acid (H_2SO_4) [6, 7]. However, the acid strength referred to in this definition is arbitrary, as the concept of acid strength is only defined in relation to a reference base. Among the various methods to measure the acid strength, the Hammett acidity function (H_0) proposed by Hammett and Deyrop is widely used [8]. H_0 is defined as below:

$$H_0 = pK_{BH} + \log \frac{[BH^+]}{[BH]} \tag{7.2}$$

Here, pK_{BH} is the acid dissociation constant of BH^+, which is the conjugate acid of B.

$$B + H^+ \longrightarrow BH^+ \tag{7.3}$$

Thus, B is the basic form of the Hammett base and BH^+ is the acidic form. The acid strength of a superacid solution is usually determined using indicator molecules B with known pK_{BH} values. To determine the H_0 value of the solution, the concentration ratio of $[BH^+]/[B]$ (*i.e.*, the ionization ratio) is typically measured using spectroscopic means. The H_0 value is equal to the pK_{BH} value of BH^+ when half the B in the solution is protonated, *i.e.*, $[B] = [BH^+]$.

Lower values of H_0 correspond to greater acid strength. Thus, the acidity function H_0 describes the acidity of a solution, meaning that H_0 is not a property of individual molecules, such as H_2SO_4. Obviously, H_0 values are solvent-dependent. The Hammett acidity function is a logarithmic scale. For example, 100% H_2SO_4 has an H_0 of -11.9 and anhydrous HF has an H_0 of -11.0; subsequently, all Brønsted acids having an $H_0 < -12$ are considered to be superacids. The H_0 values of some typical acid solutions are as follows: 100% H_2SO_4: -11.94, CF_3SO_3H: -14.1, HSO_3F: -15.07, SbF_5 (10 mol%)–HSO_3F: -18.54, HSO_3F–SbF_5 (1:1): -23, HF–SbF_5: -28.

7.2.2 *Definitions of Carbocations as Key Reaction Intermediates*

In the superacid-catalyzed conversion of hydrocarbons including CH_4, carbocations are typically reaction key intermediates. Carbocations can be generated via several different routes using homogeneous and heterogeneous catalysts. Olah proposed that the concept of carbocations should encompass all cations of carbon-containing compounds [9–11]. The carbocations in superacid-catalyzed reactions can be differentiated into two distinct classes: carbenium ions and carbonium ions.

(i) Carbenium ions: Trivalent carbenium ions, which are referred to as "classical carbenium ions," have an sp^2-hybridized electron-deficient carbon atom. The carbenium carbon has six valence electrons and thus is highly electron deficient. Their structure is adequately described using only two-electron, two-center bonds. The carbon atom easily becomes planar in the absence of constraints arising from skeletal rigidity or steric interference.

(ii) Carbonium ions: Penta-coordinate carbonium ions ("non-classical carbonium ions") have a five-coordinate (or higher) carbon atom bonded by three single bonds and a two-electron three-center bond. $^+CH_5$ carbonium ions are the parent ions of carbonium ions. The definition mentioned above can apply to higher molecular weight carbonium ions that contain a higher than five coordinate carbon atom. The structure of penta-coordinated or higher molecular weight carbonium ions cannot be expressed using only a two-electron, two-center single bond. These "carbonium ions" are electron deficient because of the sharing of two electrons between three atoms.

As discussed in Chapter 1, in addition to $\bullet CH_3$ and carbene-like species (CH_x), $^+CH_3$ is also an important reaction intermediate for the C–C bond formation in methane conversion to produce higher hydrocarbons. Throughout this chapter, trivalent carbenium ions and penta-coordinated carbonium ions are referred to simply as carbenium ions such as $^+CH_3$ and carbonium ions such as $^+CH_5$, respectively.

7.2.3 Outline of the Formation of Carbenium Ions

Carbenium ions, which are one of the key reaction intermediates for C–C bond formation, can be generated via the interaction of hydrocarbons with catalysts as follows:

(1) Addition of a proton to an alkene:

$$R-CH=CH_2 \;+\; H^+ \longrightarrow R-\overset{+}{C}H_2-CH_3 \qquad (7.4)$$

As shown in reaction (7.4), carbenium ions are generated by the addition of a proton (H^+) to an alkene. Because of their relative instability, the formation of primary carbenium ions hardly occurs.

(2) Reaction of an alkane with a proton:

This reaction can be further classified into two reaction routes. The first alkane/proton reaction occurs via the formation of both a carbenium ion and hydrogen by direct hydride abstraction from the alkane (R–H), as shown in reaction (7.5).

$$R-H \;+\; H^+ \longrightarrow \overset{+}{R} \;+\; H_2 \qquad (7.5)$$

(3) Formation of a carbenium ion from a penta-coordinated carbonium ion

In the second alkane/proton reaction route, a proton reacts with a C–H bond to form a penta-coordinated carbonium ion as shown in reaction (7.6). Subsequently, the formation of a carbenium ion and H_2 occurs, as shown in reaction (7.7).

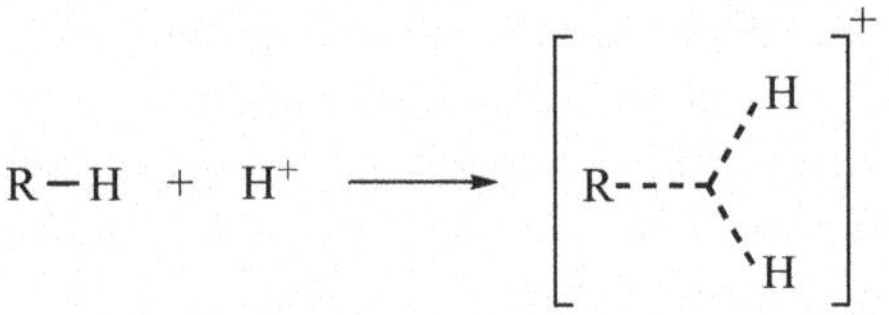

(7.6)

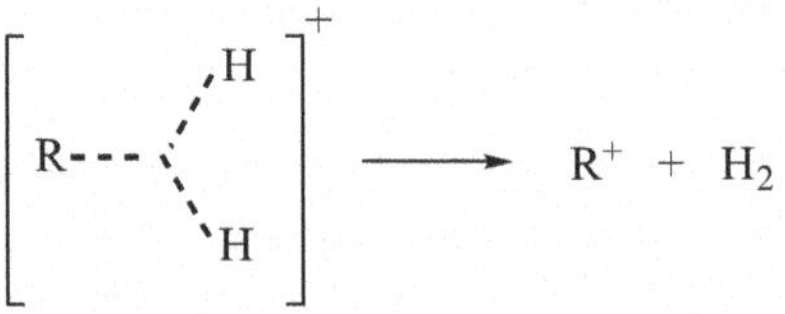

(7.7)

(4) Abstraction of hydride ions from alkanes

Carbenium ions can also be produced by the abstraction of a hydride ion (H^-) from an alkane by a Lewis acid (homogeneous catalysts) or a Lewis acid site (heterogeneous catalysts), L, as follows:

$$R{-}H + L \longrightarrow \overset{+}{R} + L{-}H \tag{7.8}$$

Here, L is usually a metal cation and/or metal cationic species, and L–H represents a metal–hydride species, such as Ag–H.

In the following sections, the formation of carbenium ions using various catalysts via reactions (7.4)–(7.8) and their reaction mechanisms are discussed.

7.3 Formation of Carbocations in Superacid Solutions

Bickel et al. reported observations related to the generation of *tert*-butyl cations (specifically, a *tert*-butyl carbenium ion) from *iso*-butane via hydride abstraction in a superacid HF–SbF_5 solution, as shown in reaction (7.9) [12].

$$(CH_3)_3C^+SbF_6^- + H_2 \rightleftharpoons (CH_3)_3CH + H^+SbF_6^- \tag{7.9}$$

Reaction (7.9) is reversible. Furthermore, the authors found that *neo*-pentane (2,2-dimethyl pentane) gave a *tert*-butyl carbenium ion together with CH_4 in HF–SbF_5 solution, as shown in reaction (7.10) [12].

$$\begin{array}{c} CH_3 \\ | \\ H_3C{-}C{-}CH_3 \\ | \\ CH_3 \end{array} + H^+SbF_6^- \longrightarrow (CH_3)_3C^+SbF_6^- + CH_4 \tag{7.10}$$

Nuclear magnetic resonance (NMR) spectroscopy provided evidence for the formation of both the *t*-butyl carbenium ion and methane. This result demonstrates that the proton (H^+) of HF–SbF_5 directly attacks the carbon–carbon bond in *neo*-pentane to produce *tert*-butyl carbenium ions. However, the reverse reaction (7.11) does not occur [13].

$$(CH_3)_3C^+SbF_6^- + CH_4 \not\longrightarrow \begin{array}{c} CH_3 \\ | \\ H_3C{-}C{-}CH_3 \\ | \\ CH_3 \end{array} + H^+SbF_6^- \tag{7.11}$$

Olah et al. also reported that *neo*-pentane formed the *t*-butyl carbenium ion (**1**) together with CH_4 in FSO_3–SbF_5 solution at 298 K [14].

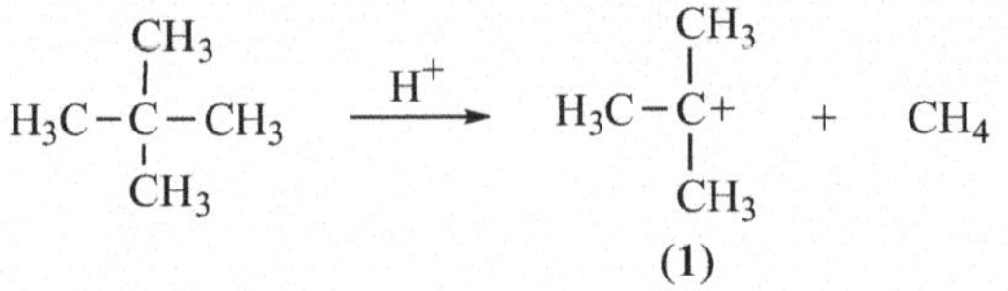

(7.12)

Furthermore, the authors demonstrated that not only (**1**), but also the *tert*-amyl carbenium ion (dimethyl ethyl carbenium ion) (**3**) was formed from *neo*-pentane in a FSO_3H–SbF_5–SO_3ClF solution at 253 K [15]. They reported that (**3**) was produced by the rearrangement of the primary carbenium ion (**2**) as shown in reaction (7.13), where (**2**) was generated by the abstraction of hydride from *neo*-pentane.

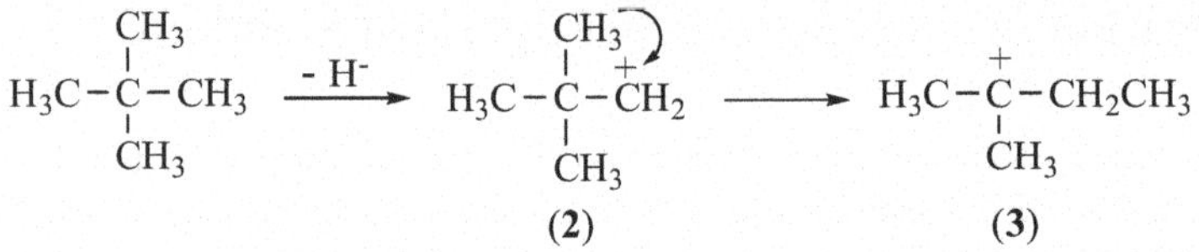

(7.13)

Thus, at 253 K, *neo*-pentane loses a hydride ion (H^-) to produce (**2**); the incident primary carbenium ion (**2**) rearranges to give the *tert*-amyl carbenium ion (**3**) (dimethyl ethyl carbenium ion), while as shown in reaction (7.12), at 298 K, the *tert*-butyl carbenium ion (**1**) was formed and CH_4 was liberated. The different reaction products of (7.12) and (7.13) indicate that at higher temperature, the abstraction of CH_3^- to give the *tert*-butyl carbenium ion (**1**) may be favored over hydride abstraction.

To explain the formation of the carbenium ions (**1**) and (**2**) from *neo*-pentane via reactions (7.12) and (7.13), respectively, Olah et al. proposed a mechanism involving penta-coordinated carbonium-ion-type intermediates (**4**) and (**5**), as shown in reaction (7.14) [16–19].

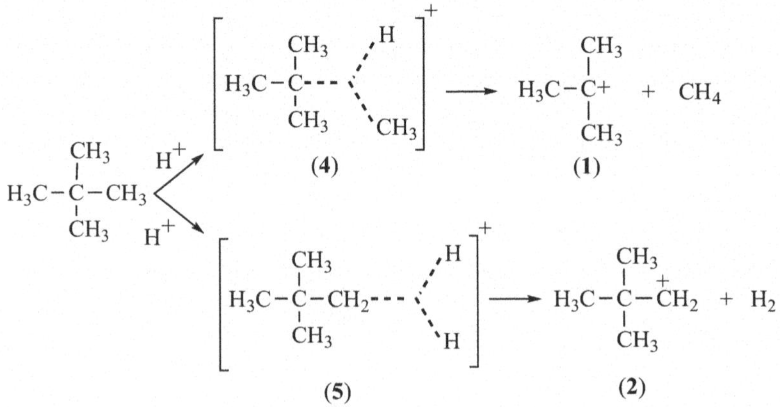

(7.14)

They also reported that in $HSO_3F–SbF_5$ and $HF–SbF_5$ solutions, the following H–D exchange reaction proceeded.

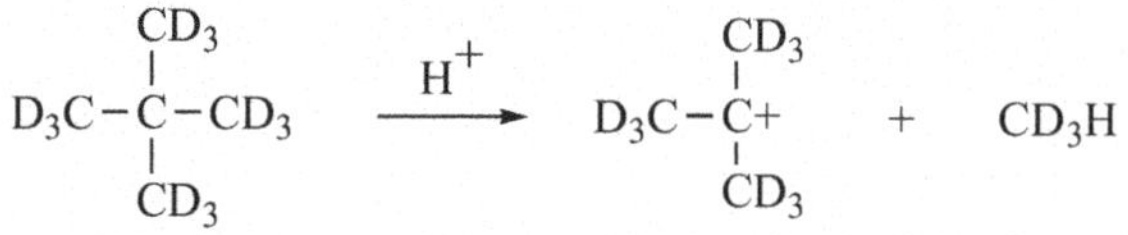
(7.15)

When *neo*-pentane was treated with deuterated superacids, such as $DSO_3F–SbF_5$, the formation of both CH_3D and deuterated *neo*-pentane was observed. These results indicate that not only the C-C bond protolysis of *neo*-pentane but also hydrogen–deuterium exchange of the methyl groups took place, as shown in reaction (7.16) [18].

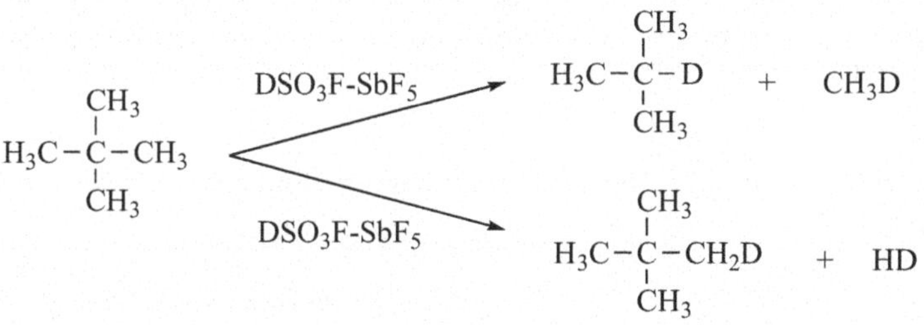
(7.16)

They also reported that the deuterated hydrocarbons $(CH_3)_3CD$, CH_3D, and $(CH_3)_2CHCH_2D$, as well as HD, were produced via the hydrogen–deuterium exchange reaction between *iso*-butane and $DSO_3F–SbF_5$ [19].

To explain the hydrogen–deuterium exchange between alkanes, such as *neo*-pentane and *iso*-butane and $DSO_3F–SbF_5$, either a C–H or C–C bond in the alkane should first be protonated to form a penta-coordinated carbonium ion intermediate and then undergo cleavage. Furthermore, the formation of alkenes is not observed when alkanes are used as reactants in superacid solutions. As an example, the reactions of the *iso*-butane–DSO_3F-SbF_5 reaction system are shown in Scheme 7.1 [18]. As shown in the scheme, the formation of methane from *neo*-pentane with simultaneous generation of tertiary carbonium ion, **(1)** can be explained without invoking the primary carbonium ion shown in (7.14) based on the reaction mechanisms involving penta-coordinated carbonium ion intermediates.

7.4 Reaction of Alkanes with Carbenium Ions to Produce C–C Bonds in Superacid Reaction Systems

As discussed in Sect. 7.2, superacids effectively generate carbenium ions from alkanes via penta-coordinated carbonium ion as key reaction intermediates. In superacid solution, the cleavage of C–H bonds usually proceeds via not only intermolecular

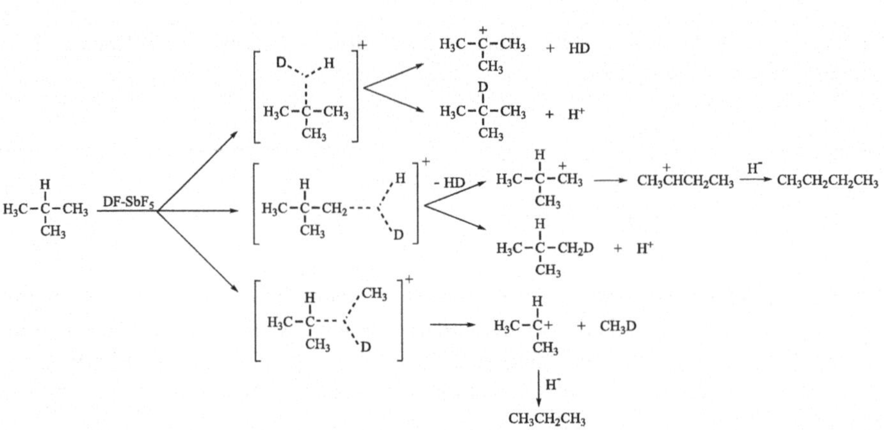

Scheme 7.1 Cleavage of the C–C and C–H bonds of *iso*-butane in DF–SbF_5 solution via penta-coordinated carbonium ion intermediates (Reprinted with permission from ref. [18]. Copyright 2020 American Chemical Society)

hydride transfer (reaction (7.17)) but also protonation–elimination in direct alkylation (reaction (7.18)).

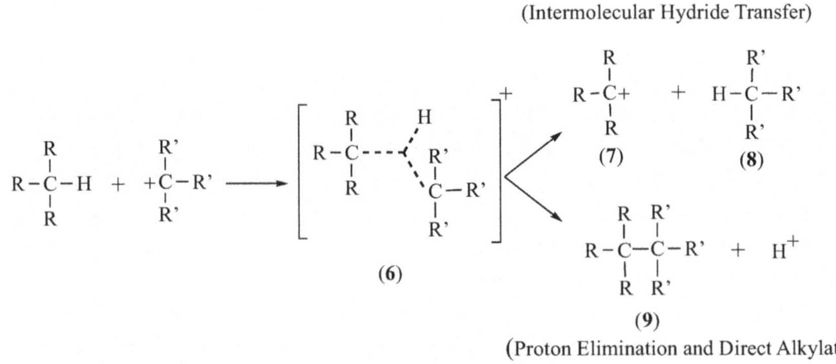

(7.17)

(7.18)

In reactions (7.17) and (7.18), R and R' are H atoms, and R_3CH and $^+CR'_3$ are CH_4 and $^+CH_3$, respectively. Thus, reaction (7.18) represents the reaction of CH_4 with $^+CH_3$ to produce C_2H_6, and also indicates that the alkylation of alkanes (both *iso*-alkane and *n*-alkane) can directly proceed using superacid catalysts. This process is explained further in Sect. 7.4.1.

7.4.1 *Conversion of Methane into Higher Hydrocarbons in FSO_3H–SbF_5 Solution*

Olah et al. reported that superacids catalyze the conversion of methane to produce hydrocarbons such as C_2H_6 together with hydrogen at around 400 K [1, 16, 18]. Thus, when methane was treated at either 423 K in an autoclave in the presence of a ten-fold excess of 1:1 FSO_3H-SbF_6 solution or at 413 K under atmospheric pressure, higher-molecular-weight hydrocarbons such as C_2H_6 and C_3H_8 were produced. According to the mechanism involving penta-coordinated carbonium-ion-type reaction intermediates, the formation of C_2H_6 and C_3H_8 can be explained by the following reaction:

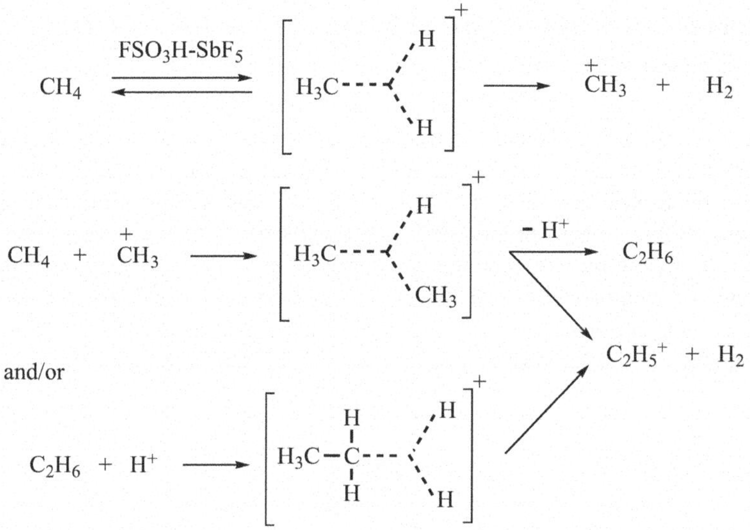

(7.19)

Thus, in superacid solution, CH_4 is protonated to afford $^+CH_5$, which then either undergoes reversible deprotonation, which accounts for the loss of hydrogen from $^+CH_5$ to form $^+CH_3$, or hydrogen exchange as described in Sect. 7.3. The $^+CH_3$ carbenium ions then react with methane to start the polycondensation to produce C_2H_6 and/or $^+C_2H_5$. C_2H_6 is further protonated to produce $^+C_2H_7$, which is then converted to $^+C_2H_5$ and H_2. $^+C_2H_5$ further reacts with CH_4 to produce C_3H_8, as shown in reaction (7.20) below:

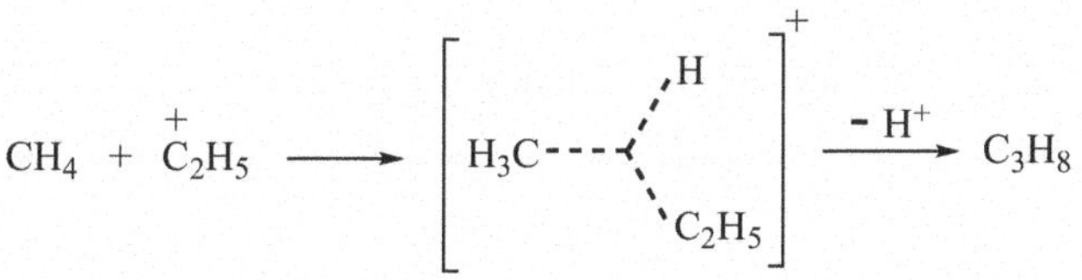

(7.20)

$^{+}CH_5$ carbonium ions have not been observed in superacid solutions, although their existence under other conditions has been confirmed using various spectroscopic methods [20–29]. Recently, Dinér et al. reported that carbenium ions, rather than carbonium ions were the key intermediate in reactions promoted by the superacid HF–SbF_5, such as the C–H bond cleavage and hydrogen–deuterium exchange reactions. In other words, the carbenium ions are directly produced without the formation of penta-coordinated carbonium ions [30]. These results indicate that the conversion of methane may proceed via $^{+}CH_3$ carbenium ions to produce the C–C bonds without using the superacid as a catalyst.

7.5 Role of Penta-Coordinated Carbonium Ion Intermediates in C–C Bond Formation in the Reaction of Alkenes with Alkanes

As shown in reactions (7.17) and (7.18), the C–C bond cleavage of *tert*-butyl carbonium ion (**6**) in superacid catalyst systems can result in not only intermolecular hydride transfer to produce $^{+}CR_3$ (**7**) and HCR'_3 (**8**), but also in proton elimination and direct alkylation to produce H^{+} and alkane (**9**), which can be expressed as $R_3CCR'_3$. For example, the reaction of *iso*-butene with the *tert*-butyl carbenium ion (**10**) and the propyl carbenium ion (**13**), which are produced by protonation of *iso*-butene and propylene, respectively, gave 2,2,3,3-tetramethylbutane (**12**) and 2,2,3-trimethylbutane (**15**), as shown in reactions (7.21) and (7.22), respectively.

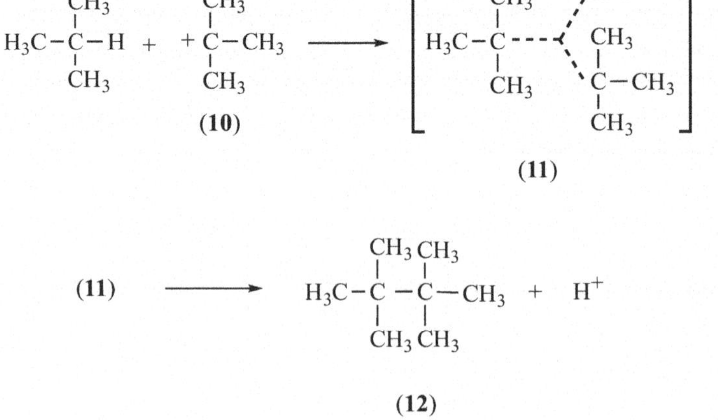

(7.21)

$$CH_3CH{=}CH_2 + H^+ \longrightarrow CH_3{-}\overset{+}{C}H_2{-}CH_3$$

(13)

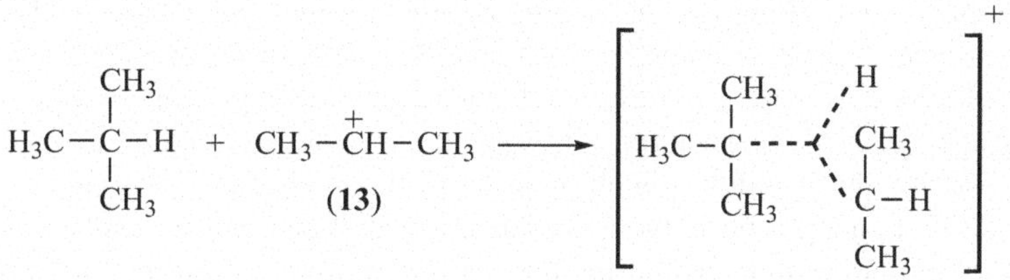

(7.22)

(14)

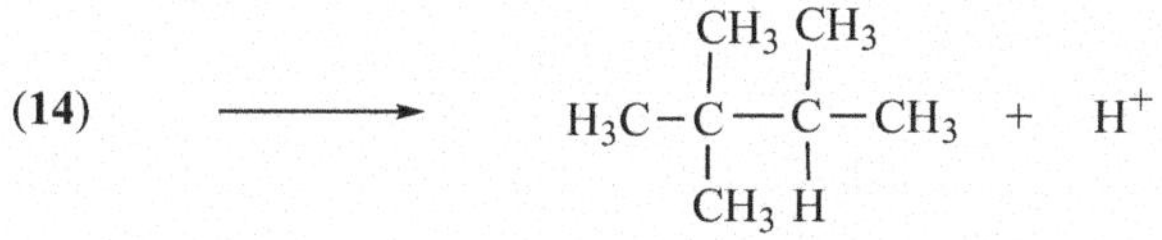

(15)

Therefore, 2,2,3,3-tetramethylbutane (**12**) and 2,2,3-trimethylbutane (**15**) are formed as the primary alkylation products via the reaction pathways suggested for superacid catalysts. Actually, Olah et al. reported the production of these alkanes in ref. [31].

On the other hand, the electrophilic alkylation of alkanes with alkenes ($R'_2CH{=}CH_2$) is usually considered to involve the alkylation of the alkane by the carbenium ion derived from the *iso*-alkane (R_3CH) via intermolecular hydride transfer [31], as shown in reactions (7.23)–(7.27) below:

$$R'_2CH{=}CH_2 + H^+ \longrightarrow R'_2\overset{+}{C}{-}CH_3 \qquad (7.23)$$

$$R'_2\overset{+}{C}{-}CH_3 + R_3CH \longrightarrow R'_2CHCH_3 + R_3C^+ \qquad (7.24)$$

$$R_3C^+ + R'_2CH{=}CH_2 \longrightarrow R'_2\overset{+}{C}{-}CH_2CR_3 \qquad (7.25)$$

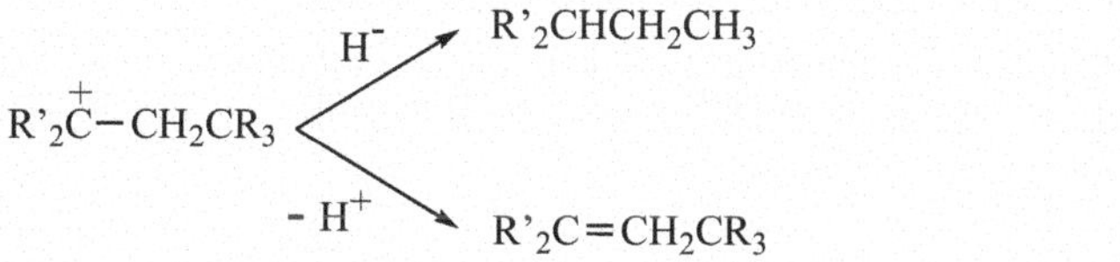

$$R'_2\overset{+}{C}{-}CH_2CR_3 \xrightarrow{H^-} R'_2CHCH_2CH_3 \qquad (7.26)$$

$$R'_2\overset{+}{C}{-}CH_2CR_3 \xrightarrow{-H^+} R'_2C{=}CH_2CR_3 \qquad (7.27)$$

According to reactions (7.23)–(7.27), the primary alkylation products of the reactions of *iso*-butene and propylene with *iso*-butane should be 2,2,4-trimethylpentane (and/or 2,2,4-trimethylpentenes) and 2,2-dimethylpentane (and/or 2,2-dimethylpentenes). However, as mentioned above, these alkanes and alkenes were not observed in superacid catalyst systems [31]. This result indicates that the reaction of alkanes with alkyl carbenium to form an alkene via proton elimination is unfavorable. However, the carbenium ions can attack both C–H and C–C bonds to produce direct alkylation and proton elimination in superacid catalyst systems.

Based on this concept, the polycondensation of methane can be explained via penta-coordinated carbonium ion intermediates. For example, in the direct conversion of CH_4 to C_2H_6, C–C bond formation proceeds via the following reaction mechanism.

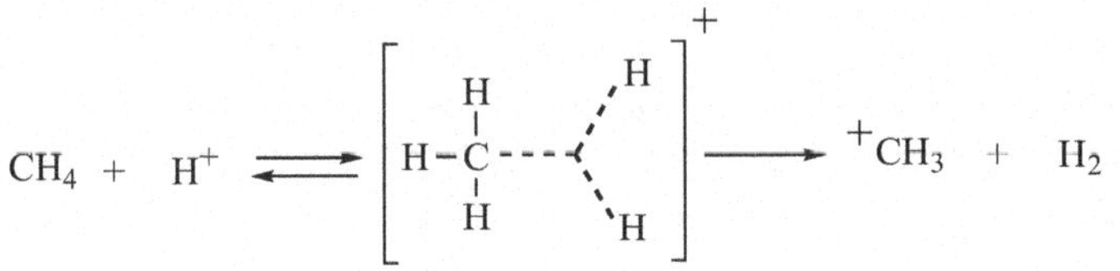

(7.28)

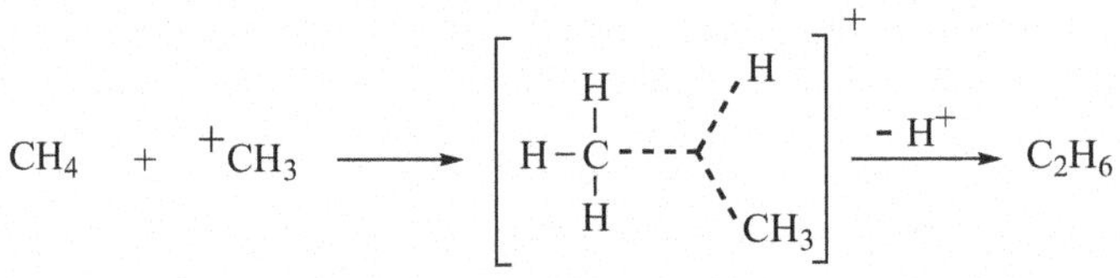

(7.29)

As shown in reaction (7.29), the reaction of CH_4 with $^+CH_3$ generated via $^+CH_5$ using superacid catalysts produces C_2H_6, not C_2H_4. Thus, reactions of alkanes with carbenium ions (^+R) afford alkanes, and no alkenes are produced whatsoever.

7.5.1 Application of the Concept of a Penta-Coordinated Carbonium Ion to Alkane Cracking Reactions Over H^+-Exchanged Zeolites

The concept of penta-coordinated carbonium-ion-type reaction intermediates can be used to explain the formation of methane in the cracking reactions of alkanes over Brønsted acid catalysts such as H^+-exchanged zeolites [32, 33]. Examples of the cracking reaction of 3-methylpentane are shown in Scheme 7.2.

As shown in reaction mechanism **a** in Scheme 7.2, the formation of CH_4 can be explained without the formation of $^+CH_3$ carbenium ion via C–C bond cleavage.

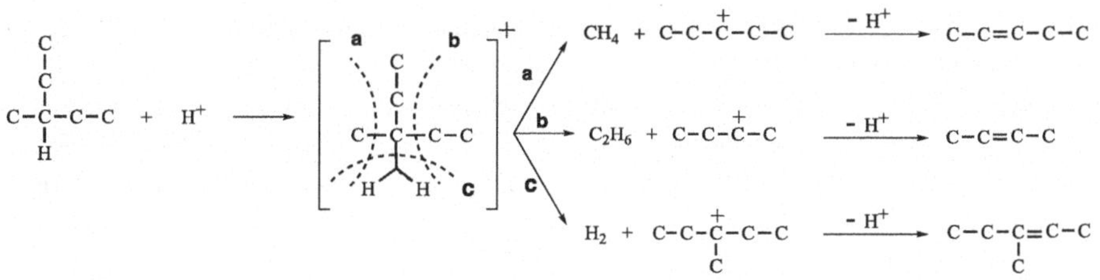

Scheme 7.2 Preferential protonation and collapse of a 3-methylpentane molecule (Reprinted from ref. [33]. Copyright 2020, with permission from Elsevier)

7.5.2 *Reaction of Methane with Ethylene to Produce Propane Using Superacid Catalysts*

The reaction of methane with lower alkenes such as ethylene and propylene is known to be thermodynamically favorable [5, 34, 35]. For example, even under non-oxidative reaction conditions, the thermodynamics of the alkylation of ethylene and propylene with methane are favorable below ~500 K, as shown in Table 7.1 [34].

Siskin reported that the reaction of CH_4 with C_2H_4 proceeded to produce C_3H_8 using HF–TaF_5 catalysts at 313 K [36]. The reaction in this catalyst system was investigated as follows: a methane–ethylene gas mixture was passed at a rate of 42 cm^3 min^{-1} through a 300 cm^3 autoclave containing 50 cm^3 of a 10:1 HF–TaF_5 (2.0 mol/0.2 mol) system stirred at 1000 rpm at 313 K, and maintained at 40 psig. To ensure the maximum protonation of ethylene and minimize possible competition from the ethylene oligomerization reaction, a 40-fold excess of acid was used and efficient mixing was maintained. Under these reaction conditions, the total C_3 hydrocarbons accounted for 58% of the reaction products after 1.5 h. Thus, C_3H_8 was produced, and the formation of propylene (C_3H_6) was not observed at all. In addition to the alkylation of CH_4 with ethylene, they also reported that the alkanes C_2H_6 and *n*-butane reacted with ethylene to produce *n*-butane and 3-methylpentane, respectively. Furthermore, the reaction of C_3H_6 with CH_4 proceeded to produce *iso*-butane [36].

After reporting these alkane–alkane alkylation reactions using superacid catalysts, Olah et al. also reported the reaction of ethylene with methane in the gas phase using heterogeneous catalysts consisting of SbF_5 intercalated into graphite, TaF_5 on AlF_3, and TaF_5, respectively, which were activated using HF [34]. In order to show

Table 7.1 Thermodynamics of methane alkylation reactions (Reprinted with permission from ref. [34]. Copyright 2020 American Chemical Society)

Reaction	ΔG/kJ mol^{-1}		
	300 K	400 K	500 K
$CH_4 + C_2H_4 \rightleftharpoons C_3H_8$	−40.5	−26.8	−13.4
$CH_4 + C_3H_6 \rightleftharpoons$ iso-C_4H_{10}	−32.6	−16.7	−1.2

Table 7.2 Ethylation of $^{13}CH_4$ with C_2H_4 (Reprinted with permission from ref [34]. Copyright 2020 American Chemical Society)

Reactant		Product distribution excluding methane/mol %				Labeled content of C_3 fraction/mol %	
$^{13}CH_4$: C_2H_4	Catalyst[a]	C_2H_6	C_3H_8	*iso*-C_4H_{10}	C_2H_5F	$^{13}CC_2H_8$	C_3H_8
98.7 : 1.3	TaF_5/AlF_3	51.9	9.9	38.2	–	31	69
99.1 : 0.9	TaF_5	–	15.5	3.0	81.5	91	9
99.1 : 0.9	SbF_5/graphite	64.1	31.5	–	4.4	96	4

[a]Catalysts pretreated with HF for 30 s

that the methane reacted with ethylene, $^{13}CH_4$ was used as the reactant, and a large excess of $^{13}CH_4$ was reacted with ethylene (C_2H_4) to minimize ethylene condensation (oligomerization) and the formation of cracking products. The reaction was carried out at ambient temperature in a pressurized fixed-bed flow reactor. The heterogeneous superacid catalysts (38% SbF_5 intercalated into graphite, TaF_5, and 1:1 w/w TaF_5–AlF_5) were treated with gaseous HF for 30 s immediately before use. The $^{13}CH_4$ and C_2H_4 gas mixtures were passed over the solid catalyst (1 g) at 5–10 ml min^{-1} and 20–60 psig at ambient temperature; the results are summarized in Table 7.2. Using the TaF_5/AlF_3, TaF_5, and SbF_5/graphite catalysts, singly ^{13}C-labeled propane ($^{13}CC_2H_8$) was produced. In particular, using the TaF_5 and SbF_5/graphite catalysts, $^{13}CC_2H_8$ was produced selectively. Thus, a significant fraction (up to 91%) of the propane was produced via the methane–ethylene condensation reaction. However, $^{13}CC_2H_6$ (^{13}C-labeled propylene) was not produced.

The authors also reported that the formation of propane was extremely sensitive towards the ratio of $^{13}CH_4$ to C_2H_4. For example, increasing the relative amount of ethylene in the reaction mixture to 2.0 mol% resulted in a significant decrease in mono-^{13}C-labeled propane. In these experiments using higher concentrations of ethylene (up to 8.0 mol%), no ^{13}C was observed in any of the products. For example, *iso*-butane contained no ^{13}C, indicating that it was derived from ethylene.

A reaction mechanism for the reaction of methane with ethylene to produce propane had previously been proposed by Siskin [36]; Olah et al. also reported the same reaction mechanism [34]. To explain the production of C_3H_8 rather than C_3H_6, the authors proposed the direct alkylation of the ethyl carbenium ion ($^+C_2H_5$), which is produced by the reaction of C_2H_4 with a proton (H^+), via a penta-coordinated carbonium ion as follows [34, 36]:

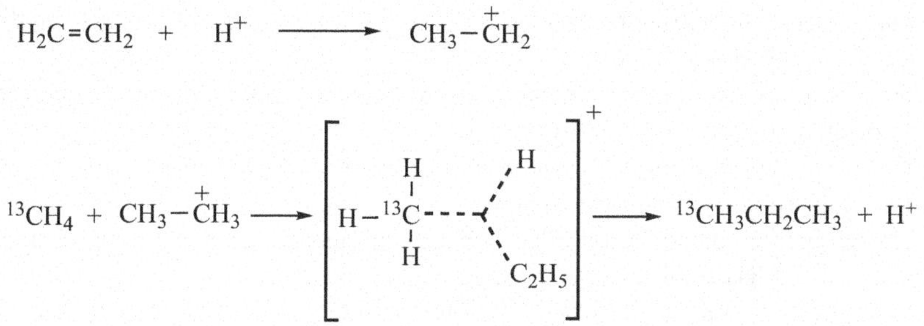

(16)

(7.30)

Superacid catalysts do not produce propylene (C_3H_6) at all, as shown in Table 7.2. Singly ^{13}C-labeled propane ($^{13}CC_2H_8$) is produced by the acid-catalyzed methane–ethylene condensation reaction [34]. Siskin stressed that the ethylation of methane is initiated by the protonation of the alkene (ethylene) to form an intermediate ethyl carbenium ion, which inserts into methane via the penta-coordinated carbonium ion (**16**) to yield singly labeled propane ($^{13}CC_2H_8$), as discussed in Sect. 7.5.1. Thus, the activation of methane is essential to produce propylene in the reaction of methane with ethylene, as described in the next section. In other words, to produce C_3H_6 via the reaction of methane with ethylene, other catalysts that can activate CH_4 to produce $^+CH_3$ carbenium ions should be developed.

7.6 Ag⁺-Exchanged Zeolite Catalysts for C–C Bond Formation in the Conversion of Methane

As discussed in Sect. 7.4, superacid catalysts activate methane to generate $^+CH_3$ carbenium ions via $^+CH_5$ carbonium ions and subsequently produce C_2H_6 via the reaction of $^+CH_3$ with CH_4 in the polycondensation of methane [1]. However, the reaction of CH_4 with ethylene, in which $^+C_2H_5$ ions are generated and react with CH_4, gave C_3H_8 [34].

Ag^+-exchanged zeolites (Ag-zeolites) can catalyze the reaction of CH_4 with C_2H_4 to produce propylene (C_3H_6) and hydrogen at around 650 K, as shown in reaction (7.31) [4, 37].

$$CH_4 + C_2H_4 \longrightarrow CH_2{=}CHCH_3 + H_2 \qquad (7.31)$$

In this section, the catalytic properties of Ag-zeolites are compared with those of superacid catalysts. To highlight the unique properties of Ag-zeolites, the following topics are discussed.

(1) The general properties of zeolites; namely, the reactivity of surface methoxy species (–OCH_3) on zeolites.
(2) The unique catalytic properties of solid-acid catalysis by Ag-zeolites: Hydrogen-induced acid catalysis.
(3) The reversible heterolytic dissociation of hydrogen molecules on Ag-zeolites and the formation of hydrogen species.
(4) The heterolytic dissociation of C–H bonds of methane over Ag-zeolites and the reaction of CH_4 with C_2H_4 to produce C_3H_6 and H_2.
(5) Activation of alkanes including CH_4 over metal cation-exchanged zeolites: The differences between the catalytic properties of Ag-zeolites and Zn^{2+}-exchanged zeolites (Zn-zeolites).

7.6.1 *Catalysis of Electrophilic Reactions by H^+-Exchanged Zeolites: Role of the Methoxy Species on the Zeolite Surface*

Methane could potentially be reacted with other hydrocarbons such as ethylene using catalysts other than superacids to generate $^+CH_3$ from CH_4. The acidic O–H groups (Brønsted acid sites) of zeolites readily react with methanol to produce surface methoxy (–OCH_3) species [38]. When $^{13}CH_3OH$ is brought into contact with H^+-exchanged zeolites, ^{13}C-labeled methoxy (–$O^{13}CH_3$) species are generated as follows:

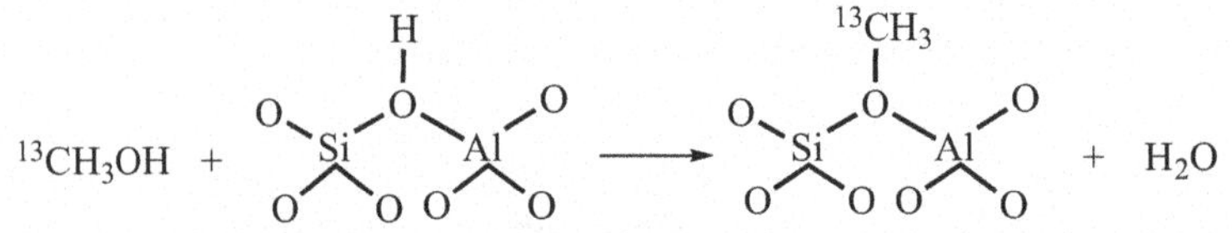

(7.32)

The surface –OCH_3 species on zeolites are well-known to act as effective methylating agents, and react electrophilically with a variety of molecules, including methanol, water, NH_3, methyl halides such as CH_3I, hydrogen chloride, aromatic hydrocarbons such as toluene, carbon monoxide, and acetonitrile [39–41]. At reaction temperatures higher than 523 K, the electrophilic reactions occur via the reaction of –OCH_3 species with various molecules, as shown in Scheme 7.3. Thus, the –OCH_3 species act as $^+CH_3$ to cause the electrophilic reactions.

Scheme 7.3 Reactions of the methoxy group with various compounds (Reprinted with permission from ref. [39]. Copyright 2020 American Chemical Society)

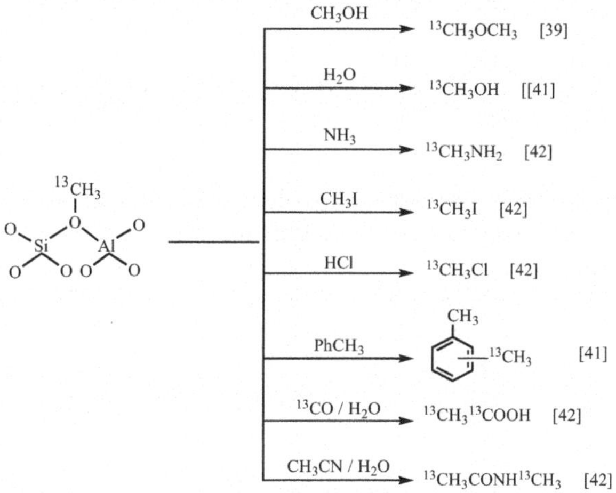

7.6.2 *Reaction of Surface Methoxy Species with Ethylene to Produce C_3H_6*

Ono et al. also reported that the protons at Brønsted acid sites on zeolites react with methanol to produce the $-OCH_3$ species [42], as shown in reaction (7.32). Furthermore, these authors demonstrated that the methyl carbenium ion ($^+CH_3$) was released from the $-OCH_3$ species. Thus, the "methoxy species" on the zeolite surface can be represented by two resonance structures.

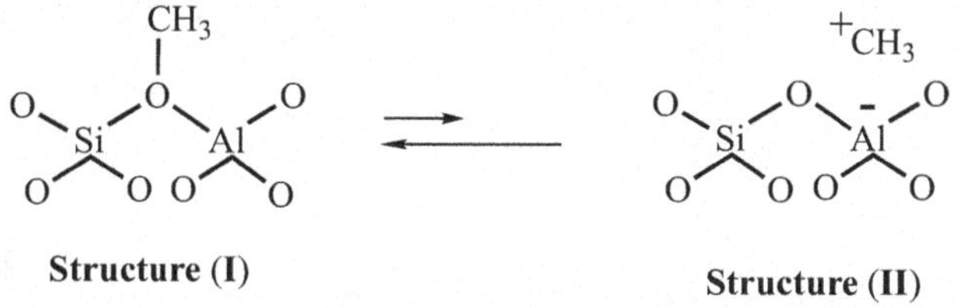

(7.33)

The surface "$-OCH_3$ species" in **Structure (I)** should be very polarized, and at temperatures above 523 K, the contribution of **Structure (II)** may become increasingly significant. Therefore, the $-OCH_3$ species on the zeolite are electrophilic and could induce a variety of electrophilic reactions. For example, as mentioned above, a reaction in which a $^+CH_3$ ion originating from the $-OCH_3$ species on zeolites reacts with alkenes such as ethylene to produce an alkene whose carbon number is one greater than that of the original alkene is shown in reaction (7.34) [42].

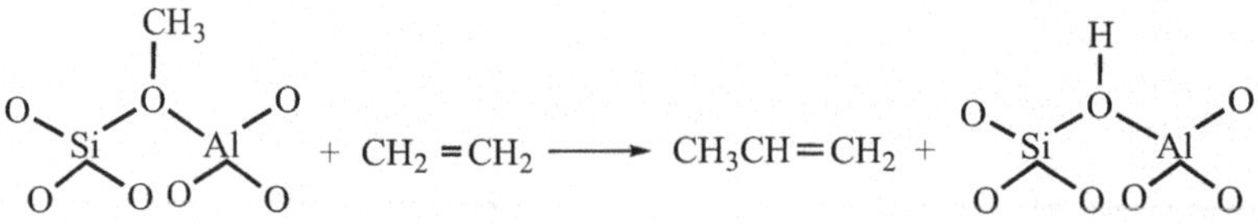

(7.34)

However, H⁺-exchanged zeolites (H-zeolites), such as H⁺-exchanged ZSM-5 (H-ZSM-5) cannot react with methane to produce –OCH_3 species on their surfaces.

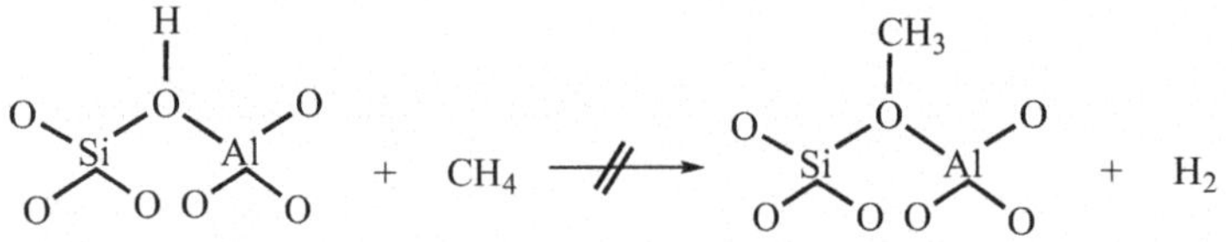

(7.35)

Furthermore, although –OCH_3 species are formed on the surface of zeolites, they cannot react with CH_4, as described in Sect. 7.7.5.

In addition to –OCH_3 species, ethoxy (–OC_2H_5) species can be formed via the reaction of surface O–H groups originating from the Brönsted acid sites on H-zeolites with ethylene (or ethanol); however these –OC_2H_5 species are also incapable of reacting with CH_4 to produce C_3H_6. This indicates that the reaction of CH_4 with C_2H_4 does not proceed over H-zeolites, and that ethyl carbenium ions ($C_2H_5^+$ ions) cannot abstract hydride ions (H^- ions) from CH_4, as shown in reaction (7.36). Further details are given in Sect. 7.7.5.

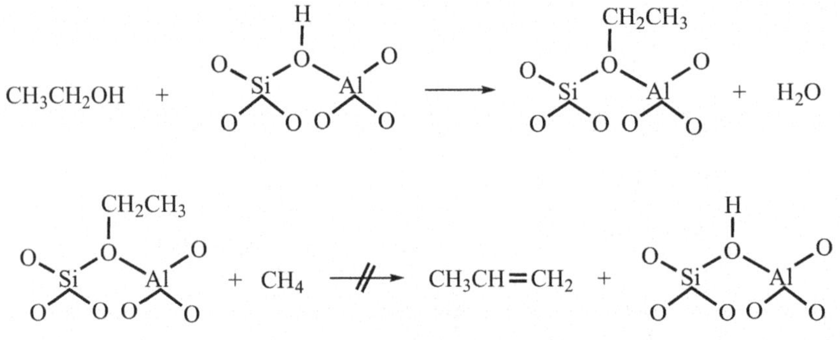

(7.36)

In terms of chemical properties, the –OCH_3 species ($^+CH_3$ carbenium ions) and ethoxy species ($^+C_2H_5$ carbenium ions) on the zeolite surface are different from the $^+CH_3$ and $^+C_2H_5$ carbenium ions in superacid solutions. According to reaction (7.34), if the –OCH_3 species due to CH_4 were generated on the surface of the zeolite, methane could react with alkenes such as ethylene to produce the corresponding alkene (*i.e.*, C_3H_6), rather than an alkane as shown in reaction (7.37) below.

$$CH_4 + C_nH_{2n} \longrightarrow C_{(n+1)}H_{2(n+1)} + H_2 \qquad (7.37)$$

7.7 Unique Properties of Silver Cations in the Zeolite-Based Catalysis of the Reaction of Methane with Ethylene

The characteristics of Ag-zeolites that lead to the activation of methane to generate $^{+}CH_3$ carbenium ions, which in turn enables the reaction of methane with alkenes such as ethylene or aromatic hydrocarbons such as benzene, are discussed in the following sections.

7.7.1 General Properties of Ag^{+}-Exchanged Zeolites: Reversible Reduction of Ag^{+} Cations and Generation of Brønsted Acid Sites

As mentioned in Chapter 6, Na^{+}-exchanged zeolites are commonly synthesized. Ag-zeolites, such as Ag^{+}-exchanged Y-type zeolite (Ag-Y), are often prepared via the ion-exchange method. For example, Na^{+}-exchanged Y-type zeolite (Na-Y) can be converted to Ag-Y using an aqueous solution of $AgNO_3$, as shown in reaction (7.38).

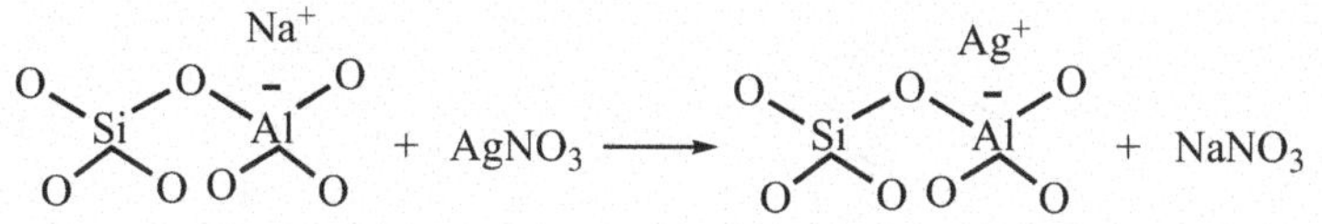

(7.38)

At this stage, Brønsted acid sites (acidic O–H groups) do not exist in the Ag-zeolite.

When Ag-zeolites are brought into contact with hydrogen and oxygen, unique silver species are generated. A number of studies of the reduction and oxidation of silver cations have been conducted; these processes are conventionally expressed as shown in reactions (7.39) and (7.40), respectively [43, 44].

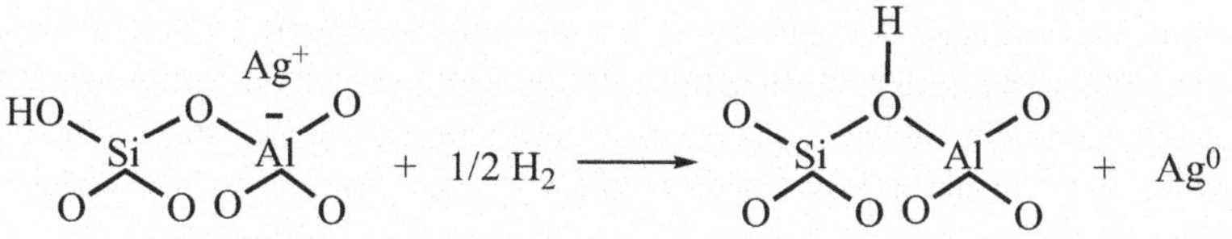

(7.39)

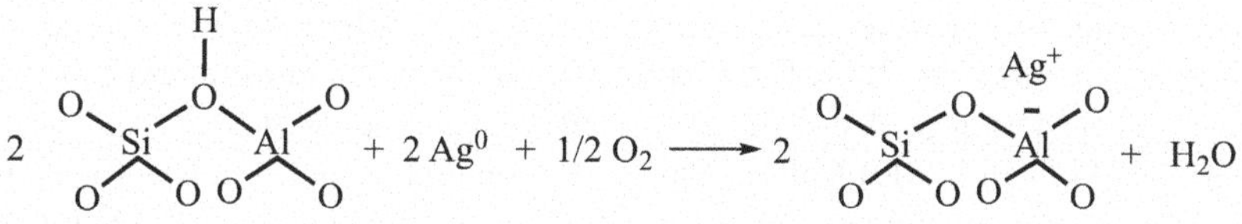

(7.40)

Thus, the metallic silver (Ag^0), which is formed by the reduction of Ag^+ cations with hydrogen, is re-oxidized with oxygen to form both Ag^+ cations and H_2O as shown in reaction (7.40). The disappearance of the Ag^+ cations due to reduction with H_2 and the reappearance of the Ag^+ cations upon oxidation have been confirmed in the infrared spectra of CO adsorbed on Ag-zeolites, such as Ag-Y [45, 46]. The formation of acidic protons (H^+) has also been confirmed by the appearance of O–H groups in infrared spectroscopy [43]. When Ag-Y was placed in contact with D_2, two deuteroxyl bands (O–D bands) were observed. These O-D bands were similar to the high and low frequency bands of D-Y (D^+-exchanged Y-type zeolite) [47].

However, the chemistry of the reduction–oxidation process is more complex than the simplified version shown in reactions (7.39) and (7.40), and a great deal of effort has been devoted to elucidating the silver species involved [43, 47–60]. In fact, reduction at low temperatures has been found to result in the formation of highly dispersed silver cationic clusters (Ag_n^{q+}), such as Ag_n^+, which contain unreduced Ag^+ ions (Ag_3^+ on average) [43].

$$(n - 1)\,Ag^0 \longrightarrow Ag^0_{(n-1)} \tag{7.41}$$

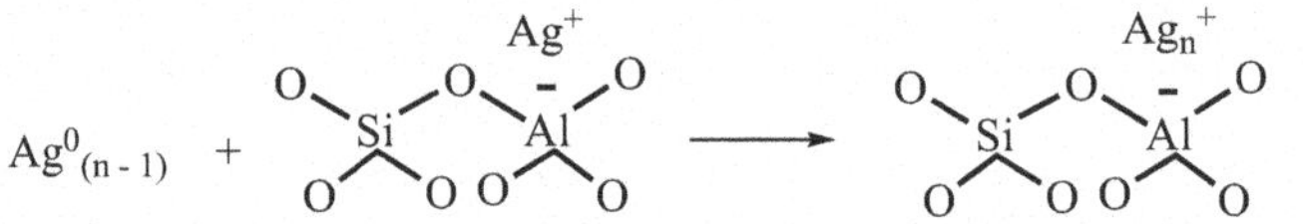

(7.42)

The silver cationic clusters (Ag_n^{q+}) can be easily re-oxidized with oxygen.

Hydrogen reduction at high temperature results in the formation of metallic Ag particles on the external surface of the zeolite. These large metallic particles can only be re-oxidized at temperatures higher than 500 K. Furthermore, the formation of small Ag cationic clusters such as Ag_n^+ has been reported for various zeolites, including A, X, and Y type zeolites.

Jacobs et al. later found that reaction (7.39) was partially reversible [46]. Hydrogen was desorbed from previously reduced Ag-Y or Ag^+-exchanged mordenite (Ag-mordenite) by evacuating these Ag-zeolites at temperatures higher than 573 K. At the same time, metallic silver was re-oxidized to Ag^+ ions. Baba et al. also reported the reversible interconversion of Ag^+ ions and silver metal particles [61, 62].

The reversibility between Ag^+ cations and Ag^0 metal was also observed via X-ray diffraction (XRD) spectra. Figure 7.1 shows the formation and disappearance of silver metal during a reduction–evacuation cycle [61, 62]. When Ag-Y was reduced with hydrogen (13.3 kPa) at 303 K for 30 min, 20% of the Ag ions were reduced [61, 62]. The main diffraction peaks of the hydrogen-treated sample at 37.6° and 44.4° correspond the (111) and (200) Miller indices of Ag metal particles, respectively, in Fig. 7.1a. The average particle size was *ca.* 17 nm, as calculated using the Scherrer equation. The other diffraction peaks originate from the Y zeolite. After hydrogen

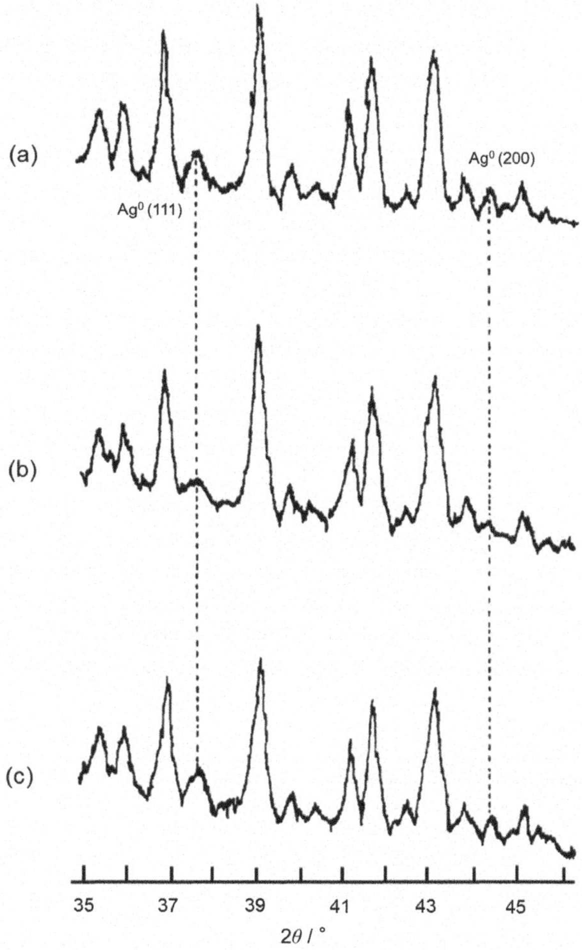

Fig. 7.1 XRD patterns of Ag metal particles in Ag-Y (Reproduced from Ref. [61] and [62] with permission from The Royal Society of Chemistry). **a** Ag-Y after being heated under vacuum at 303 K for 4 h followed by reduction with H_2 (13.3 kPa) at 303 K for 25 min. **b** The same Ag-Y sample after subsequent evacuation at 303 K for 3 h. **c** The Ag-Y sample after being re-exposed to H_2 (16.6 kPa) at 303 K for 25 min

reduction, the sample was evacuated at 303 K for 3 h and then subjected to XRD measurement under vacuum. As shown in Fig. 7.1b, the metallic Ag^0 peaks almost disappeared, indicating that the metal Ag particles had transformed into Ag^+ cations. When the sample was re-exposed to hydrogen (13.3 kPa) at 303 K for 30 min, the diffraction peaks of the Ag metal particles reappeared (Fig. 7.1c). During this second reduction stage, almost the same amount of hydrogen was consumed as during the initial reduction of Ag-Y. The reversibility depended on the hydrogen reduction conditions, such as the reduction temperature and the length of the reduction stage, which determined the degree of reduction of the Ag^+ cations [46, 61, 62].

7.7.2 *Generation of Hydrogen Species Over Ag^+-Exchanged Zeolites in the Presence of Hydrogen: Reversible Heterolytic Dissociation of Hydrogen Molecules*

As discussed in Sect. 7.6.1, the Ag^+ cations in Ag-zeolites are reduced by H_2 to produce acidic protons (Brønsted acid sites) and Ag^0 metal, as shown in reaction (7.39); this reaction is partially reversible. The acidic protons originating from the surface O–H groups in the zeolites were observed using infrared spectroscopy [58].

Baba et al. investigated the surface hydrogen species of Ag-zeolites using ^1H magic angle spinning NMR (^1H MAS NMR) spectroscopy. The Ag-zeolites were treated with hydrogen, and their ^1H MAS NMR spectra were then measured in the presence of hydrogen [63–65]. Figure 7.2 shows the ^1H MAS NMR spectra of Ag-A. The Ag-A was placed in contact with H_2 (40 kPa) at 313 K for 30 min (first hydrogen reduction). The amount of hydrogen consumed was 3.02×10^{-4} mol g^{-1}, which corresponded to the reduction of 21% of the Ag^+ cations in the zeolite. The ^1H MAS NMR spectrum of the reduced Ag-A was measured in the presence of hydrogen (Fig. 7.2a), and two peaks were observed at 4.0 and −1.8 ppm. The peak at 4.0 ppm was unambiguously attributed to the acidic protons (Brønsted acid sites), because such protons have been observed at chemical shifts of 3.9–4.4 ppm in various H^+-exchanged zeolites [66–68]. The peak at −1.8 ppm was a quartet with an intensity ratio of 1:3:3:1 and a coupling constant of (131 ± 1) Hz, indicating that

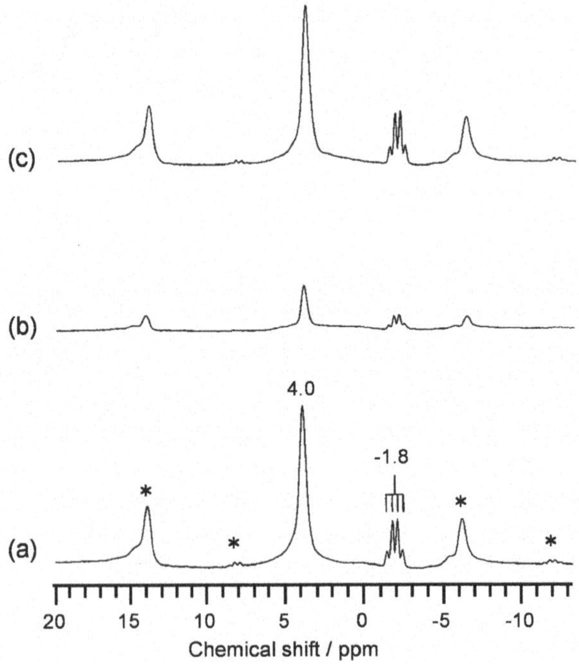

Fig. 7.2 ^1H MAS NMR spectra of AgA recorded at 298 K in the presence of hydrogen (Reprinted with permission from ref [63]. Copyright 2020 American Chemical Society). **a** AgA reduced with H_2 (40 kPa) at 313 K for 30 min. The spectrum was recorded under 40 kPa of H_2. **b** The same sample after being evacuated at 313 K for 2 h. **c** The sample after being re-exposed to H_2 (40 kPa) at 313 K for 30 min. The asterisks (*) show the spinning sidebands

three equivalent Ag atoms or ions interacted with a hydrogen nucleus to generate the Ag–hydride species, Ag_3–H. The formation of this Ag–hydride species indicated that the heterolytic dissociation of hydrogen molecules proceeds over cationic silver clusters, i.e., Ag_3^+, as shown in the reactions below.

$$2\,Ag^+ + H_2 \longrightarrow 2\,Ag^o + 2\,H^+ \quad (7.43)$$

$$2\,Ag^o + Ag^+ \longrightarrow Ag_3{}^+ \quad (7.44)$$

$$Ag_3{}^+ + H_2 \longrightarrow Ag_3\text{–}H + H^+ \quad (7.45)$$

The reactions (7.43) and (7.45) show that acidic H^+ (O–H groups) is generated not only by the reduction of Ag^+ cations with hydrogen, but also by the heterolytic dissociation of the H–H bond in H_2 molecules on Ag_3^+ cationic clusters. The formation of a triangular Ag_3^+ cationic cluster had previously been proposed as the result of treatment of Ag-Y with hydrogen at low-temperature [58] and γ-irradiation of Ag-A [69].

Baba et al. also found that reactions (7.43), (7.44), and (7.45) proceeded reversibly, as shown in Fig. 7.2. They elucidated the reversibility of Ag_3–H and acidic proton formation by measuring ^{1}H MAS NMR spectra of Ag-A. When the reduced sample was evacuated at 313 K for 2 h, the intensities of the signals associated with the acidic protons and Ag_3–H were reduced to 13% of their intensities in the spectrum of the sample before evacuation (Fig. 7.2b). However, re-exposing the sample to hydrogen (40 kPa) at 313 K for 30 min (second hydrogen reduction) caused the signals at 4.0 ppm and −1.8 ppm to reappear, as shown in Fig. 7.2c. The intensities of these signals returned to almost their original levels, which are shown in Fig. 7.2a. Here, the hydrogen consumption in the second hydrogen reduction was almost equal to that in the first hydrogen reduction in Fig. 7.2a.

The results shown in Fig. 7.2 clearly indicate that reactions (7.43), (7.44), and (7.45) are reversible when the degree of Ag^+ cation reduction in the Ag-A zeolite is low. However, the extent of reversibility decreased at Ag^+ cation reduction degrees higher than ~30%.

Furthermore, they reported the reversibility of heterolytic dissociation on Ag-Y zeolites, as shown in Fig. 7.3 [4, 64]. When Ag-Y (46% Ag^+ exchange) was placed in contact with H_2 at 373 K for 12 min, the degree of Ag^+ cation reduction was 21%. In this first hydrogen reduction, three peaks were observed in the presence of hydrogen at 4.7, 3.9, and −0.1 ppm, as shown in Fig. 7.3a. The peaks at 4.7 and 3.9 ppm were attributed to the acidic O–H groups formed in the sodalite cage and the supercage of the Y zeolite, respectively, because these chemical shift values were in agreement with those previously observed for H-Y [70]. The peak at −0.1 ppm was assigned

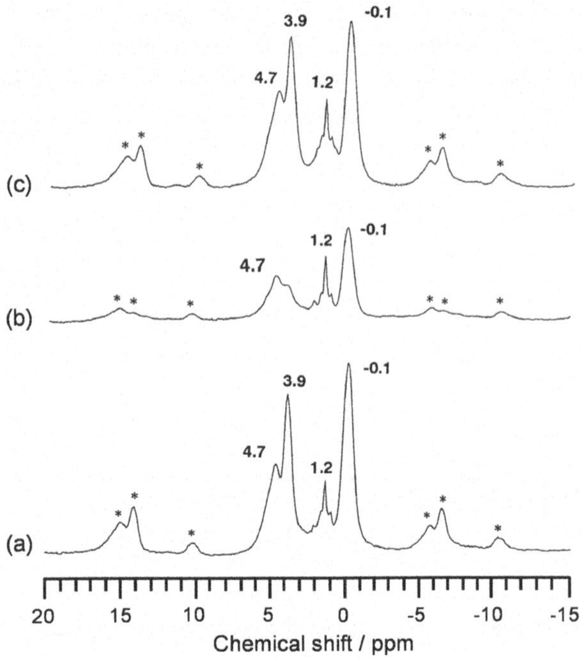

Fig. 7.3 Changes in the ^{1}H MAS NMR spectrum of AgY during a hydrogen adsorption–desorption cycle (Reprinted from ref. [4] and [64]. Copyright 2020, with permission from Elsevier). **a** AgY reduced with H_2 (40 kPa) at 313 K for 60 min. **b** The same sample after being evacuated at 313 K for 2 h. **c** The sample after being re-exposed to H_2 (40 kPa) at 313 K for 1 h. The asterisks (*) show the spinning sidebands

to a silver hydride (Ag_n–H) species. The silver cationic cluster (Ag_n^+) of Ag-Y may be larger (n > 3) than the Ag_3^+ cationic clusters of Ag-A.

$$Ag_n^+ + H_2 \longrightarrow Ag_n\text{–}H + H^+ \quad (n > 3) \tag{7.46}$$

The sample from the first hydrogen reduction was further evacuated at 373 K for 1 h. This operation greatly diminished the intensities of the three peaks, as shown in Fig. 7.3b. The peak at 3.9 ppm due to the bridged hydroxyls (O–H groups) in the supercage nearly disappeared, even though the intensity of the peak at −0.1 ppm (silver hydride) decreased by half. Re-exposing the sample to H_2 (40 kPa) at 333 K (the second hydrogen reduction), mostly recovered the intensities of the three peaks, as shown in Fig. 7.3c. These results mentioned above demonstrate that the reduction of the Ag^+ cations and the heterolytic dissociation of H_2 molecules on Ag-Y proceed reversibly. However, ^{1}H MAS NMR studies of the zeolites Ag-Y and Ag-A have shown that the reversibility depends on the degree of reduction of the Ag^+ cations [4, 64], as the value of n in silver cationic clusters Ag_n^{q+} decreases with increasing number of silver atoms per cationic cluster at higher degrees of Ag^+ cations reduction.

The phenomenon of reversible inter-conversion between Ag^+ cations and Ag metallic clusters was also investigated in detail for the zeolites Ag-ZSM-5 [71] and Ag-mordenite [72]. It was concluded that the heterolytic dissociation of hydrogen molecules in Ag-zeolites is reversible, with the reversibility depending on the degree of Ag^+ ion reduction. Thus, reaction (7.46) should be expressed as the following reversible reaction:

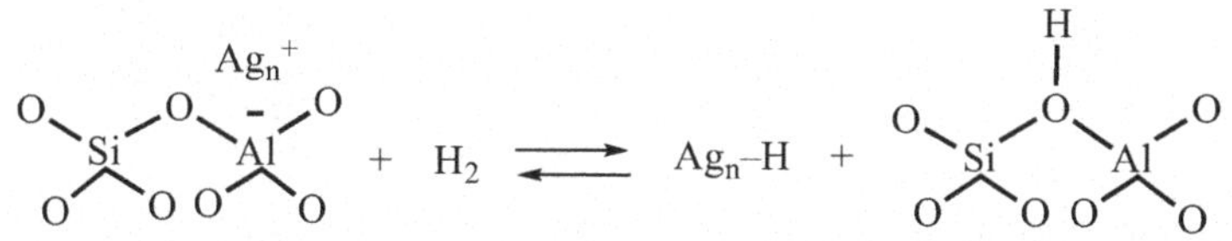

(7.47)

Reversible reaction (7.47) is crucial to regenerate Ag_n^+ in Ag-zeolites, which allows the reaction of methane with ethylene or benzene to proceed catalytically, as described in Sect. 7.5. Thus, the reversible inter-conversion between Ag_n^+ and Ag_n–H is essential to generate the catalytic activity of the Ag-zeolite for the conversion of methane following the activation of methane.

7.7.3 Cleavage of the C–H Bond of CH_4 Over Ag^+-Exchanged Zeolites

As described above, hydrogen molecules undergo reversible heterolytic dissociation over zeolites Ag-A and Ag-Y, as shown in Figs. 7.2 and 7.3, respectively. The H–H bond energy of the H_2 molecule is nearly equal to that of the C–H bond of the CH_4 molecule at ~ 420 kJ mol^{-1} [73]. On the basis of these bond energies, the C–H bond of CH_4 should also be cleaved over Ag-zeolites. Baba et al. found that silver hydride species are generated when CH_4 is placed in contact with the zeolite Ag-Y [63]. Figure 7.4 shows the ^{1}H MAS NMR spectrum of Ag-Y in the presence of CH_4 after being exposed to CH_4 (14 kPa) for 1 h at 393 K. Two peaks were observed at 0.4 and −0.1 ppm. The peak at 0.4 ppm was attributed to CH_4 adsorbed on Ag-Y and/or gas-phase CH_4. The peak at −0.1 ppm originated from Ag-hydride species (Ag_n–H). This peak was also observed for Ag-Y placed in contact with hydrogen, as shown in Fig. 7.3. However, the peak due to acidic O-H groups (acidic protons) at around 4 ppm was not observed. These results demonstrate that Ag_n–H is generated by the cleavage of the C–H bond of CH_4 molecules as follows:

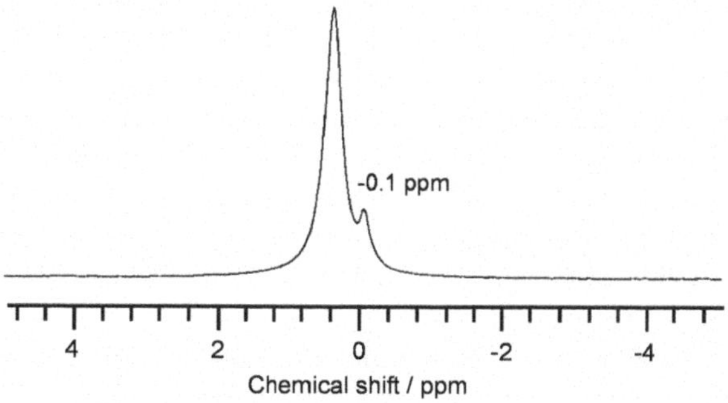

Fig. 7.4 ^{1}H MAS NMR spectrum of Ag-Y exposed to CH_4 (14 kPa) at 393 K for 1 h. The spectrum was recorded at 298 K (Reprinted with permission from ref [63]. Copyright 2020 American Chemical Society)

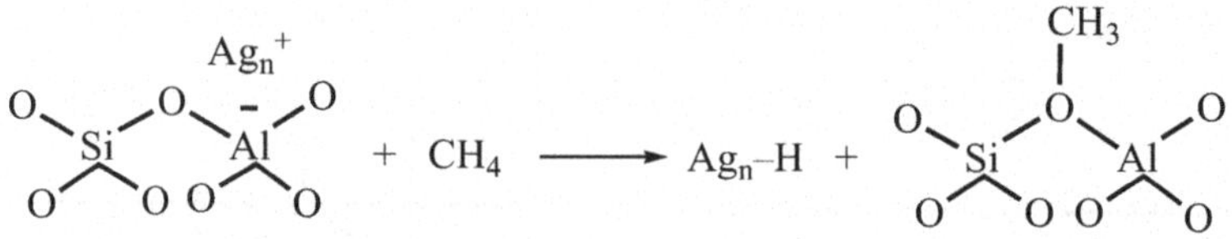

(7.48)

However, the signal of the protons of the methoxy (–OCH_3) groups was possibly overlapped by the signal of physically adsorbed and/or gas-phase methane.

As mentioned above, the signal corresponding to the acidic protons of O–H groups, which are usually observed at around 4 ppm, was not observed. This result indicated that neither Ag–CH_3 nor acidic protons (O–H groups) were produced by the cleavage of the C–H bond of CH_4 over Ag-zeolites. Therefore, reaction (7.48) proceeds selectively, while reaction (7.49) does not proceed [63].

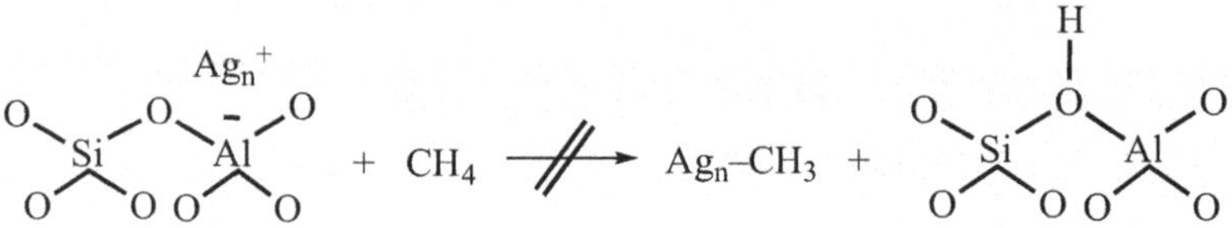

(7.49)

7.7.4 ^{13}C MAS NMR Evidence for the Formation of Methoxy Species on Ag^+-Exchanged Zeolites

In 2013, Gabrienko et al. reported the experimental observation of the formation of methoxy (–OCH_3) groups on Ag-ZSM-5 using ^{13}C cross-polarized (CP)/MAS NMR at 508–623 K [74]. $^{13}CH_4$ was placed in contact with Ag-ZSM-5 at room temperature, and the adsorption and subsequent reaction of $^{13}CH_4$ over Ag/H-ZSM-5 was monitored via ^{13}C CP/MAS NMR spectroscopy over the range 298–823 K to investigate the surface –OCH_3 species. At 298 K, only a signal at −16 ppm attributed to gaseous $^{13}CH_4$ was observed. An additional signal at 59 ppm appeared at 578 K, and was still observable at 623 K. The chemical shift of this signal was typical of –$O^{13}CH_3$ species generated on the zeolite surface. Therefore, the observation of the –$O^{13}CH_3$ species, as well as of Ag_n–H, provided clear evidence for the cleavage of the C–H bond of methane on Ag-zeolites, as shown in reaction (7.48). However, silver-methyl species, such as Ag_n–CH_3, were not observed in the ^{13}C MAS NMR spectra, nor was a signal corresponding to the acidic O–H group protons found at ~ 4 ppm in the ^{1}H MAS NMR spectra. Based on these results, methane undergoes cleavage of a C–H bond over Ag-zeolites to simultaneously generate –OCH_3 species, which are the source of the $^{+}CH_3$ moieties and silver hydrides such as Ag_n-H, as shown in reaction (7.48).

Furthermore, they reported that the –$O^{13}CH_3$ species peak at 59 ppm disappeared when the temperature was increased from 623 K to 673 K, while new peaks appeared at 4 ppm and 109 ppm. These signals were assigned to ethane and ethylene (π-complex with Ag^+ cations), respectively. When the system was further heated to 723 K, a new broad signal at 128 ppm was observed, while a broad peak at 128–134 ppm and a peak at ~ 20 ppm were evident at 823 K. These two peaks were attributed to methyl-substituted aromatics such as toluene and xylenes. The results described above clearly show that methane is converted into aromatic hydrocarbons via ethane and ethylene.

Gabrienko et al. also examined the reaction of ethane and ethylene over Ag-ZSM-5 via ^{13}C MAS NMR spectroscopy and proposed that ethane was the primary product formed in methane conversion [74]. Thus, ethylene was formed stepwise from ethane. Further experiments demonstrated that methane was more effectively incorporated into aromatic hydrocarbons in the presence of ethylene. They also investigated the H/D exchange reaction between CD_4 and the acidic O–H groups on H-ZSM-5 and Ag-ZSM-5 zeolites by measuring in situ ^1H MAS NMR spectra. The H/D exchange rate over Ag-ZSM-5 was two orders of magnitude faster than that over H-ZSM-5. This result indicates that the Ag^+ cation facilitated the reversible dissociation of C–H bond of methane, which provides further evidence that the $-OCH_3$ species are the key reaction intermediates of the H/D exchange of CD_4 and $-OCH_3$ groups.

7.7.5 Reaction of Methane with Ethylene Over Ag^+-Exchanged Zeolites

The $-OCH_3$ species on zeolites can react with ethylene to produce propylene, and regenerate the acidic O–H groups, as shown in reaction (7.34). The $-OCH_3$ groups generated on Ag-zeolites act as $^+CH_3$ donors, allowing the reaction of methane with ethylene, propylene, and benzene to proceed [37, 75–78]. In the following sections, the characteristics of Ag-zeolite catalysts relevant to C–C bond formation in the reaction of methane with ethylene are discussed. In particular, the activation mechanism of CH_4 over Ag-zeolites is emphasized and compared with that of other metal-cation-exchanged zeolites (metal-cation zeolites). Additionally, the differences between the catalytic properties of Ag-zeolites and those of superacid catalysts in the conversion of methane in the presence of ethylene are focused upon.

7.7.5.1 Reaction of $^{13}CH_4$ with Ethylene Over Ag^+-Exchanged Zeolites

In order to examine whether CH_4 reacts with C_2H_4 to produce propylene (C_3H_6) and H_2 (reaction (7.31)), the reaction of ^{13}C-labeled methane ($^{13}CH_4$) with ethylene was carried out using Ag-zeolites (Ag-Y, Ag-A, and Ag-ZSM-5) in a closed gas-circulation reactor at 673 K [75]. To minimize side reactions such as the oligomerization–cracking of ethylene, a large excess of $^{13}CH_4$ relative to ethylene was used and the reaction time was shortened to 1 min. Propylene was formed over all three of the tested Ag-zeolites, as shown in Table 7.3. In all cases, a significant proportion of the propylene was singly ^{13}C-labeled, *i.e.*, $^{13}CC_2H_6$. This indicated that the propylene was produced by the reaction of $^{13}CH_4$ with C_2H_4, in other words, by the addition of a $^{13}CH_3{}^+$ carbenium ion to an ethylene molecule. Although not quantified, the formation of H_2 was confirmed. Unlabeled propylene (C_3H_6) was also produced, probably because of the oligomerization/cracking of ethylene. Over Ag-Y and Ag-A, ethane was also formed as a gaseous product in addition to propylene. However, the ethane was not labeled with a ^{13}C atom, showing that the ethane did not originate from methane.

As shown in Table 7.3, the ethylene conversion of H-ZSM-5 was very low, and the mole fraction of ^{13}C-labeled propylene was only 6%, which is close to the natural abundance of the ^{13}C atom in propylene. Butanes and ethane were also produced, and had mole fractions of ^{13}C atoms close to those expected based on the natural abundance of ^{13}C. Therefore, methane was not converted to butanes and ethane, which were produced from ethylene, indicating that the Brønsted acid sites (acidic protons) on H-zeolites such as H-ZSM-5 cannot cleave the C–H bond of CH_4 to generate $^+CH_3$ carbenium ions (–OCH_3 species), and instead only catalyze the conversion of ethylene to C_3H_6 and higher hydrocarbons such as butenes. The latter result indicates that $^+C_2H_5$ carbenium ions are easily produced by the addition of H^+ to C_2H_4; however, these $^+C_2H_5$ carbenium ions cannot react with CH_4, as shown in reaction (7.36). Thus, propylene is produced solely from C_2H_4 over H-zeolites. The conversion of C_2H_4 to C_3H_6 over H^+-exchanged zeolites is well known. For example, SAPO-34, which has a pore entrance diameter approximately equal to the kinetic diameter of the C_3H_6 molecule, can selectively produce C_3H_6 from C_2H_4 [79] via $^+C_2H_5$ carbenium ions as a reaction intermediate.

Table 7.3 Reaction of $^{13}CH_4$ with ethylene over Ag^+-exchanged zeolites and H-ZSM-5 in a closed circulation system^a (Reprinted from ref. [75], Copyright 2020, with permission from Elsevier)

Catalyst		Ag (51%)-Y	Ag (69%) -A	Ag (17%)-ZSM-5	H (100%)-ZSM-5
Pressure/kPa	$^{13}CH_4$	39.4	38.8	39.5	39.5
	C_2H_4	1.21	1.12	0.412	0.399
Conversion/%	C_2H_4	10	37	10	0.8
Selectivity/mol%	C_2H_6	35	72	0	10
	C_3H_6	65	28	100	71
	C_4H_8	0	0	0	19
Fraction of singly ^{13}C labeled propene in total propene/%	$^{13}CCH_6$	86	80	87	6

^{a}Catalyst: 0.1 g; temperature: 673 K, reaction time: 1 min, reactor volume: 408 cm^3
Values in parentheses denote the degree of Ag^+ or H^+ ion exchange

However, in the case of superacid catalysts, reaction (7.30) proceeds via the formation of $^{+}C_2H_5$ carbenium ions; C_3H_6 is not produced in this reaction and C_3H_8 is a product, as described in Sect. 7.5.2.

7.7.5.2 Reaction of CH_4 with Ethylene or Propylene Over the Zeolite Ag-ZSM-5 in a Flow Reactor

The reaction of methane with ethylene over the zeolite Ag-ZSM-5 was also carried out in a flow reactor [37, 75]. An equimolar mixture of CH_4 (33.8 kPa) and C_2H_4 (33.8 kPa) was passed over the catalyst at 598 K with a W/F of 3.6 g h mol^{-1}, where W (g) is the weight of the catalyst and F (mol h^{-1}) is the total flow rate. The time courses of the methane and ethylene conversion over Ag-ZSM-5 are shown in Fig. 7.5. Propylene and butenes were the main products. Under these reaction conditions, the conversion of CH_4 remained constant (8%), however the C_2H_4 conversion gradually decreased with increasing time on stream.

The product distributions obtained using the same gas feed mixture and W/F at a temperature of 673 K over Ag-Y, H-Y, Ag-A, Ag-ZSM-5, and H-ZSM-5 are summarized in Table 7.4. The conversions of methane over Ag-A, Ag-Y, and Ag-ZSM-5 were 2.1%, 6.5%, and 13.2%, respectively. Appreciable quantities of aromatic hydrocarbons were formed over Ag-ZSM-5 at this temperature, while Ag-A produced C_2H_4, C_3H_6, and C_3H_8 via a molecular sieving effect. On the other hand, H-zeolites (H-Y and H-ZSM-5) were not able to convert methane. This shows that H^+-exchanged zeolites cannot cleave the C–H bond of CH_4.

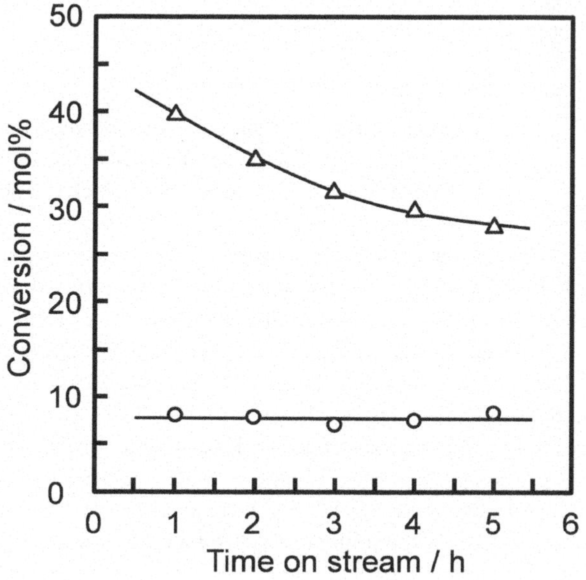

Fig. 7.5 Conversion of CH_4 (○) and C_2H_4 (△) in the reaction of a CH_4–C_2H_4 mixture over Ag-ZSM-5 as a function of the time on stream (Reproduced from Ref. [37] with permission from the PCCP Owner Societies). Reaction conditions: 598 K, CH_4 = C_2H_4 = 33.8 kPa, W/F = 3.8 g h mol^{-1}

Table 7.4 Conversion of CH_4 in the presence of C_2H_4 over silver-exchanged zeolites (Reproduced from Ref. [37] with permission from the PCCP Owner Societies, and reprinted from ref. [75], Copyright 2020, with permission from Elsevier)

Catalyst		Ag (46%)-Y	H (68%)-Y	Ag (60%)-A	Ag (17%)-ZSM-5	H (100%)-ZSM-5
Reaction temperature/K		673	673	623	673	673
Conversion/mol%	CH_4	6.5	0	2.1	13.2	0
	C_2H_4	15.9	11.8	3.2	86.3	93.9
Selectivity/mol%	C_2H_6	41.0	62.6	33.6	1.8	2.0
	C_3H_6	29.7	10.3	38.9	20.5	10.3
	C_3H_8	4.2	4.5	27.5	11.6	26.5
	C_4H_8	7.5	4.2	0	9.8	9.0
	C_4H_{10}	11.7	13.5	0	13.6	18.2
	C_{5+} Aliphatics	5.9	4.0	0	12.6	11.9
	Aromatics	0	0	0	30.1	22.1

Pressure: $CH_4 = C_2H_4 = 33.8$ kPa, temperature: 673 K, running time: 1 h, *W/F*: 3.6 g h mol^{-1}
Helium was used as a carrier gas and internal standard to determine the amounts of hydrocarbons
Values in parentheses denote the degree of Ag^+ or H^+ exchange

The reaction of CH_4 with C_3H_6 proceeded at lower reaction temperatures than its reaction with C_2H_4 over Ag-zeolites, because C_3H_6 is more nucleophilic than C_2H_4. The reaction was carried out at 523 K and 623 K using Ag-ZSM-5 and Ag-A, respectively. The conversion and hydrocarbon product distribution of these reactions at a running time of 1 h are shown in Table 7.5 [37, 75]. Using Ag-ZSM-5, mainly butenes were produced, while Ag-A gave only 1-butene, with other butenes (2-butenes and *iso*-butene) not being observed.

7.7.5.3 Experimental Evidence for the Formation of $^+CH_3$ Carbenium Ions Over Ag^+-Exchanged Zeolites: Reaction of $^{13}CH_4$ with Benzene to Produce ^{13}C–Labeled Toluene

The formation of toluene by the reaction of CH_4 with benzene on Ag-zeolites provides clear evidence for the formation of $^+CH_3$ carbenium ions (–OCH_3 species) [39]. Thus, $^+CH_3$ carbenium ions are essential to the production of toluene in the reaction of methane with benzene. To demonstrate this phenomenon, the reaction of $^{13}CH_4$ with benzene was carried out at 473 K using Ag-ZSM-5 as a catalyst in a closed gas-circulation reactor [37]. The pressure of $^{13}CH_4$ was 39.8 kPa, which was much higher than that of benzene (1.6 kPa). Toluene was formed in 5.8% yield in 2 min at 6.1% conversion of benzene. Toluene with a ^{13}C atom-labelled methyl group was produced. The proportion of ^{13}C-labeled toluene (**17**) ($^{13}CH_3$–C_6H_5) was 93%, indicating that the majority of the toluene was produced by the methylation of benzene by reaction (7.50) over Ag-ZAM-5.

Table 7.5 Conversion of CH_4 in the presence of C_3H_6 using Ag-ZSM-5 and Ag-A zeolites (Reproduced from Ref. [37] with permission from the PCCP Owner Societies, and reprinted from ref. [75], Copyright 2020, with permission from Elsevier)

Ag^+-exchanged zeolite		Ag (17%)-ZSM-5	Ag (100%)-A
Reaction temperature/K		523	623
Conversion/%	CH_4	4.5	2.1
	C_3H_6	42.6	3.2
Hydrocarbon distribution/mol%	C_2H_4	0	0
	C_2H_6	0	33.6
	C_3H_8	0	38.9
	C_4H_8	38.6	11.2 (1-Butene)
	C_4H_{10}	3.4	0
	C_5H_{10}	8.3	0
	C_5H_{12}	3.6	0
	C_6	26.5	0
	C_7	11.1	0
	Aromatic hydrocarbons	8.5	0

$CH_4 = C_3H_6 = 33.8$ kPa, running time: 1 h, $W/F = 3.6$ g h mol^{-1}

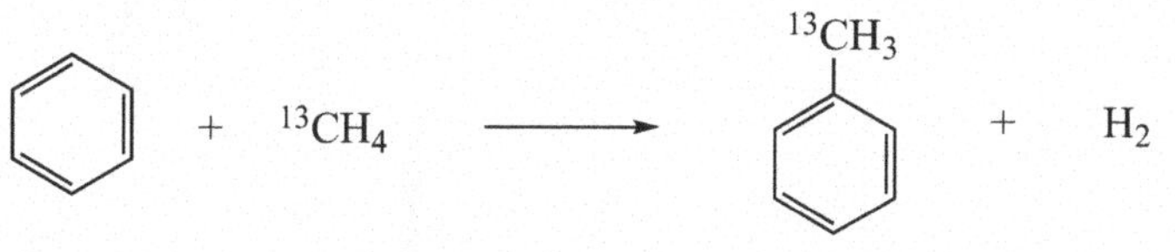

(17)

(7.50)

The results described above provide further support for the formation of $^+CH_3$ carbenium ions over Ag-zeolites and the electrophilic attack of the $^+CH_3$ carbenium ions on the benzene molecule to produce toluene. However, H-ZSM-5 cannot catalyze the reaction of methane with benzene to produce toluene and H_2, meaning that reaction (7.50) does not proceed over H-zeolites. Thus, they cannot cleave the C–H bond of methane to produce $^+CH_3$ carbenium ions.

The reaction of benzene with CH_4 was also examined at 598 K in a flow reactor using both Ag-ZSM-5 and H-ZSM-5 [75]. The contact time (W/F) was 8.2 g h mol^{-1} and the partial pressures of CH_4 and benzene were 31.5 and 38.2 kPa, respectively. Under these reaction conditions, Ag-ZSM-5 gave 89.4% toluene and 10.6% benzene at a running time of 1 h, at which the conversions of CH_4 and benzene were 2.2 and 1.6%, respectively. The xylenes were presumably formed by the disproportionation of toluene on the acidic sites of Ag-ZSM-5 and/or by the methylation of toluene with CH_4. However, H-ZSM-5 did not show any catalytic activity in this reaction.

As described above, superacid catalysts, such as $HSO_3F–SbF_5$ and $HF–SbF_5$, can produce $^+CH_3$ carbenium ions from CH_4, as shown in reaction (7.28). However, in mixtures of methane and benzene, superacids do not catalyze the reaction of methane with benzene to produce toluene or other aromatic hydrocarbons. This indicates that superacid catalysis is essentially different from Ag-zeolite catalysis.

7.7.5.4 Mechanism of C–C Bond Formation in the Conversion of Methane Over Ag^+-Exchanged Zeolites

As discussed in Sect. 7.7.5, the formation of propylene and toluene by the reaction of methane with ethylene and benzene, respectively, over Ag-zeolites proceeds via $^+CH_3$ carbenium ions. Based on reactions (7.47) and (7.48), the reaction mechanism for the methane conversion is shown in Scheme 7.4. Ag_n^+ cationic clusters activate methane to generate Ag_n–H. The Ag_n^+ cationic clusters are then regenerated by the reaction of Ag_n–H with the acidic protons, which is the reverse reaction shown in reaction (7.47).

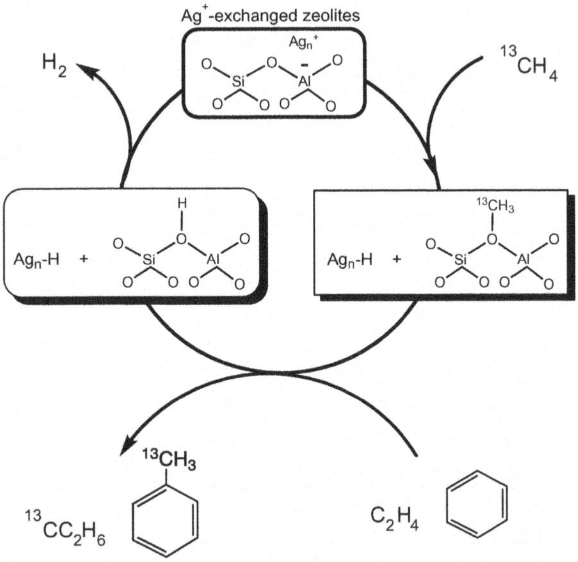

Scheme 7.4 Mechanism of methane conversion over Ag^+-exchanged zeolites

7.7.6 Reaction of Methane with Ethylene Over Metal-Cations-Exchanged Zeolites

7.7.6.1 Mechanism of the C–H Bond Cleavage of CH_4 Over Metal-Cation-Exchanged Zeolites

As mentioned above, the C–H bond cleavage of CH_4 proceeds over Ag-zeolites to form silver hydrides (Ag_n–H) and $-OCH_3$ species, as shown in reaction (7.48). However, neither Ag-methyl (Ag–CH_3) species nor acidic protons are generated. On the other hand, for metal-cation-exchanged zeolites (M-cation zeolites), such as Zn^{2+}-exchanged zeolites (Zn-zeolites), two types of methane activation mechanisms (**Mechanism 1** and **Mechanism 2**) have been proposed, as described below.

(1) **Mechanism 1**: Heterolytic adsorption of CH_4 to form metal–CH_3 (M–CH_3) as the metal-alkyl species, along with acidic O–H groups (reaction (7.51)).

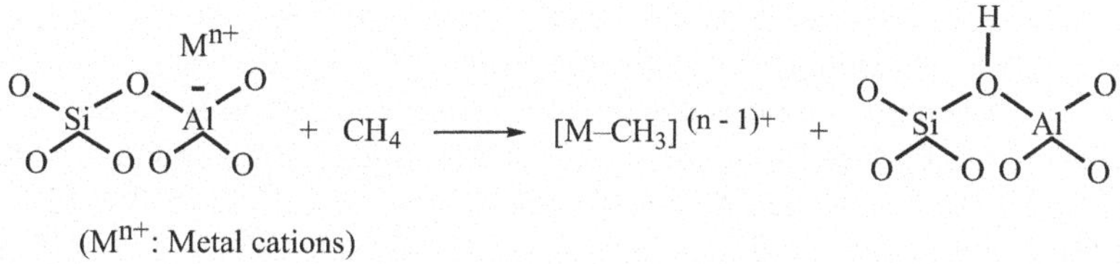

(7.51)

Thus, CH_3^- and H^+ are formally generated.

(2) **Mechanism 2**: Heterolytic dissociation of CH_4 to form metal hydrides (M–H) and methoxy ($-OCH_3$) species on the zeolites, as in the case of Ag-zeolites.

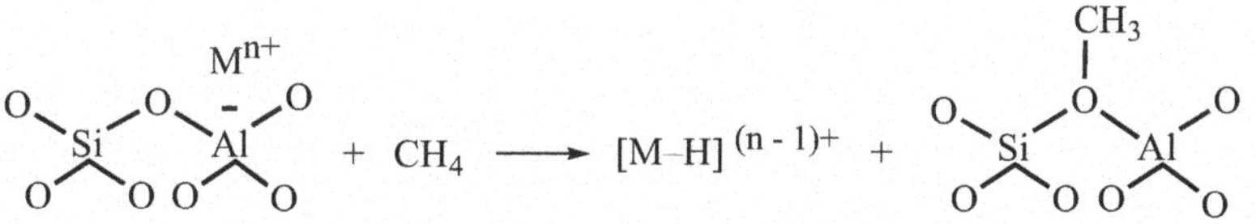

(7.52)

Thus, $^+CH_3$ carbenium ions and H^- ions are formally generated.

Theoretical studies of the activation of methane over Ag-zeolites and Zn-zeolites usually propose that **Mechanism 1** is much more favorable than **Mechanism 2** [79, 80]. However, **Mechanism 2** has not been proposed to occur over Ag-zeolites. For example, Ag-ZSM-5 gave only Ag–CH_3 and acid protons via **Mechanism 1**. However, as mentioned in Sect. 7.7.3, Ag-hydride species such as Ag_n–H were observed on Ag-zeolites by ^1H MAS NMR spectroscopy, while acid protons were not observed at all. Furthermore, ^{13}C MAS NMR spectra showed $-OCH_3$ groups after the adsorption of methane on zeolite Ag-ZSM-5, while the Ag-CH_3 species was not generated [74]. These results indicate that the activation of methane over Ag-zeolites proceeds selectively via **Mechanism 2**.

In the case of Zn-zeolites, the C–H bond cleavage reaction proceeds via both **Mechanism 1** and **Mechanism 2**. For example, Zn^{2+}-exchanged β-zeolite (Zn-β zeolite) [81–84], and Zn-ZSM-5 zeolite [85–87] activate CH_4 to generate both Zn–CH_3 species and –OCH_3 groups, which were observed via ^{13}C MAS NMR spectroscopy.

Zn^{2+} and In^{3+} (InO^+) cation-exchanged zeolites have also been reported to form zinc–methyl (Zn–CH_3) species and indium–methyl species (In–CH_3), which were further transformed to zinc–methoxy (Zn–OCH_3) [88] and In-methoxy (In–OCH_3) species [89], respectively. These metal–methoxy (M–OCH_3) species reacted with ethylene and benzene to produce propylene and toluene, respectively.

7.7.6.2 Reaction of $^{13}CH_4$ with C_2H_4 Using Metal-Cation-Exchanged Zeolites: Roles of Metal Cations and/or Metal Cationic Species

Because the –OCH_3 species react with ethylene to produce C_3H_6 over M-cation zeolites as shown in reaction (7.34), the reaction of CH_4 with ethylene should proceed via $^+CH_3$ carbenium ion intermediates (**Mechanism 2**). Baba et al. reported that the reaction of $^{13}CH_4$ with C_2H_4 using various M-cation ZSM-5 zeolites proceeded to produce singly ^{13}C-labeled hydrocarbons such as $^{13}CC_2H_6$ at 673 K [75, 76, 90]. The results for In-, Fe-, and Mo-ZSM-5 zeolites, as well as for Ag-ZSM-5, are summarized in Table 7.6. These M-cation zeolites produced mainly propylene, while

Table 7.6 Reaction of $^{13}CH_4$ with C_2H_4 over metal cation-exchanged ZSM-5 zeolites (Reprinted from ref. [75], Copyright 2020, with permission from Elsevier)

Catalyst[a]		In-ZSM-5	Fe-ZSM-5	Mo-ZSM-5	Ag-ZSM-5
Pressure/kPa	$^{13}CH_4$	39.4	39.5	39.6	39.5
	C_2H_4	0.346	0.359	0.346	0.412
Conversion/mol%	C_2H_4	13.0	9.2	4.0	10.0
Selectivity/mol%	C_2H_6	17	33	37	0
	C_3H_6	23	67	63	100
	Benzene (C_6H_6)	41	0	0	0
	Toluene (C_7H_8)	19	0	0	0
(Singly ^{13}C labeled hydrocarbon/hydrocarbon) × 100	$^{13}C^{12}C_2H_6$ in C_3H_6	33	62	24	87
	$^{13}C^{12}C_5H_6$ in C_6H_6	9	–	–	–
	$^{13}C^{12}C_6H_8$ in C_7H_8	54	–	–	–

Catalyst weight: 0.1 g, reaction temperature: 673 K, reaction time: 1 min, reactor volume: 408 cm^3
[a]The molar ratio of metal cations to Al^{3+} was 0.17

In-ZSM-5 also gave benzene and toluene. The main product was propylene, with singly ^{13}C-labeled propylene ($^{13}CC_2H_6$) being formed to varying extents. However, no multi ^{13}C-labeled propylene or other hydrocarbons were observed. The $^{13}CC_2H_6$ fraction of the produced propylene depended on the kind of metal cations, clearly indicating that methane was activated on metal cations and/or metal cationic species via **Mechanism 2**, not **Mechanism 1**. Furthermore, propylene was produced by both the reaction of CH_4 with C_2H_4 and the direct conversion of C_2H_4 via oligomerization/cracking reactions.

When the zeolite In-ZSM-5 was used as a catalyst, singly ^{13}C-labeled toluene ($^{13}CC_6H_8$) was also produced. The fraction of ^{13}C-labeled toluene ($^{13}CC_6H_8$) among the total toluene was 54%. However, the $^{13}CC_5H_6$ fraction of benzene was 9%, which was close to the percentage expected based on the natural abundance of ^{13}C atoms, indicating that the benzene originated solely from C_2H_4, and was not produced by the conversion of propylene.

To investigate the position of ^{13}C atom in the $^{13}CC_6H_8$, its fragmentation MAS spectrum was measured [90]. Two kinds of ^{13}C-labeled toluene were observed: $^{13}CH_3$–C_6H_5 and CH_3–$^{13}CC_5H_5$. The fraction of $^{13}CH_3$–C_6H_5 in the produced toluene was 17%, while that of CH_3-$^{13}CC_5H_5$ was 37% (54 – 17 = 37%). $^{13}CH_3$-C_6H_5 was produced by the reaction of $^{13}CH_4$ with benzene. CH_3-$^{13}CC_5H_5$ was presumably formed by the reaction of $^{13}CC_2H_6$ with butanes (C_4H_8) produced by the dimerization of C_2H_4 and/or by isomerization of $^{13}CH_3$–C_6H_5.

7.7.6.3 Reaction of $^{13}CH_4$ with Benzene Using in-ZSM-5

According to **Mechanism 2**, $^+CH_3$ carbenium ions should be also generated over M-cation zeolites in addition to Ag-zeolites. To provide experimental evidence for this, the reaction of $^{13}CH_4$ with benzene was carried out for 1 min using In-ZSM-5 in a closed gas-circulation reactor. The reaction temperature was 673 K, and the pressures of $^{13}CH_4$ and benzene were 39.8 and 1.63 kPa, respectively. Under the above reaction conditions, the conversion of benzene was 5.6%, and both toluene and H_2 were reaction products. The mole fraction of singly ^{13}C-labeled toluene in the total produced toluene was 97%, indicating that the $^+CH_3$ carbenium ions generated on In-ZSM-5 subsequently reacted with benzene to produce toluene. Therefore, methane can be activated not only by Ag_n^+ cationic clusters but also metal cations, such as In cations, to generate $^+CH_3$ carbenium ions via **Mechanism 2** (reaction (7.52)).

As described in Sect. 7.7.7.2, both $^{13}CH_3$–C_6H_5 and CH_3–$^{13}CC_5H_5$ were produced in the reaction of $^{13}CH_4$ with C_2H_4 over In-ZSM-5. To further investigate the position of the ^{13}C atom in the singly ^{13}C-labeled toluene, the fragmentation of toluene was analyzed using mass spectroscopy. The fragmentation spectrum of the ^{13}C-labeled toluene is shown in Fig. 7.6 as Spectrum A, while that of ^{13}C-free toluene (CH_3–C_6H_5) is shown as Spectrum B. In spectrum A, the main peak observed

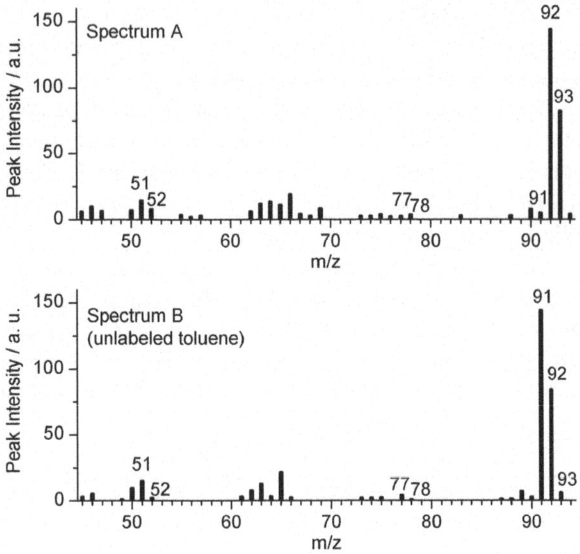

Fig. 7.6 (Spectrum A) Mass spectrum of the toluene formed by the reaction of $^{13}CH_4$ over In-ZSM-5 in the presence of benzene at 673 K (Reprinted with permission from ref [90]. Copyright 2020 American Chemical Society) (Spectrum B) Spectrum of unlabeled toluene shown for comparison

at *m/z* 92 was attributed to $^{13}CC_6H_7^+$, while that located at *m/z* 91 in Spectrum B corresponded to $^+C_7H_7^+$. Toluene fragmentation occurs via two paths [91], as shown in Scheme 7.5.

Path A: The methyl group is ejected from toluene to give a molecular ion with *m/z* 77, which then gives other molecular ions with *m/z* 51.

Path B: Protons (H^+) are ejected to give molecular ions with *m/z* 91, which then produce the other molecular ions with *m/z* 65.

In Spectrum A, the mole fractions of $^{13}CH_3$–C_6H_5 and CH_3–$^{13}CC_5H_5$ can be determined by comparing the ion abundance of the peaks at *m/z* 51 and *m/z* 52, respectively. Thus, the mole fraction of $^{13}CH_3$-C_6H_5 was estimated to be 56% of the total mole fraction of ^{13}C-labeled toluene ($^{13}CC_6H_8$) of 97%, while that of CH_3–$^{13}CC_5H_5$

Scheme 7.5 Fragmentation of toluene in MS

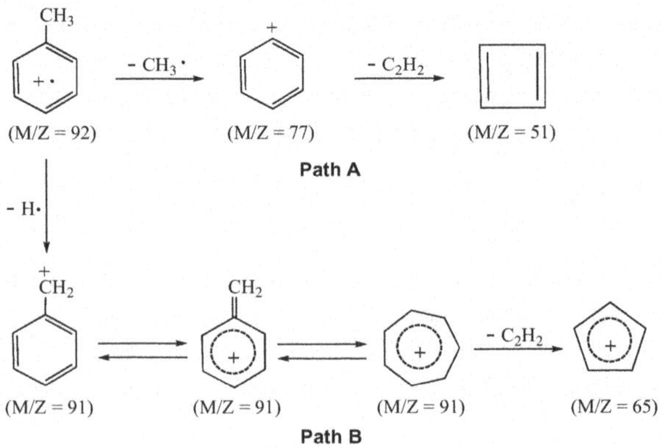

was 41% (97−56 = 41%). CH_3–$^{13}CC_5H_5$ was produced by the isomerization of the $^{13}CH_3$-C_6H_5 initially produced by the reaction of $^{13}CH_4$ with benzene.

The reaction of CH_4 with benzene over In-ZSM-5 was carried out at 623 K in a flow reactor [90]. The conversions of CH_4 (33.8 kPa) and benzene (33.8 kPa) were 2.0% and 1.8%, respectively, at 1 h time on stream and a W/F of 8.2 g h mol^{-1}. The main product was toluene (87% selectivity), while xylenes (13% selectivity) were also produced. Xylenes were presumably formed by the methylation of toluene with CH_4, even though the disproportionation of toluene proceeded to produce xylenes and benzene.

7.7.6.4 Catalytic Performance of Metal-Cation-Exchanged ZSM-5 Zeolites for the Reaction of Methane with Ethylene

M-cation zeolites with metals other than Ag also showed catalytic activity in the reaction of methane with ethylene [75]. Their catalytic activities at 1 h time on stream are summarized in Table 7.7. The reactions were carried out at 673 K using various M-cation ZSM-5 zeolites with metal-to-Al molar ratios of 0.17. The conversion of CH_4 strongly depended on the metal cations, showing that metal cations, as well as cationic silver clusters such as Ag_n^+, can activate CH_4. However, H-ZSM-5 cannot activate CH_4, as discussed above.

The conversions of CH_4 over almost all of the M-cation ZSM-5 zeolites decreased with time on stream. However, In-ZSM-5 showed much more stable catalytic activity, with the conversion actually increasing from 8.1% at 1 h to 8.6% after 5 h, due to the fact that In cations are more resistant to reduction than Ag^+ cations. Furthermore, the catalytic activity of In-ZSM-5 strongly depended on its preparation conditions, *i.e.*, the temperature used for calcination before hydrogen treatment and the pretreatment of In cations with hydrogen [90]. The highest CH_4 conversion was 11.8% for In-ZSM-5 pretreated with hydrogen at 723 K after calcination at 903 K. The CH_4 conversion remained constant, as shown in Fig. 7.7 [90].

7.7.7 *Differences Among the Catalytic Properties of Ag^+-, Zn^{2+}-, and H^+-Exchanged Zeolites in the Activation of Lower Alkanes Including CH_4*

As discussed in Sect. 7.7.6, M-cation zeolites with metals other than Ag can activate methane, as shown in reactions (7.51) and (7.52). The generation of $[M–CH_3]^{(n-1)+}$ and $[M–H]^{(n-1)+}$ leads to the formation of CH_4 and H_2, respectively, as reported in references 93 and 94. However Ag-zeolites selectively produce silver hydride species, such as Ag_n–H, as well as $[M–H]^{(n-1)+}$, which are converted to H_2.

The generation of $[M–H]^{(n-1)+}$ and $[M–CH_3]^{(n-1)+}$ can be applied to activate higher alkanes than CH_4, such as butenes, to produce many kinds of carbenium ions.

Table 7.7 Catalytic activities for the conversion of methane in the presence of ethylene over ZSM-5 exchanged with various metal cations (Reprinted from ref. [75], Copyright 2020, with permission from Elsevier)

Metal ion[a]		Ag	Mo	In	Fe	V	Ga	Pd	Pb	La	Zn	Cu	H[b]
Conversion/%	CH_4	13.2	9.5	8.1	7.9	7.6	7.0	5.5	5.4	2.5	1.8	1.7	0
	C_2H_4	86.3	80.2	97.6	88.8	94.0	97.2	95.5	78.1	93.2	88.8	97.0	93.9
Selectivity towards hydrocarbons/%[c]		68.9	94.6	98.6	76.3	75.7	92.1	82.5	80.5	96.7	99.8	99.2	91.3
Composition/mol%	C_2H_6	1.8	0.0	1.7	1.0	2.3	3.7	3.7	0.0	1.7	1.2	1.7	2.0
	C_3H_6	20.6	20.3	5.5	20.6	11.2	5.9	9.2	24.3	11.5	16.7	6.0	10.3
	C_3H_8	11.7	13.7	25.0	18.2	24.8	28.6	22.4	13.0	23.0	7.9	24.6	26.5
	C_4H_8	9.9	14.7	3.6	15.7	8.0	4.0	6.3	17.8	8.5	10.8	4.3	9.0
	C_4H_{10}	13.7	16.0	24.9	14.8	22.1	23.5	18.5	11.3	18.9	11.9	36.5	18.2
C_5–C_7 hydrocarbons		12.7	14.2	13.4	13.9	11.9	7.4	8.7	14.2	10.6	12.4	8.1	11.9
Aromatic hydrocarbons		30.3	21.1	25.9	15.8	20.1	26.9	31.2	19.4	25.8	39.1	18.8	22.1

$CH_4 = C_2H_4 = 33.8$ kPa, running time: 1 h, *W/F*: 3.6 g h mol^{-1}, reaction temperature = 673 K
[a]The molar ratio of metal cations to Al^{3+} was 0.17
[b]H (100%)-ZSM-5
[c]Moles carbon in hydrocarbon products per 100 mol ($CH_4 + C_2H_4$) carbon converted

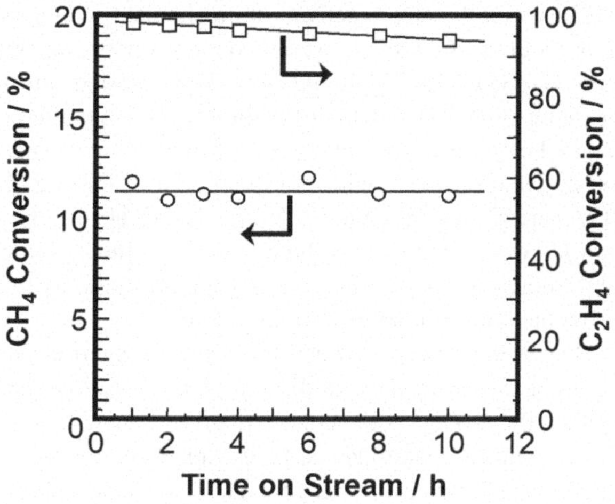

Fig. 7.7 Conversion of CH_4 (□) and C_2H_4 (○) in the reaction of a CH_4–C_2H_4 mixture as a function of the running time over In-ZSM-5 (Reprinted with permission from ref [90]. Copyright 2020 American Chemical Society). Reaction conditions: 673 K, CH_4 = C_2H_4 = 33.8 kPa, *W*/*F*: 3.8 g h mol^{-1}

As mentioned before, the carbenium ions also gored over H-zeolites and superacid catalysts.

In this section, the differences between the catalytic properties of Ag-zeolites and Zn-zeolites are focused upon and compared with the catalytic properties of H-zeolites.

7.7.7.1 Primary Reactions of Lower Alkanes Over H-Zeolites

Based on the mechanism of alkane activation in superacid solutions, the Haag-Dessau mechanism was proposed in 1994. The mechanism involves penta-coordinated carbonium ions as a reaction intermediate in the reactions of lower alkanes, such as *iso*-butane, over H-zeolites [32, 33] as shown in reaction (7.53).

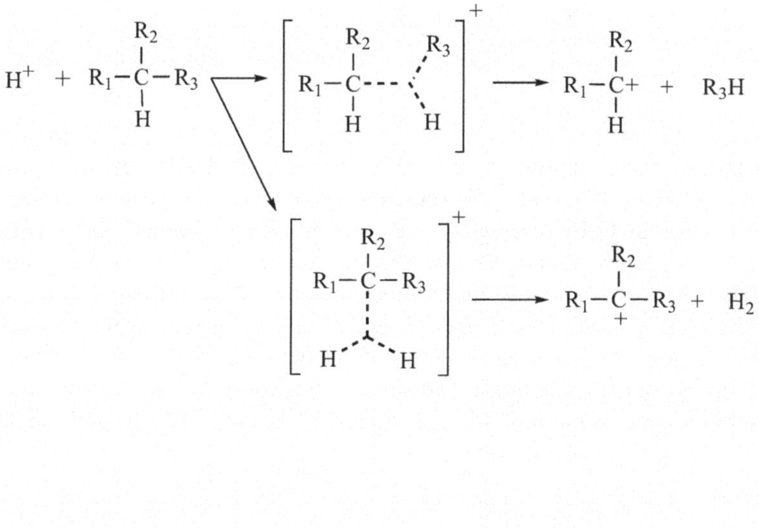

(7.53)

Based on the above activation mechanism, which is valid for lower alkanes other than methane, Ono et al. examined the primary reactions in the cracking of *n*-butane over H-ZSM-5 at 773 K [92, 93]. They carried out the reaction at a low *n*-butane pressure of 8.0 kPa, which resulted in an *n*-butane conversion of less than 5%, to avoid the hydride-transfer reaction. Some of their experimental results (*n*-butane conversion and the selectivities of the produced hydrocarbons at different contact times *W*/*F*) are shown in Table 7.8 [92]. CH_4, C_2H_6, C_2H_4, C_3H_8, C_3H_6, C_4H_8 (butanes), and H_2 were produced as the reaction products, while $C_5{}^+$ (aliphatic hydrocarbons with more than five carbon atoms) and aromatic hydrocarbons were not produced at *n*-butane conversions lower than ~ 5.4%.

The authors proposed that the primary reaction products in *n*-butane conversion could be determined by plotting the selectivities of the hydrocarbon products and H_2 against the *n*-butane conversion. In this manner, the relative rates of reaction products in the primary reactions can be estimated by extrapolating the product selectivities at zero conversion of *n*-butane. The estimated values are also shown in Table 7.8. According to these results, the pairs (CH_4 and C_3H_8), (C_2H_6 and C_2H_4), and (H_2 and C_4H_8) are the primary products [94–96]. They concluded that the primary reactions can be expressed by three reactions (7.54–7.56), which involve the activation of *n*-butane via penta-coordinated carbonium ion intermediates, as follows.

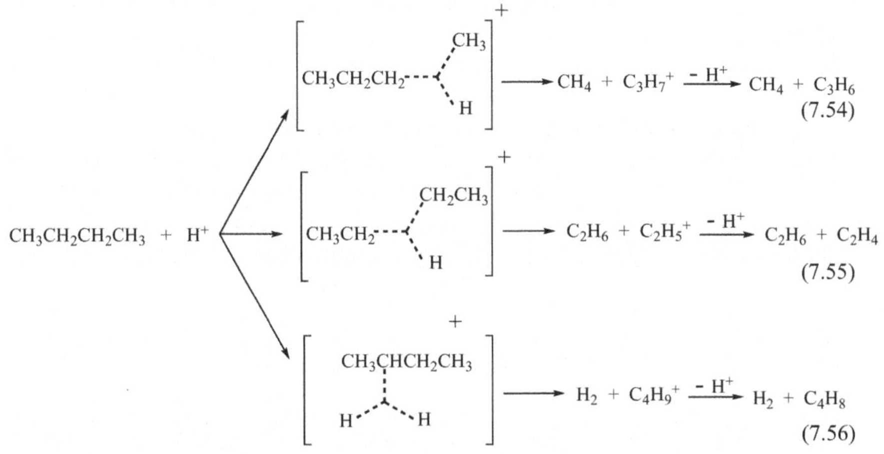

In these reactions, the carbenium ions $^+C_3H_7$, $^+C_2H_5$, and $^+C_4H_9$, release a proton (H^+), and then C_3H_6, C_2H_4, and C_4H_8 (butanes) are formed, respectively. However, these alkenes, such as C_4H_8, are easily converted to other hydrocarbons, and their selectivities at zero conversion on *n*-butane can not be considered as the real primary products. Therefore, in Table 7.8, the number of moles for each pair of products, (CH_4 and C_3H_6), (C_2H_6 and C_2H_4), and (H_2 and C_4H_8), do not completely coincide. This means that the alkenes or carbenium ions in the reactions (7.54), (7.55), and (7.56) undergo secondary reactions. The selectivities towards CH_4, C_2H_6, and H_2 extrapolated to zero conversion of *n*-butane represent the contributions of these three

Table 7.8 Dependence of the product selectivity on the contact time in *n*-butane cracking (8.0 kPa) over H-ZSM-5 at 773 K (Reproduced from Ref. [92] with permission from The Royal Society of Chemistry)

W/F/g h mol^{-1}	Conversion/%	Selectivity/mol in 100 mol *n*-butane converted							
		CH_4	C_2H_6	C_3H_8	*iso*-C_4H_{10}	C_2H_4	C_3H_6	C_4H_8	H_2
0.075	0.57	49	39	2.4	2.3	36	37	18	28
0.085	0.71	40	36	2.3	2.4	33	33	27	–
0.1	1.82	42	36	2.6	1.6	34	33	26	–
0.2	1.61	42	39	4.8	1.5	38	34	21	–
0.4	3.7	33	33	8.6	1.4	35	31	27	21
0.6	5.4	32	32	13	1.7	37	30	24	20
–	0[a]	42	38	0	0	33	34	27	20

[a]Extrapolated

Table 7.9 Absolute reaction rates for the primary products in n-butane cracking at 773 K (Reprinted from ref. [97], Copyright 2020, with permission from Elsevier)

Reaction	Reaction Rate
Reaction (7.54)	0.5×10^{-2} mol h^{-1} g^{-1}
Reaction (7.55)	0.5×10^{-2} mol h^{-1} g^{-1}
Reaction (7.56)	0.3×10^{-2} mol h^{-1} g^{-1}

reactions. Thus, the ratio of the extrapolated selectivity shows that reactions (7.54), (7.55), and (7.56) occur in a ratio of 42:38:20, *i.e.*, cleavage of C–C bonds dominates over that of C–H bonds in n-butane.

Furthermore, the absolute rates of the three reactions can be determined from the yields of CH_4, C_2H_6, and H_2 and the contact time (*W/F*). The details of the method were reported in ref. [97, 98]. The absolute reaction rates for the formation of the primary products over H-ZSM-5 are shown in Table 7.9 [97]. They also indicated that the primary reactions not only in n-butane cracking, but also in *iso*-butane, n-pentane, *iso*-pentane, and *neo*-pentane cracking, could be explained by the activation of these alkanes rather than by penta-coordinated carbenium ions over H-ZSM-5 [92, 97].

7.7.7.2 Catalytic Properties of Ag- and Zn-Zeolites Relevant to the Primary Reactions of Lower Alkanes: C–H and C–C Bond Cleavage Reactions

As mentioned in Sect. 7.7.7.1, in the n-butane cracking reaction, the initial selectivities towards the primary products H_2, CH_4, and C_2H_6 were estimated by extrapolating the their selectivities to zero conversion of n-butane using H-ZSM-5 zeolite as a catalyst. Additionally, the absolute formation rates of these primary products were determined. These treatment methods can be also applied to the cracking of lower alkanes over M-cation zeolites, even though the activation mechanisms of lower alkanes over M-cation zeolites are different from the mechanism over H-zeolites.

Ono et al. reported that the C–C and C–H bond cleavage reactions of lower alkanes (n-butane, *iso*-butane, n-pentane, *iso*-pentane, and *neo*-pentane) over M-cation zeolites such as Zn-zeolite proceed via **Mechanism 1** and/or **Mechanism 2**, as shown in reactions (7.51) and (7.52), respectively [97, 98]. They examined the conversion of lower alkanes at low reaction pressures and conversions of less than ~ 3% at 773 K using Ag-, Zn- and H-ZSM-5 zeolites to determine the primary reactions in the monomolecular cracking of these alkanes. The absolute formation rates of the primary products of lower alkane conversion over Ag- and Zn-ZSM-5 zeolites are summarized in Table 7.10 [97]. To allow comparison of the catalytic properties of Ag- and Zn-ZSM-5 catalysts with those of H-ZSM-5, the data over H-ZSM-5 are also included in Table 7.10.

(1) n-Butane cracking

When the Ag- and Zn-ZSM-5 zeolites are used as catalysts, the cracking of n-butane proceeds through the carbenium ion mechanism, and the primary reactions below take place.

Table 7.10 Rates of formation of the primary products in alkane conversion over H-ZSM-5, Ag-ZSM-5, and Zn-ZSM-5 (Reprinted from ref. [97], Copyright 2020, with permission from Elsevier)

Reactant	Catalyst	Formation rate/10^{-2} mol h^{-1} g^{-1}				
		Total Rate	H_2	CH_4	C_2H_6	C_3H_8
n-Butane	H-ZSM-5	1.3	0.3	0.5	0.5	–
	Ag-ZSM-5	4.8	4.5	0.2	0.1	–
	Zn-ZSM-5	7.0	5.8	0.7	0.4	–
iso-Butane	H-ZSM-5	1.0	0.5	0.5	–	–
	Ag-ZSM-5	27	27	0.2	–	–
	Zn-ZSM-5	44	36	8.1	–	–
n-Pentane	H-ZSM-5	1.5	0.4	0.3	0.6	0.2
	Ag-ZSM-5	8.8	7.8	0.3	0.5	0.2
	Zn-ZSM-5	11	8.9	0.9	1.0	0.5
iso-Pentane	H-ZSM-5	2.2	1.0	0.9	0.3	–
	Ag-ZSM-5	20	20	0.0	0.0	–
	Zn-ZSM-5	23	19	4.2	0.2	–
neo-Pentane	H-ZSM-5	1.0	0.0	1.0	–	–
	Ag-ZSM-5	0.7	0.0	0.4	–	–
	Zn-ZSM-5	28	0.0	28	–	–

Reaction conditions: 773 K, Reactant pressure = 8 kPa

$$CH_3CH_2CH_2CH_3 \rightarrow H_2 + C_4H_8 \quad (7.57)$$

$$CH_3CH_2CH_2CH_3 \rightarrow CH_4 + C_3H_6 \quad (7.58)$$

$$CH_3CH_2CH_2CH_3 \rightarrow C_2H_6 + C_2H_4 \quad (7.59)$$

Table 7.10 shows that over H-ZSM-5, the C–C bond cleavage reactions (7.58) and (7.59) predominate over the C–H bond cleavage reaction (7.57) to produce H_2. However, over Ag- and Zn-ZSM-5 zeolites, the C–H bond cleavage reaction (dehydrogenation reaction) is predominant over the C–C bond cleavage reaction. The rates of dehydrogenation (reaction (7.57)) over both Ag- and Zn-ZSM-5 zeolites are 15–20 times as fast as that over H-ZSM-5. Furthermore, the formation rates of CH_4 and C_2H_6 originating from the C–C bond cleavage reaction over Ag-ZSM-5 are lower than those over H-ZSM-5. The relative rates of reactions (7.57), (7.58), and (7.59) were 94:4:2 over Ag-ZSM-5, which was different from the ratio observed over H-ZSM-5 (3:5:5). Thus, *n*-butane cracking over H-ZSM-5 proceeds via the penta-coordinated carbonium-ion mechanism shown in the reactions (7.54), (7.55) and (7.56), which hardly proceed over the Ag- and Zn-ZSM-5 zeolites.

(2) *iso*-Butane cracking

The primary reactions of *iso*-butane cracking are reactions (7.60) and (7.61).

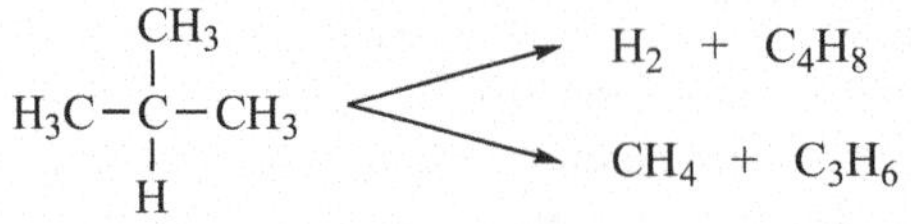

(7.60)

(7.61)

Thus, the formation rates of H_2 and CH_4 are the rates of the C–H and C–C bond cleavage reactions, respectively. The primary reaction rates over Ag- and Zn-ZSM-5 zeolites, together with those for H-ZSM-5, are summarized in Table 7.10. The rates of the C–H bond cleavage reaction over Ag- and Zn-ZSM-5 zeolites are much faster than that over H-ZSM-5. The rate of the C–C bond cleavage reaction (reaction (7.61)) over Ag-ZSM-5 is evidently different from that over Zn-ZSM-5. Ag-ZSM-5 suppressed this reaction (the formation of CH_4), as compared with H-ZSM-5, while Zn-ZSM-5 greatly enhanced it. The ratio of the CH_4 formation rates is Zn-ZSM-5/H-ZSM-5 = 81:5.

(3) *n*-Pentane cracking

In *n*-pentane cracking, the following four primary reactions occur.

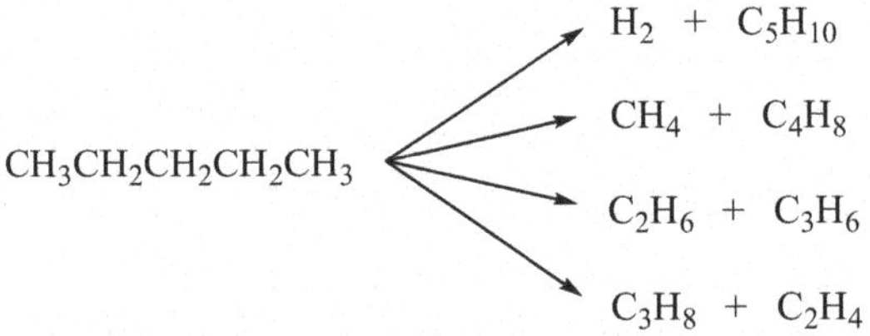

(7.62)

(7.63)

(7.64)

(7.65)

The C–H bond cleavage reaction (reaction (7.62)) gives H_2 and C_5H_{10}. C_5H_{10} is readily decomposed to lower alkenes and alkanes. The C–C bond cleavage reaction gives CH_4, C_2H_6, and C_3H_8 as primary reaction products through reactions (7.63), (7.64), and (7.65), respectively. The reaction rate for each of the above reactions is summarized in Table 7.10. Over H-ZSM-5, the ratio of the C–H bond cleavage reaction rate to that of the C–C bond cleavage reaction was 4:11. The rates of the C–H bond cleavage reaction (reaction (7.62)) over the zeolites Ag- and Zn-ZSM-5 were about 20 times faster than the rate over H-ZSM-5. The formation rates of CH_4, C_2H_6, and C_3H_8 over Ag-ZSM-5 were almost comparable to those over H-ZSM-5. However, the formation rates of these alkanes over Zn-ZSM-5 were almost three times faster than over H-ZSM-5.

(4) *iso*-Pentane cracking

In *iso*-pentane cracking, the primary reactions can be accounted for by reactions (7.66), (7.67), and (7.68) below.

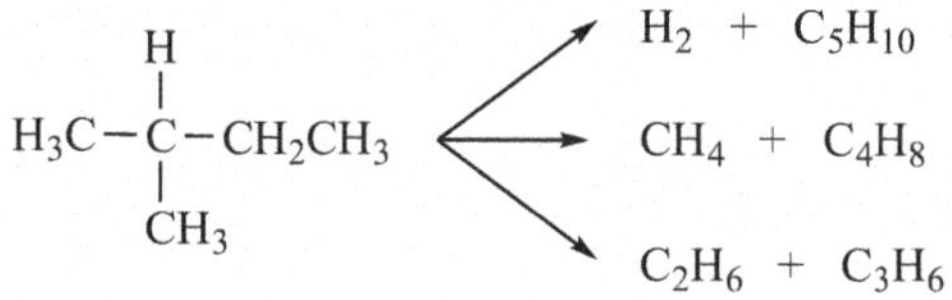

(7.66)

(7.67)

(7.68)

When H-ZSM-5 was used as a catalyst, the ratio of these reaction rates was 10:9:3. The C–H bond cleavage reaction rates over both Ag- and Zn-ZSM-5 were almost 20 times faster than that over H-ZSM-5, with dehydrogenation being the most predominant among the three primary reactions. Notably, the formation rate of CH_4 (reaction (7.67)) was also enhanced over Zn-ZSM-5. However, Ag-ZSM-5 completely suppressed both the reactions (7.67) and (7.68), which give CH_4 and C_2H_6 via C–C bond cleavage, indicating that Ag-ZSM-5 selectively cleaved the C–H bond of *iso*-pentane. On the other hand, Zn-ZSM-5 simultaneously dissociated both the C–H and C–C bonds of *iso*-pentane

(5) *neo*-Pentane cracking

The cracking of *neo*-pentane proceeds via reaction (7.69) below.

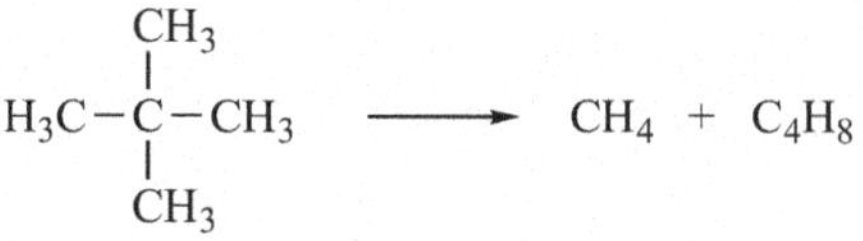

(7.69)

The primary products in this reaction can be accounted for using only reaction (7.69). No C–H bond cleavage reaction occurs because the formation of the primary carbenium ions $^+CH_2C(CH_3)_3$ by the abstraction of H^- ions via **Mechanism 2** hardly proceeds. Therefore, as shown in Table 7.10, Ag-ZSM-5 depressed *neo*-pentane cracking. However, this cracking reaction easily proceeds over Zn-ZSM-5 because the Zn^{2+} cations can cleave the C–C bond of *neo*-pentane via **Mechanism 1**. The formation rate of CH_4 over Zn-ZSM-5 was 28 times as fast as that over H-ZSM-5. This result shows that Zn^{2+} cations directly cleave the C–C bonds of alkanes.

Based on the results shown in Table 7.10, the absolute rates of dehydrogenation and those of C–C bond cleavage in lower alkanes were determined, and are shown in Tables 7.11 and 7.12, respectively. These results can be summarized as follows [98].

(1) Ag-zeolites selectively cleave the alkane C–H bond to produce H_2 more effectively than H-zeolites. For example, in *iso*-butane cracking, the hydrogen formation rate over Ag-ZSM-5 was 54 times greater than that over H-ZSM-5. Furthermore, Ag-ZSM-5 hardly catalyzed *neo*-pentane cracking.
(2) Zn-zeolites directly cleave both the C–H and C–C bonds of alkanes, and enhance C–C bond cleavage to a greater extent than C–H bond cleavage. For example,

Table 7.11 Absolute rates (10^{-2} mol h^{-1} g^{-1}) of dehydrogenation (hydride abstraction) of lower alkanes over Ag-ZSM-5 and Zn-ZSM-5 (Reprinted from ref. [97], Copyright 2020, with permission from Elsevier)

	$H_3C-\overset{CH_3}{\underset{H}{C}}-CH_3$ >	$H_3C-\overset{CH_3}{\underset{H}{C}}-CH_2CH_3$ >	$CH_3CH_2CH_2CH_2CH_3$ >	$CH_3CH_2CH_2CH_3$ >	$H_3C-\overset{CH_3}{\underset{CH_3}{C}}-CH_3$
Ag-ZSM-5	27	20	7.8	4.3	0
Zn-ZSM-5	36	19	8.9	5.8	0

Table 7.12 Absolute rates (10^{-2} mol h^{-1} g^{-1}) of CH_4 formation by the C–C bond cleavage of lower alkanes over H-ZSM-5 and Zn-ZSM-5 (Reprinted from ref. [97], Copyright 2020, with permission from Elsevier)

	CH_3 $H_3C-C-CH_3$ > CH_3	CH_3 $H_3C-C-CH_3$ > H	CH_3 $H_3C-C-CH_2CH_3$ > H	$CH_3CH_2CH_2CH_2CH_3$ >	$CH_3CH_2CH_2CH_3$
H-ZSM-5	1.0	0.5	0.9	0.5	0.5
Zn-ZSM-5	28	8.1	4.2	0.9	0.7

these effects were clearly observed in the cracking of *iso*-butene, *n*-pentane, and *iso*-pentane. The rates of these reactions over Zn-ZSM-5 were much faster than those over H-ZSM-5.

Results (1) and (2) clearly show that the activation mechanism of lower alkanes over M-cation zeolites is different from that over H-zeolites. The reaction proceeds via carbenium ion reaction intermediates over M-cation zeolites, while the reactions of lower alkanes over H-ZSM-5 proceed via penta-coordinated carbonium ion intermediates.

7.7.7.3 Different Catalytic Properties of Ag-Zeolites and Zn-Zeolites in the Activation of Methane: Characteristics of Ag-Zeolites Relevant to C–C Bond Cleavage

The order of the C–H bond cleavage reaction rates over Ag-ZSM-5 and Zn-ZSM-5 are listed in Table 7.11 [97, 98]. The hydrogen formation rates over Ag-ZSM-5 in lower alkane cracking follow the same order as those of Zn-ZSM-5. However, these formation rates are much faster than those observed over H-ZSM-5, showing that the reaction mechanism of C–H bond cleavage over Ag- and Zn-ZSM-5 zeolites is different from that over H-ZSM-5. Hydride abstraction occurs over Ag- and Zn-Zeolites via **Mechanism 2**, as mentioned in Sect. 7.7.6. The order reflects the stability of the secondary and/or tertiary carbenium ions formed by the abstraction of hydride ions by the metal cations, such as Zn^{2+}. Therefore, *iso*-butane and *iso*-pentane showed the highest reactivities, because these alkanes produce tertiary carbenium ions. *n*-Pentane and *n*-butane have reactivities an order of magnitude smaller than those of *iso*-butane or *iso*-pentane because they give secondary carbenium ions via hydride abstraction. Ag-ZSM-5 shows little activity for the conversion of *neo*-pentane, because hydride abstraction produces a primary carbenium ion. However, Ag-zeolites abstract hydride ions (H^-) from all the lower alkanes (*n*-butane, *iso*-butane, *n*-pentane, and *iso*-pentane), except for *neo*-pentane to generate secondary or tertiary carbenium ions. This clearly indicates that the C–H bond cleavage reaction occurs over Ag-zeolites via **Mechanism 2**. For example, the dehydrogenation step of *n*-butane over Ag-zeolites can be expressed as follows:

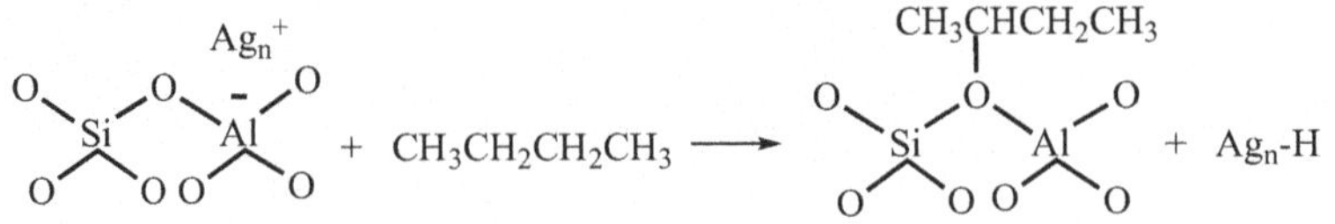

(7.70)

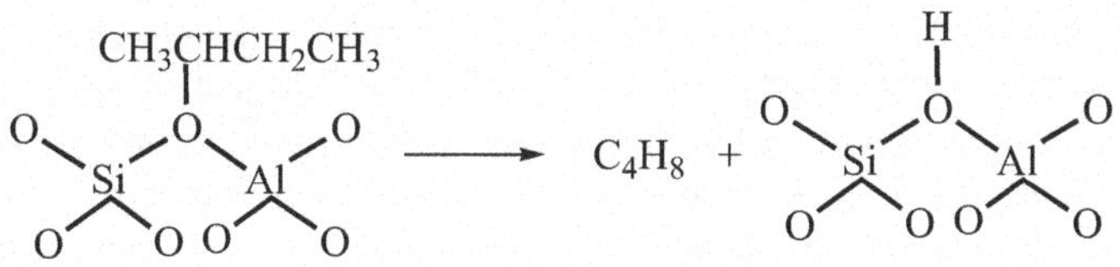

(7.71)

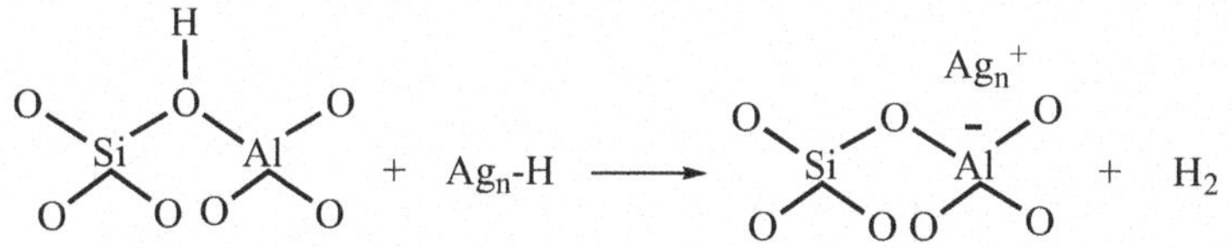

(7.72)

The initial step of hydride abstraction proceeds via the heterolytic dissociation of a C–H bond of *n*-butane due to the synergetic action of Ag_n^+ cationic clusters and oxygen anions (O^{2-}) in the zeolite framework, as depicted in reaction (7.70). **Mechanism 2** is in agreement with the activation mechanism of CH_4 over Ag-zeolites. As mentioned above, the rate of hydride abstraction via **Mechanism 2** depends on the stability of the carbenium ion intermediates, with the C–H bond cleavage reaction of alkanes including CH_4 being the slowest step, *i.e.*, the rate-determining step of alkane activation. In other words, the subsequent steps (reactions (7.71) and (7.72)) are faster than reaction (7.70).

Notably, the C–C bond cleavage reaction barely proceeds over Ag-zeolites, as shown in the cracking of *neo*-pentane. This reaction may occur on the Brönsted acid sites generated on Ag-zeolites as shown in reaction (7.39) because the conversion of *neo*-pentane proceeds slightly to produce CH_4 and *iso*-butene. Therefore, cationic Ag species, such as Ag_n^+ cationic clusters, do not have the ability to cleave the C–C bonds of alkanes. Thus, it can be concluded that Ag-zeolites selectively cleave the C–H bond in CH_4 to give only $-OCH_3$ species and silver hydride (Ag_n–H) as shown in reaction (7.48), indicating that the CH_3^+ carbenium ion is the real intermediate in the reaction of CH_4 with C_2H_4 to produce propylene and hydrogen. This also shows that the reaction of CH_4 with benzene proceeds to produce toluene and hydrogen via $^+CH_3$ carbenium ions.

7.7.7.4 Characteristics of Zn-Zeolites Relevant to the Activation of Alkanes Including CH_4

As shown in Table 7.10, Zn-ZSM-5 greatly enhances the C–C bond cleavage reactions in lower alkanes as compared with H-ZSM-5. The absolute CH_4 formation rates are listed in Table 7.12. Their relative magnitudes correspond to the stability of the carbenium ions produced by the abstraction of methyl carbanions ($^-CH_3$) from the alkanes. For example, *neo*-pentane gives the tertiary carbenium ion $^+C(CH_3)_3$ after

$^{-}CH_3$ abstraction by Zn^{2+} cations, while *iso*-butane and *iso*-pentane give the secondary carbenium ions $^{+}C(CH_3)_2H$ and $^{+}C(CH_3)(C_2H_5)H$, respectively. However, in the first step, the abstraction of $^{-}CH_3$ from *n*-butane and *n*-pentane should give primary carbenium ions. In Table 7.12, the CH_4 formation rates for reactions involving tertiary and secondary carbenium ions are faster than that involve primary carbenium ions, indicating that the cracking of lower alkanes proceeds via the carbenium ion mechanism over Zn-zeolites. Based on this, a plausible reaction mechanism for the production of CH_4 from *n*-butane over Zn-zeolites is shown below.

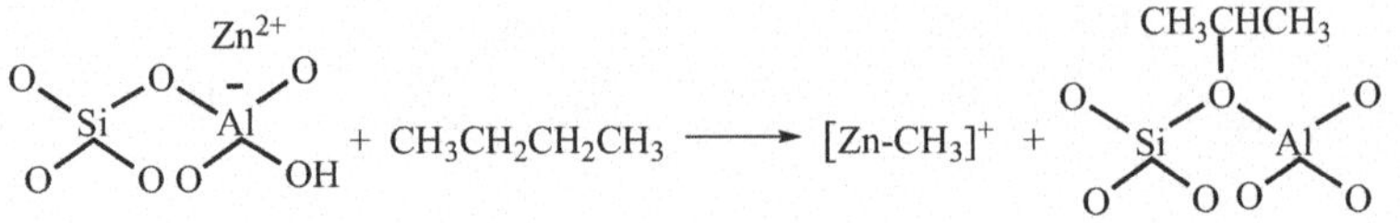

(7.73)

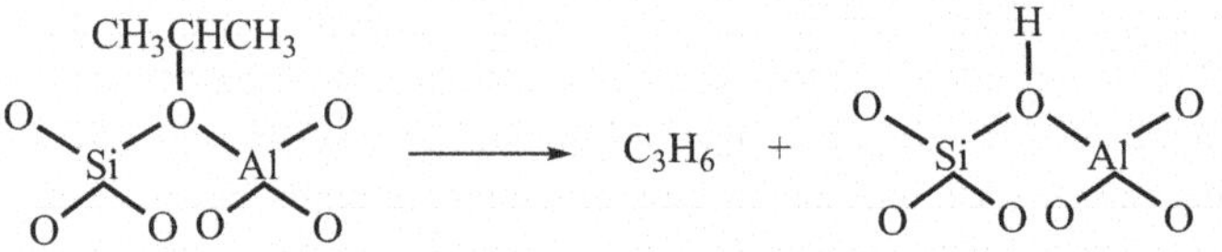

(7.74)

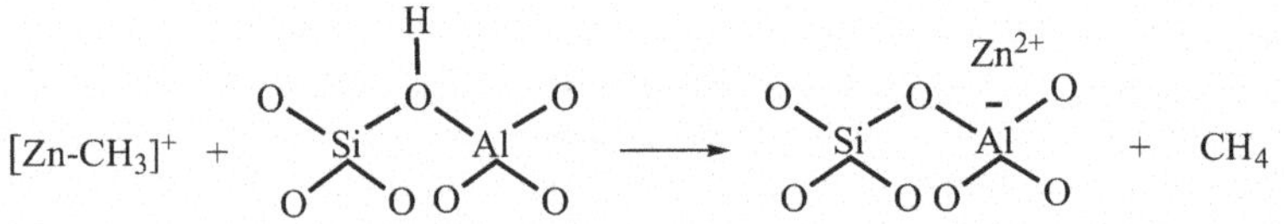

(7.75)

The same reaction mechanism is seen in the formation of C_2H_6 (reaction (7.59)) and CH_4 (reactions (7.73), (7.74), and (7.75)).

The formation of Zn–alkyl species, such as Zn–CH_3, from lower alkanes over the zeolites Zn-ZSM-5 and Zn-β was reported based on ^{1}H and ^{13}C MAS NMR spectroscopy of Zn–alkyl species [99–102]. The formation of Zn–C_2H_5 species via the adsorption of C_2H_6 on Zn-ZSM-5 zeolites was also supported by quantum mechanical studies [103, 104]. These results indicate that the C–C bond cleavage reaction proceeds by the abstraction of an alkyl carbanions, as well as $^{-}CH_3$ via **Mechanism-1**, as shown in reaction (7.51). However, the results described above seem to conflict with the observation of Zn–CH_3 and –OCH_3 species on Zn-zeolites by ^{13}C MAS NMR spectroscopy [101]. M–CH_3 species were also observed in In-, Fe-, and Mo-ZSM-5 zeolites [89]. However, $^{+}CH_3$ carbenium ions are the real reaction intermediates in the reaction of methane with benzene or ethylene to produce new C–C bonds. Even though the formation of M-CH_3 is observed when CH_4 is placed in contact with M-cation zeolites, they are completely unable to react with benzene or ethylene. Therefore, Ag-zeolites can selectively produce $^{13}CC_2H_6$ in the reaction of $^{13}CH_4$ with C_2H_4, as shown in Table 7.3.

7.8 Summary of the Features of Carbocation-Mediated C–C Bond Formation in Methane Conversion

As discussed in this chapter, the key intermediate in carbocation-mediated methane conversion is $^{+}CH_3$, which is essential for selective C–C bond formation. Thus, the selective formation of $^{+}CH_3$ is an important factor. Superacids and Ag^{+}-exchanged zeolites can selectively generate $^{+}CH_3$ through the heterolytic dissociation of methane C–H bonds. Carbenium ions such as $^{+}CH_3$ are effective alkylating agents and can react electrophilically with electron-rich molecules, except for ethylene and benzene, to form non-C–C bonds (*e.g.*, C–O and C–N bonds). Furthermore, superacids and Ag-zeolites can produce carbenium ions through the heterolytic C–H bond dissociation of alkanes other than CH_4. This indicates that carbenium ion generation methods should be extended to other molecules to develop new reactions, *i.e.*, catalysts that do not resemble superacids or Ag-zeolites should be designed to promote the development of new reactions, catalyst characterization methods, and chemical processes in methane conversion.

References

1. Olah GA, Schlosberg RH (1968) Chemistry in super acids. I. Hydrogen exchange and polycondensation of methane and alkanes in FSO_3H-SbF_5 ("magic acid") solution. Protonation of alkanes and the intermediacy of CH_5^{+} and related hydrocarbon ions. The high chemical reactivity of "paraffins" in ionic solution reactions. J Am Chem Soc 90:2726–2727
2. Sommer J, Muller M, Loadi K (1982) Alkene-alkane alkylation catalyzed by the HF-TaF_5 acid system. Nou J Chim 6:3–6
3. Choudhary VR, Kinage AK, Choudhary TV (1997) Low-temperature nonoxidative activation of methane over H-galloaluminosilicate (MFI) zeolite. Science 275:1286–1288
4. Baba T, Sawada H, Takahashi T, Abe M (2002) Chemisorption study of hydrogen and methane by ^{1}H MAS NMR and conversion of methane in the presence of ethylene on Ag-Y zeolite. Appl Catal A 231:55–63
5. Guo J, Lou H, Zhang X (2009) Energy-efficient coaromatization of methane and propane. J Nat Gas Chem 18:260–272
6. Olah GA, Surya Prakash GK, Sommer J (1979) Superacids. Science 206:13–20
7. Olah GA, Surya Prakash GK, Goeppert A (2006) Fluorinated superacidic systems. Actulite Chimique 301–302:68–72
8. Hammett LP, Deyrup AJ (1932) A series of simple basic indicators. I. The acidity functions of mixtures of sulfuric and perchloric acids with water. J Am Chem Soc 54:2721–2739
9. Olah GA (1973) Carbocations and electrophilic reactions. Angew Chem Int Ed 12:173–211
10. Olah GA (1987) Electrophilic methane conversion. Acc Chem Res 20:422–428
11. Olah GA (2001) 100 years of carbocations and their significance in chemistry. J Org Chem 18:5943–5957
12. Bickel AF, Gaasbeek CJ, Hogeveen H, Oelderik JM, Platteeuw JC (1967) Chemistry and spectroscopy in strong acidic solutions: reversible reaction between aliphatic carbenium ions and hydrogen. Chem Commun 634–635
13. Hogeveen H, Bickel AF (1967) Chemistry and spectroscopy in strong acidic solutions: Electrophilic substitution at alkane-carbon by protons. Chem Commun 635–636

14. Olah GA, Lukas J (1967) Stable carbonium ions. XXXIX. Formation of alkylcarbonium ions *via* hydride ion abstraction from alkanes in fluorosulfonic acid-antimony pentafluoride solution. Isolation of some crystalline alkylcarbonium ion salts. J Am Chem Soc 89:2227–2228
15. Olah GA, Lukas J (1967) Stable carbonium ions. XLVII. Alkylcarbonium ion formation from alkanes *via* hydride (alkide) ion abstraction in fluorosulfonic acid-antimony pentafluoride-sulfuryl chlorofluoride solution. J Am Chem Soc 89:4739–4744
16. Olah GA, Klopman G, Schlosberg RH (1969) Chemistry in super acids. III. Protonation of alkanes and the intermediacy of alkanonium ions, pentacoordinated carbon cation of the CH_5^+ type, hydrogen exchange, protolytic cleavage, hydrogen abstraction, and polycondensation of methane, ethane, 2,2-dimetylpropane and 2,3,3,3-tetramethylbutane in FSO_3H-SbF_5. J Am Chem Soc 91:3261–3268
17. Olah GA, Halpern Y, Shen J, Mo YK (1971) Electrophilic reactions at single bonds. III. H-D exchange and protolysis (deuterolysis) of alkanes with superacids. The mechanism of acid-catalyzed hydrocarbon transformation reactions involving the σ electron pair donor ability of single bonds via three-center bond formation. J Am Chem Soc 93:1251–1256
18. Olah GA, Halpern Y, Shen J, Mo YK (1973) Electrophilic reactions at single bonds. XII. Hydrogen-deuterium exchange, protolysis (deuterolysis), and oligocondensation of alkanes with superacids. J Am Chem Soc 95:4960–4970
19. Olah GA, DeMember JR, Shen J (1973) Electrophilic reactions at single bonds. X. Hydrogen transfer, alkylation, and alkylolysis of alkanes with methyl and ethyl fluoroantimonate. J Am Chem Soc 95:4952–4956
20. White ET, Tang J, Oka T (1999) CH_5^+: The infrared spectrum observed. Science 284:135–137
21. Kramer GM (1999) CH_5^+ stability and mass spectroscopy. Science 286:1051
22. Hiraoka K, Kebarle P (1975) Energetics, stabilities, and possible structures of $CH_5^+(CH_4)_n$ clusters from gas phase study of equilibria $CH_5^+(CH_4)_{n-1} + CH_4 = CH_5^+(CH_4)_n$ for $n = 1$–5. J Am Chem Soc 97:4179–4183
23. Tal'roze VL, Ljubimova AK (1999) Secondary processes in the ion source of a mass spectrometer (presented by academician N. N. Semenov 27 VIII (1952)—reprinted from report of the Soviet Academy of Science volume LXXXVI-n5 (1952). J Mass Spectrom 33:502–504
24. Ahlberg P, Karlsson A, Goeppert A, Lill SON, Dinér P, Sommer J (2001) Solvated CH_5^+ in liquid superacid. Chem—Eur J 7:1936–1943
25. Boo DW, Lee YT (1993) Infrared spectra of CH_5^+ core in $CH_5^+(H_2)$. Chem Phys Lett 211:358–363
26. Marx D, Parrinello M (1995) Structure quantum effects and three-centre two-electron bonding in CH_5^+. Nature 375:216–218
27. Esser A, Forbert H, Marx D (2018) Tagging effects on the mid-infrared spectrum of microsolvated protonated methane. Chem Sci 9:1560–1573
28. Ivanov SD, Asvany O, Witt A, Hugo E, Mathias G, Redlich B, Marx D, Schlemmer S (2010) Quantum-induced symmetry breaking explains infrared spectra of CH_5^+ isotopologues. Nature Chem 2:298–302
29. Asvany O, Yamada KMT, Brünken S, Potapov A, Schlemmer S (2015) Experimental ground-state combination differences of CH_5^+. Science 347:1346–1349
30. Dinér P (2012) Superacid-protonated ionization of alkanes without carbonium ion formation: a density function theory study. J Phys Chem A 116:9979–9984
31. Olah GA, Olah JA (1971) Electrophilic reactions at single bonds. IV. Hydrogen transfer from, alkylation of, and alkyloysis of alkanes by alkylcarbenium fluoroantimonates. J Am Chem Soc 93:1256–1259
32. Haag WO, Dessau RM (1984) Duality of mechanism of acid-catalyzed paraffin cracking. In: Proceedings of the 8th International Congress on Catalysis, Berlin, vol 2, Dechema, Frankfurt am Main, pp 305–316
33. Kotrel S, Knözinger H, Gates BC (2000) The Haag-Gessau mechanism of protolytic cracking of alkanes. Microporous Mesoporous Mater 35–36:11–20

34. Olah GA, Felberg JD, Lammertsma K (1983) Electrophilic reactions at single bonds. 18. Heterogeneous gas-phase ethylation of methane with ethylene over solid superacids. A ^{13}C isotope tracer study. J Am Chem Soc 105:6529–6531
35. Ng FTT, Rourke CR, Lynn J (1992) Solid superacid catalysts for methane conversion and alkylation. Stud Surf Sci Catal 73:91–97
36. Siskin M (1976) Strong acid chemistry. 3. Alkane-alkene alkylations in hydrofluoric acid-tantalum pentafluoride. Evidence for the presence of ethyl(1+) ion in solution. J Am Chem Soc 98:5413–5414
37. Baba T, Sawada H (2002) Conversion of methane into higher hydrocarbons in the presence of ethylene over H-ZSM-5 loaded with silver cations. Phys Chem Chem Phys 4:3919–3923
38. Wang W, Seiler M, Hunger M (2001) Role of surface methoxy species in the conversion of methanol to dimethyl ether on acidic zeolites investigated by in situ stopped-flow MAS NMR spectroscopy. J Phys Chem B 105:12553–12558
39. Wang W, Hunger M (2008) Reactivity of surface alkoxy species on acidic zeolite catalysts. Acc Chem Res 41:895–904
40. Wang W, Buchholz A, Seiler M, Hunger M (2003) Evidence for an initiation of the methanol-to-olefin process by reactive surface methoxy group on acidic zeolite catalysts. J Am Chem Soc 125:15260–15267
41. Jiang Y, Hunger M, Wang W (2006) On the reactivity of surface methoxy species in acidic zeolites. J Am Chem Soc 128:11679–11692
42. Ono Y, Mori T (1981) Mechanism of methanol conversion into hydrocarbons over ZSM-5 zeolite. J Chem Soc Faraday Trans 1(77):2209–2221
43. Beyer H, Jacobs PA, Uytterhoeven JB (1976) Redox behaviour of transition metal ions in zeolites. Part 2—Kinetic study of the reduction and reoxidation of silver-Y zeolites. J Chem Soc Faraday Trans 1 72:674–685
44. Reikert L (1969) Redox equilibria in zeolites. Ber Bunsen-Ges 73:331–338
45. Gellens LR, Mortier WJ, Uytterhoeven JB (1981) On the nature of the changed silver clusters in zeolites of type A, X and Y. Zeolites 1:11–18
46. Jacobs PA, Uytterhoeven JB, Beyer HK (1977) Redox behaviour of transition metal ions in zeolites. Part 6—Reversibility of the reduction reaction in silver zeolites. J Chem Soc Faraday Trans 1 73:1755–1762
47. White JL, Jelli AN, André JM, Fripiat JJ (1967) Perturbation of OH groups in decationated Y-zeolites by physically adsorbed gases. Trans Faraday Soc 1(63):461–475
48. Gellens LR, Mortier WJ, Schoonheydt RA, Uytterhoeven JB (1981) The nature of the changed silver clusters in dehydrated zeolites of type A. J Phys Chem 85:2783–2788
49. Xu B, Kevan L (1991) Formation of silver ionic clusters and silver metal particles in zeolite rho studied by electron spin resonance and far-infrared spectroscopies. J Phys Chem 95:1147–1151
50. Sun T, Seff K (1994) Silver clusters and chemistry in zeolites. Chem Rev 94:857–870
51. Baker MD, Ozin GA, Godber J (1985) Direct probe Fourier transform far-infrared spectroscopy of metal atoms, metal ions, and metal clusters in zeolites. Catal Rev—Sci Eng 27:591–651
52. Jacobs PA, Uytterhoeven JB, Beyer HK (1979) Some unusual properties of activated and reduced AgNaA zeolites. J Chem Soc Faraday Trans 1(75):56–64
53. Gellens LR, Mortier WJ, Uytterhoeven JB (1981) Oxidation and reduction of silver in zeolite Y: a structural study. Zeolites 1:85–90
54. Kim Y, Seff K (1987) Silver clusters formation in large cavity of zeolite A. Crystal structure of Ag4.6Na7.4Si12Al12O48 dehydrated and treated with hydrogen at 350 °C. J Phys Chem 91:668–671
55. Kim Y, Seff K (1987) Crystal structure of fully dehydrated partially Ag^{+}-exchanged zeolite 4A, $Ag_{7.6}Na_{4.4}$-A. Silver ions prefer 6-ring sites. One Ag^{+} ion is reduced. J Phys Chem 91:671–674
56. Tsutsumi K, Takahashi H (1972) The formation of metallic silver in silver-form zeolites. Bull Chem Soc Jpn 45:2332–2337

57. Ozin GA, Hugues F (1983) Silver atoms and small silver clusters stabilized in zeolite Y: Optical spectroscopy. J Phys Chem 87:94–97
58. Baker MD, Ozin GA, Godber J (1984) Direct observation of the reversible redox couple trisilver(2+) (Ag_3^{2+}) Ag_3^o in silver zeolite A by Fourier transform far-infrared spectroscopy. J Phys Chem 88:4902–4904
59. Baker MD, Ozin GA, Godber J (1985) Far-infrared studies of silver atoms, silver ions, and silver clusters in zeolites A and Y. J Phys Chem 89:305–311
60. Baker MD, Godber J, Ozin GA (1985) Watching silver clusters grow in zeolites: direct probe Fourier transform far-infrared spectroscopy of the red form of fully silver ion exchanged zeolite A. J Phys Chem 89:2299–2304
61. Baba T, Akinaka N, Nomura M, Ono Y (1993) Reversible redox behaviour between silver cations and silver metal particles in hydrated Ag-Y zeolite. J Chem Soc Faraday Trans 89:595–599
62. Baba T, Akinaka N, Nomura M, Ono Y (1992) Reversible interconversion of silver cations and silver metal particles in Ag-Y. J Chem Soc Chem Commun 339–340
63. Baba T, Komatsu N, Sawada H, Yamaguchi Y, Takahashi T, Sugisawa H, Ono Y (1999) ^{1}H Magic Angle Spinning NMR evidence of dissociate adsorption of hydrogen on Ag^+-exchanged A- and Y-zeolites. Langmuir 15:7894–7896
64. Baba T, Komatsu N, Takahashi T, Sugisawa H, Ono Y (1999) Chemisorption of hydrogen on Ag-Y studied by ^{1}H MAS NMR. Stud Surf Sci Catal 125:269–276
65. Baba T, Tohjo Y, Takahashi T, Sawada H, Ono Y (2001) Properties of chemisorbed hydrogen species on Ag-A zeolite partially reduced with hydrogen as studies by ^{1}H MAS NMR. Catal Today 66:81–89
66. Pfeifer H, Freude D, Kärger J (1991) Basic principle and recent results of H magic-angle-spinning and pulsed field gradient nuclear magnetic resonance studies on zeolites. Stud Surf Sci Catal 65:89–115
67. Brunner E, Karge HG, Pfeifer H (1992) On the correlation between the ^{1}H NMR chemical shift and the stretching vibration frequency of hydroxyl groups in solids. Z Phys Chem 176:173–183
68. Brunner E (1995) Solid state NMR—a powerful tool for the investigation of surface hydroxyl groups in zeolites and their interactions with adsorbed probe molecules. J Mol Struct 355:61–85
69. Morton JR, Preston KF (1986) Paramagnetic silver clusters in irradiated Ag-A molecular sieves. J Magn Reson 68:121–128
70. Freude D, Ernst H, Wolf I (1994) Solid-state nuclear magnetic resonance studies of acid sites in zeolites. Solid State Nucl Magn Reson 3:271–286
71. Ausavasukhi A, Suwannaran S, Limtrakul J, Sooknoi T (2008) Reversible interconversion behavior of Ag species in AgHZSM-5: XRD, ^{1}H MAS NMR, TPR, TPHE, and catalytic studies. Appl Catal A 345:89–96
72. Berndt H, Richter M, Gerlach T, Baerns M (1998) Influence of Brönsted acidity on the redox properties of silver species in zeolite mordenite. J Chem Soc Faraday Trans 94:2043–2046
73. Musaev DG, Morokuma K (1995) Ab initio molecular orbital study of the mechanism of H–H, C–H, N–H, O–H, and Si–H bond activation on transient cyclopentadienylcarbonylrhodium. J Am Chem Soc 117:799–805
74. Garbrienko AA, Arzumanov SS, Morzo IB, Toktarev AV, Wang W, Stepanov AG (2013) Methane activation and transformation on Ag/H-ZSM-5 zeolite studied with solid-state NMR. J Phys Chem C 117:7690–7702
75. Baba T, Abe Y (2003) Metal cation-acidic proton bifunctional catalysts for methane activation: Conversion of $^{13}CH_4$ in the presence of ethylene over metal cations-loaded H-ZSM-5. Appl Catal A 250:265–270
76. Baba T, Inazu K (2006) Heterolytic dissociation of C-H bond of methane over Ag^+-exchanged zeolites and conversion of methane into higher hydrocarbons in the presence of ethene or benzene. Chem Lett 35:142–147
77. Baba T (2005) Conversion of methane over Ag^+-exchanged zeolite in the presence of ethene. Catal Surv Asia 9:147–154

78. Baba T, Iwase Y, Inazu K, Masih D, Matsumoto A (2007) Catalytic properties of silver-exchanged zeolites for propene production by conversion of methane in the presence of ethene. Microporous Mesoporous Mater 101:142–147
79. Ding B, Huang H, Wang W (2008) Methane activation over Ag-exchanged ZSM-5 zeolites: a theoretical study. Appl Surf Sci 254:4944–4948
80. Oda A, Torigoe H, Itadani A, Ohkubo T, Yumura T, Kobayashi H, Kuroda Y (2013) Mechanism of CH_4 activation on a monomeric Zn^{2+}-ion exchanged in MFI-type zeolite with a specific Al arrangement: similarity to the activation site for H_2. J Phys Chem C 117:19525–19534
81. Luzgin MV, Rogov VA, Arzumanov SS, Toktarev AV, Stepanov AG, Parmon VN (2008) Understanding methane aromaticzation on Zn-modified high-silica zeolite. Angew Chem Int Ed 47:4559–4562
82. Luzgin MV, Gabrienko AA, Rogov VA, Toktarev AV, Parmon VN, Stepanov AG (2010) The "alkyl" and "carbenium" pathways of methane activation on Ga-modified zeolite BEA: ^{13}C solid-state NMR and GC-MS study of methane aromatization in the presence of higher alkane. J Phys Chem C 114:21555–21561
83. Luzgin MV, Rogov VA, Arzumanov SS, Toktarev AV, Stepanov AG, Parmon VN (2009) Methane aromatization on Zn-modified zeolite in the presence of a co-reactant higher alkane: how does it occur? Catal Today 144:265–272
84. Arzumanov SS, Gabrienko AA, Freude D, Stepanov AG (2016) Competitive pathways of methane activation on Zn^{2+}-modified ZSM-5 zeolite: H/D hydrogen exchange with Brønsted acid sites versus dissociate adsorption to form Zn-methyl species. Catal Sci Technol 6:6381–6388
85. Wang X, Xu J, Qi G, Li B, Wang C, Deng F (2013) Alkylation of benzene with methane over ZnZSM-5 zeolites studied with solid-state NMR spectroscopy. J Phys Chem C 117:4018–4023
86. Xu J, Zheng A, Wang X, Qi G, Su J, Du J, Gan Z, Wu J, Wang W, Deng F (2012) Room temperature activation of methane over Zn modified H-ZSM-5 zeolite: insight from solid-state NMR and theoretical calculations. Chem Sci 3:2932–2940
87. Liu BS, Zhang Y, Liu JF, Tian M, Zhang FM, Au CT, Cheung ASC (2011) Characterization and mechanism of methane dehydroaromatization over Zn-based/H-ZSM-5 catalyst under conditions of atmospheric pressure and supersonic jet expansion. J Phys Chem C 115:16954–16962
88. Gabrienko AA, Arzumanov SS, Toktarev AV, Danilova IG, Prosvirin IP, Kriventsov VV, Zaikovskii VI, Fruede D, Stepanov AG (2017) Different efficiency of Zn^{2+} and ZnO species for methane activation on Zn-modified zeolite. ACS Catal 7:1818–1830
89. Gabrienko AA, Arzumanov SS, Moroz IB, Prosvirin IP, Toktarev AV, Wang W, Stepanov AG (2014) Methane activation on In-modified ZSM-5: The state of indium in zeolite and pathways of methane transformation to surface species. J Phys Chem C 118:8034–8043
90. Baba T, Abe Y, Nomoto K, Inazu K, Echizen T, Ishikawa A, Murai K (2005) Catalytic transformation of methane over In-loaded ZSM-5 zeolite in the presence of ethylene. J Phys Chem B 109:4263–4268
91. Messe M, Beier H, Zeeh B (1996) Spectroskopiche Methoden in der Organischen Chemie, Georg Thieme Verlag 1996, Chapter 4
92. Ono Y, Kanae K (1991) Transformation of butanes over ZSM-5 zeolites. Part-1.—Mechanism of cracking of butanes over H-ZSM-5. J Chem Soc Faraday Trans 87:663–667
93. Ono Y, Kanae K (1991) Transformation of butanes over ZSM-5 zeolites. Part-2—Formation of aromatic hydrocarbons over Zn-ZSM-5 and Ga-ZSM-5. J Chem Soc Faraday Trans 87:669–675
94. Shigeishi R, Harris GI, Dyer J (1991) The conversion of butane in HZSM-5. J Catal 130:423–439
95. Krannila H, Haag WO, Gates BC (1992) Monomolecular and bimolecular mechanisms of paraffin cracking: *n*-Butane cracking catalyzed by HZSM-5. J Catal 135:115–124
96. Narbeshuber TF, Vinek H, Lercher JA (1995) Monomolecular conversion of light alkanes over H-ZSM-5. J Catal 157:388–395

97. Ono Y, Osako K, Yamawaki M, Nakashiro K (1994) Mechanism of the activation of butanes and pentanes over ZSM-5 zeolites. Stud Surf Sci Catal 83:303–312
98. Ono Y, Baba T (2015) Unique properties of silver cations in solid-acid catalysts by zeolites and heteropolyacids. Phys Chem Chem Phys 17:15637–15654
99. Kolyagin YG, Ordomsky VV, Khimyak YZ, Rebrov AI, Fajula F, Ivanova II (2006) Initial stages of propane activation over Zn/MFI catalyst studied by in situ NMR and IR spectroscopic techniques. J Catal 238:122–133
100. Kolyagin YG, Ivanova II, Pirogov YA (2009) ^{1}H and ^{13}C MAS NMR, studies of light alkanes activation over MFI zeolite modified by Zn vapor. Solid State Nucl Magn Reson 35:104–112
101. Arzumanov SS, Gabrienko AA, Freude D, Stepanov AG (2009) In situ high temperature MAS NMR study of the mechanisms of catalysis. Ethane aromatization on Zn-modified BEA. Solid State Nucl Magn Reson 35:113–119
102. Ivanova II, Kolyagin YG, Ordomsky VV, Asachenko EV, Pasynkova EM, Pirogov YA (2009) Surface species formed during propane aromatization over Zn/MFI catalysts as determined by in situ spectroscopic techniques. J Mol Catal A 305:47–53
103. Frash MV, van Santen RA (2000) Activation of ethane in Zn-exchanged zeolites: a theoretical study. Phys Chem Chem Phys 2:1085–1089
104. Frash MV, van Santen RA (2000) Activation of small alkanes in Ga-exchanged zeolites: a quantum chemical study of ethane dehydrogenation. J Phys Chem A 104:2468–2475

The manufacturer's authorised representative in the EU is Springer Nature Customer Service Centre GmbH, Europaplatz 3, 69115 Heidelberg, Germany. If you have any concerns regarding our products, please contact ProductSafety@springernature.com

Printed and bound by CPI Group (UK) Ltd, Croydon, CR0 4YY
05/10/2026
02228120-0002